I0051579

Artificial Intelligence and Knowledge Processing

Artificial intelligence (AI) and knowledge processing play a vital role in various automation industries and their functioning in converting traditional industries to AI-based factories. This book acts as a guide and blends the basics of AI in various domains, which include machine learning, deep learning, artificial neural networks, and expert systems, and extends their application in all sectors.

The book discusses the designing of new AI algorithms used to convert general applications to AI-based applications. It highlights different machine learning and deep learning models for various applications used in healthcare and wellness, agriculture, and automobiles. The book offers an overview of the rapidly growing and developing field of AI applications, along with knowledge of engineering and business analytics. Real-time case studies are included across several different fields such as image processing, text mining, healthcare, finance, digital marketing, and HR analytics. The book also introduces a statistical background and probabilistic framework to enhance the understanding of continuous distributions. Topics such as ensemble models, deep learning models, artificial neural networks, expert systems, and decision-based systems round out the offerings of this book.

This multicontributed book is a valuable source for researchers, academics, technologists, industrialists, practitioners, and all those who wish to explore the applications of AI, knowledge processing, deep learning, and machine learning.

Artificial Intelligence and Knowledge Processing
Improved Decision-Making and Prediction

Edited by
Hemachandran K, Raul V. Rodriguez,
Umashankar Subramaniam and
Valentina Emilia Balas

CRC Press
Taylor & Francis Group
Boca Raton London New York

CRC Press is an imprint of the
Taylor & Francis Group, an **informa** business

Design cover image: © Shutterstock

First edition published 2024
by CRC Press
2385 Executive Center Drive, Suite 320, Boca Raton, FL 33431

and by CRC Press
4 Park Square, Milton Park, Abingdon, Oxon, OX14 4RN

CRC Press is an imprint of Taylor & Francis Group, LLC

© 2024 selection and editorial matter, Hemachandran K., Raul V. Rodriguez, Umashankar Subramaniam, and Valentina Emilia Balas; individual chapters, the contributors

Reasonable efforts have been made to publish reliable data and information, but the author and publisher cannot assume responsibility for the validity of all materials or the consequences of their use. The authors and publishers have attempted to trace the copyright holders of all material reproduced in this publication and apologize to copyright holders if permission to publish in this form has not been obtained. If any copyright material has not been acknowledged please write and let us know so we may rectify in any future reprint.

Except as permitted under U.S. Copyright Law, no part of this book may be reprinted, reproduced, transmitted, or utilized in any form by any electronic, mechanical, or other means, now known or hereafter invented, including photocopying, microfilming, and recording, or in any information storage or retrieval system, without written permission from the publishers.

For permission to photocopy or use material electronically from this work, access www.copyright.com or contact the Copyright Clearance Center, Inc. (CCC), 222 Rosewood Drive, Danvers, MA 01923, 978–750–8400. For works that are not available on CCC please contact mpkbookspermissions@tandf.co.uk

Trademark notice: Product or corporate names may be trademarks or registered trademarks and are used only for identification and explanation without intent to infringe.

Library of Congress Cataloging-in-Publication Data
Names: K., Hemachandran, editor. I Rodriguez, Raul Villamarin, editor. I
    Subramaniam, Umashankar, editor. I Balas, Valentina Emilia, editor.
Title: Artificial intelligence and knowledge processing : improved decision-making
    and prediction / Hemachandran K., Raul V. Rodriguez, Umashankar Subramaniam,
    and Valentina Emilia Bas.
Description: Boca Raton : CRC Press, 2023. I Includes bibliographical references
    and index.
Identifiers: LCCN 2023010710 (print) I LCCN 2023010711 (ebook) I
    ISBN 9781032354163 (hardback) I ISBN 9781032357577 (paperback) I
    ISBN 9781003328414 (ebook)
Subjects: LCSH: Artificial intelligence—Case studies. I Decision making—Data processing.
Classification: LCC TA347.A78 A7889 2023 (print) I LCC TA347.A78 (ebook) I
    DDC 006.3—dc23/eng/20230609
LC record available at https://lccn.loc.gov/2023010710
LC ebook record available at https://lccn.loc.gov/2023010711

ISBN: 978-1-032-35416-3 (hbk)
ISBN: 978-1-032-35757-7 (pbk)
ISBN: 978-1-003-32841-4 (ebk)

DOI: 10.1201/9781003328414

Typeset in Times
by Apex CoVantage, LLC

Contents

Preface

Artificial intelligence (AI) is essential because it gives computers the ability to plan, comprehend, reason, communicate, and perceive. AI technology is effective because it efficiently processes vast amounts of data. Knowledge processing representing a variety of distinct ways of creating knowledge, including socialization, externalization, combination, and internalization, is fundamental to the numerous automation industries and their operations. On the other hand, computers must enhance their capacity for prediction to become more proficient at making decisions.

According to experts, artificial intelligence is a component of production that can open up new avenues for growth and transform how labour is performed across sectors. For instance, the PricewaterhouseCooper (PWC) article estimates that by 2035, AI might contribute $15.7 trillion to the world economy. With approximately 70% of the worldwide influence, China and the United States are well positioned to reap the greatest rewards from the impending AI boom. According to a Deloitte poll, 75% of firms believe that sharing and maintaining knowledge across their changing workforces are crucial to their success. Only 9% of businesses claim to be prepared to confront this trend, while 55% of enterprise data remains inactive.

This book covers applications of analytics techniques and their limits, statistics and probability, the incorporation of robotics and AI in the medical and health sectors, AI and Internet of Things (IOT) integration for smart systems, polarity in native language comments on the Internet, edge and fog computing techniques, denoising using autoencoders, prediction of terrorist attacks, breast cancer prediction using machine learning algorithms, and machine learning techniques for detecting and analysing online fake reviews.

We believe that this book will help students, businesspeople, academics, mentors, and anyone else interested in learning more about the uses of AI. We are grateful to our contributors, who come from prestigious institutions and businesses and who made a significant contribution by sharing their expertise for societal benefit. We would like to extend our sincere gratitude to our editing and production staff for their tireless efforts and unwavering support in helping us publish this book on schedule.

Editors' Biographies

Hemachandran K is currently working as a professor in the Department of Analytics and Artificial Intelligence at the School of Business, Woxsen University, Hyderabad, Telangana, India. He is a passionate teacher with 14 years of teaching experience and 5 years of research experience. His research interests are machine learning, deep learning, computer vision, natural language processing (NLP), knowledge engineering, and decision support systems. He has three patents to his credentials. He has more than 20 journals and international conference publications to his credit and has served as a resource person at various national and international scientific conferences.

Raul V. Rodriguez is a professor at the School of Business, Woxsen University, Hyderabad, Telangana, India. He holds a PhD in artificial intelligence and robotics process automation applications in human resources. He was a former co-CEO at Irians Research Institute. His areas of expertise and interest are machine learning, deep learning, natural language processing, computer vision, robotic process automation, multiagent systems, knowledge engineering, and quantum artificial intelligence. He has co-authored two reference books and has more than 70 publications to his credit.

Umashankar Subramaniam is an associate professor in electrical engineering at the College of Engineering, Prince Sultan University, Riyadh, Saudi Arabia. Previously he worked as an associate professor and head of the Department of Energy, VIT, Vellore. He has more than 15 years of teaching, research, and industrial research and development (R&D) experience. He has published more than 250 research papers in national and international journals and prestigious conferences. He is an editor of *IEEE Access*, *Heliyon*, and other high-impact journals.

Valentina Emilia Balas is a professor in the Department of Automatics and Applied Software at the Faculty of Engineering, "Aurel Vlaicu" University of Arad, Romania. She holds a PhD cum laude in applied electronics and telecommunications from the Polytechnic University of Timisoara. Dr. Balas is the author of more than 350 research papers in refereed journals and international conferences. Her research interests are in intelligent systems, fuzzy control, soft computing, smart sensors, information fusion, modeling, and simulation. She is the editor-in-chief of the *International Journal of Advanced Intelligence Paradigms* (IJAIP) and the *International Journal of Computational Systems Engineering* (IJCSysE).

Contributors

Balajee A
Assistant Professor
Department of Computer Science and
    Engineering
Srinivasa Ramanujan Centre
SASTRA (Deemed to be University)
Kumbakonam, India

Mir Aadil
Assistant Professor
School of Computer Science and IT
Jain (Deemed to be University)
Bangalore, India

Tejaswini Aala
Woxsen School of Business
Woxsen University
Hyderabad, India

Kirti Aija
Woxsen University
Hyderabad, India

Reham Alahmadi
Department of Basic Science
College of Science and Theoretical
    Studies
Medinah-Female Branch
Saudi Electronic University
Riyad, Saudi Arabia

Sirisha Alamanda
Chaitnya Bharathi Institute of
    Technology
Hyderabad, India

Dheeraj Anchuri
Student
Woxsen University
Hyderabad, India

Sunitha Purushottam Ashtikar
Research Scholar
School of Business
SR University
Warangal, India

Gogineni Venkata Ashwith
Woxsen University
Hyderabad, India

D. Suresh Babu
VR Siddhartha Engineering College
    Vijayawada
India

M. Balamurugan
Associate Professor
Department of CSE
CHRIST (Deemed to be University)

Divya Batra
National Institute of Fashion
    Technology
New Delhi, India

K. Bharath
Assistant Professor of Computer
    Science & Engineering
Malla Reddy College of Engineering
Hyderabad, India

J. Bhuvana
Associate Professor
Department of CSIT
Jain (Deemed to be University)

Tanushree Biswas
St. Xavier's University
Kolkata

J. Gladson Maria Britto
Professor of Computer Science &
  Engineering
Malla Reddy College of Engineering
Hyderabad, India

Varanasi Chandradhar
Woxsen University
Hyderabad, India

Channabasava Chola
Department of Electronics and
  Information Convergence Engineering
College of Electronics and Information
Kyung Hee University
Republic of Korea

P. Srihitha Chowdary
VR Siddhartha Engineering College
Vijayawada, India

Jk. Dhivya
Assistant Professor
Department of English
Arunachala College of Engineering for
  Women

B. Dinesh
Sree Vidyanikethan Engineering College
India

Subhashini Durai
Assistant Professor
GRD Institute of Management
Coimbatore, India

Mahmoud El Samad
Lebanese International University
Lebanon

Sravani Elaprolu
PGDM
School of Business
Woxsen University
Hyderabad, India

Megha Gada
Woxsen University
Hyderabad, India

Shahid Mohammad Ganie
Assistant Professor
Department of Analytics
School of Business
Woxsen University
Hyderabad, India

Tanvi Gorantla
School of Business
Woxsen University
Hyderabad, India

Happy
Lingayas Vidyapeeth
Faridabad, India

Dr. Harishree
Assistant Professor of English
SRM University

Arshia Jabeen
Department of ECE
Vignan's Institute of Management
  and Technology for Women,
  Kondapur (V), Ghatkesar (M),
  Medchal-Malkajgiri (D)
Telangana, India

Swetha Jaladi
Sree Vidyanikethan Engineering
  College
India

Jegan Jayapal
Assistant Professor
Woxsen University
Hyderabad, India

Ashwin Kumaar K
Woxsen University
Hyderabad, India

Hemachandran K
Professor
Woxsen School of Business
Woxsen University
Hyderabad, India

Paul Nirmal Kumar K
PGDM
Woxsen School of Business
Hyderabad, India

Rajesh Kannan K
Chaitnya Bharathi Institute of
    Technology
Hyderabad, India

Gabriel Kabanda
Adjunct Professor of Machine
    Learning
Woxsen School of Business
Woxsen University
Hyderabad, India

Vishwa KD
Woxsen University
Hyderabad, India

Ahmad Kheir
Coventry University
Cairo Branch in TKH Universities
Egypt

C. Naga Sai Kiran
Woxsen University
Hyderabad, India

S. Ravi Kishan
VR Siddhartha Engineering College
    Vijayawada
India

Varadaraja Krishna
Woxsen University
Hyderabad, India

Raja Krishnamoorthy
Professor
Department of ECE
Vignan's Institute of Management and
    Technology for Women, Kondapur
    (V), Ghatkesar (M), Medchal-
    Malkajgiri (D)
Telangana, India

J.R. Arun Kumar
Sree Vidyanikethan Engineering
    College
India

Pokala Pranay Kumar
MPS Data Science
University of Maryland
Baltimore, MD, USA

Neelam Kumari
Dublin Business School
Dublin, Ireland

Mursal Furqan Kumbhar
Department of Computer
    and Information Systems
    Engineering
NED University of Engineering and
    Technology
Karachi, Pakistan

Senbagamalar L
Research Scholar
Department of Information
    Technology
Karpagam College of Engineering
Coimbatore, India

B. Leander
Undergraduate Scholar
Department of Electronics and
    Communication Science
Agurchand Manmull Jain College
Chennai, India

Satheesh Kumar M
PGDM
School of Business
Woxsen University
Hyderabad, India

Geetha Manoharan
Assistant Professor
School of Business
SR University
Warangal, India

Harshitha Methukula
Department of ECE
Vignan's Institute of Management and
    Technology for Women, Kondapur
    (V), Ghatkesar (M), Medchal-
    Malkajgiri (D)
Telangana, India

Dipra Mitra
Amity University
Jharkhand

Kaja Bantha Navas Raja Mohamed
National Institute of Fashion Technology
Gandhinagar, India

R. Raj Mohan
Assistant Professor
Department of Electronics and
    Communication Science
Agurchand Manmull Jain College
Chennai, India

Vikor Molnar
Institute of Manufacturing Science
University of Miskolc
Hungary

Soumen Mondal
PDF
IIT Guwahati

T. Monish
VR Siddhartha Engineering College
    Vijayawada
India

Abdullah Y. Muaad
Department of Studies in Computer
    Science
University of Mysore

Tanmai Sree Musalamadugu
School of Business
Woxsen University
Hyderabad, India

Ghalia Nasserddine
Lebanese International University
Lebanon

Tuhin Utsab Paul
St. Xavier's University
Kolkata

Laxminarayan Pimpdae
PGDM
School of Business
Woxsen University
Hyderabad, India

Anil Audumbar Pise
University of the Witwatersrand
Johannesburg, South Africa

Shivani Prabhu
Woxsen University
Hyderabad, India

Ch. Prathima
Assistant Professor
School of Computing (IT)
MohanBabu University
India

T. Prathima
Chaitnya Bharathi Institute of
    Technology
Hyderabad, India

Murugan R
Associate Professor
School of Computer Science and IT
Jain (Deemed to be University)
Bangalore, India

Sunny Raj
National Institute of Fashion Technology
Gandhinagar, India

Gunaseelan Alex Rajesh
Sri Venkateswara Institute of Information
    Technology and Management
Coimbatore, Tamil Nadu, India

Mitta Chaitanya Kumar Reddy
Woxsen University
Hyderabad, India

P. Supreeth Reddy
Chaitnya Bharathi Institute of
    Technology
Hyderabad, India

Yashwitha Buchi Reddy
Sree Vidyanikethan Engineering College
India

Manjeet Rege
Professor and Chair
Department of Software Engineering
    and Data Science
University of St. Thomas
MN

Raul V. Rodriguez
Professor
School of Business
Woxsen University
Hyderabad, India

Sudeshna Sani
Koneru Lakshmaiah Education Foundation

P. Sanjna
Woxsen University
Hyderabad, India

N.P. Saravanan
Assistant Professor (SG)
Department of Computer Science and
    Engineering
Kongu Engineering College
Perundurai, India

Pusarla Bhuvan Sathvik
Student
Woxsen University
Hyderabad, India

Abhishek Asodu Shetty
Woxsen University
Hyderabad, India

Shivani
National Institute of Fashion
    Technology
Gandhinagar, India

Gaddam Venkat Shobika
Student
Woxsen School of Business
Woxsen University
Hyderabad, India

S. Pavan Siddharth
Woxsen University
Hyderabad, India

Abha Singh
Department of Basic Science College of
    Science and Theoretical Studies
Dammam-Female Branch
Saudi Electronic University
Riyad, Saudi Arabia

B.R.S.S. Sowjanya
Woxsen University
Hyderabad, India

Pingili Sravya
Woxsen University
Hyderabad, India

D. Sujith
Woxsen University
Hyderabad, India

T. Sunil
Professor of Computer Science &
    Engineering
Malla Reddy College of Engineering
Hyderabad, India

Skanda S. Tallam
Woxsen University
Hyderabad, India

Krishna Sai Talupula
Woxsen University
Hyderabad, India

B. Harsha Vardhan
VR Siddhartha Engineering College
   Vijayawada
India

B. Vasavi
Woxsen University
Hyderabad, India

K. Vinodh
Woxsen University
Hyderabad, India

Manoj Yadav
Lingayas Vidyapeeth
Faridabad, India

1 Introduction to Artificial Intelligence

Mahmoud El Samad, Ghalia Nasserddine, and Ahmad Kheir

1.1 INTRODUCTION

Since the creation of the first computer, humans have concentrated on developing various approaches to decrease the computer size and increase its operational capacity. During the evolution of computer systems, researchers were interested in creating machines that think, work, and act like humans [1]. This enthusiasm induced the development of artificial intelligence theory (AI) and gave rise to the creation of computer-based machines (e.g., robots) that have intelligence almost like humans [2]. More precisely, AI is a set of algorithms and techniques that are mutually and widely used nowadays to create machines and software solutions emulating the capabilities of a human being. These solutions perform tasks that used to be performed solely by a human, with few additional advantages compared to their human counterparts, like the ability to perform these tasks with a minimal margin of error and with a significantly decreased required time to find the results.

According to John McCarthy, the father of AI, artificial intelligence is defined as "the science and engineering of making intelligent machines, especially intelligent computer programs" [3]. Additionally, the word artificial in AI stands for human-created; the word intelligence represents the power of thinking. Therefore, AI is a human-made machine with thinking power [3, 4].

In the literature, AI is divided into two main types: AI type 1, based on capabilities, and AI type 2, based on functionality. AI type 1 includes three subtypes: (i) narrow AI, (ii) general AI, and (iii) super AI. Narrow AI is a type of AI which is able to perform a dedicated task with intelligence. Narrow AI (also called weak AI) is only trained for one specific task. General AI is a type of intelligence that can process human tasks. The main idea is to let these systems carry out daily tasks without human intervention. Many research efforts are now focused on implementing machines with general AI, but it is still an active research area. Super AI is a high level of intelligent systems that can perform tasks better than humans with cognitive properties such as the ability to think, plan, learn, and communicate. Super AI is still a theoretical concept of AI. The real development of these systems is still in a very early stage.

DOI: 10.1201/9781003328414-1

In AI type 2, we can extract four subtypes [5]: (i) reactive machines, (ii) limited memory machines, (iii) theory of mind, and (iv) self-aware AI. In the first type, reactive machines have no memory and do not use past experiences to determine the best actions. They simply perceive the world and react to it. In the second type, limited memory, the machines hold data for a short period. However, they cannot add any new information to the library of their experiences. The third type, theory of mind, is where the researcher hopes to create a machine that imitates human mental models. Finally, the self-aware AI has not been developed until now. In this type, machines are conscious of themselves. They can perceive their internal states and others' emotions and act accordingly.

Nowadays, AI is integrated into our daily activity in many forms, including computer gaming, Alexa, Google Assistant, etc. Recently, AI has also become part of many fields like healthcare, social media, education, banking, and finance [6]. Recent years have witnessed a significant expansion in digitized financial services. AI has been considered a robust tool in the financial field [7]. Many analytical tools, including machine learning, are used by firms to analyze data collected over time. AI improves the pattern recognition step by the use of modern statistical methods and large volumes of data to provide the best solution to any defined problem set [8].

In the context of AI, machine learning is now gaining importance. Machine learning is a trendy concept that goes back many decades [9, 10]; it is considered a subfield of AI [9]. The idea of the machine learning concept is to develop programming models that can process human activities by using a self-learning approach without any human interaction. In the classical programming models, the human role is crucial, while machine learning is based on the automation of analytical models where the system can learn from given data sets and proceed with decisions with minimal human interference [9–12].

The rest of this chapter is organized as follows. In Section 1.2, we present different types of AI. In Section 1.3, we describe AI systems and subsets. Section 1.4 discuss some relevant AI applications in the finance domains and other domains. Finally, Section 1.5 concludes the chapter and presents the current challenges and opportunities in this area of research.

1.2   TYPES OF ARTIFICIAL INTELLIGENCE

AI is one of the most complex human creations. Indeed, this field stands largely unexplored. Actually, the outstanding AI applications that are explored today represent only the top of the AI iceberg [1]. As the main goal of AI systems is to imitate human functionality, the degree by which these systems can perform human tasks is used to determine their types. Hence, based on how a system can be compared to a human in terms of capabilities and functionalities, AI can be classified under one of seven types as shown in Figure 1.1 [13].

According to Figure 1.1, AI systems based on their capabilities can be classified into three subtypes [13, 14]: narrow AI, general AI, and Super AI. Next, we will detail each subtype.

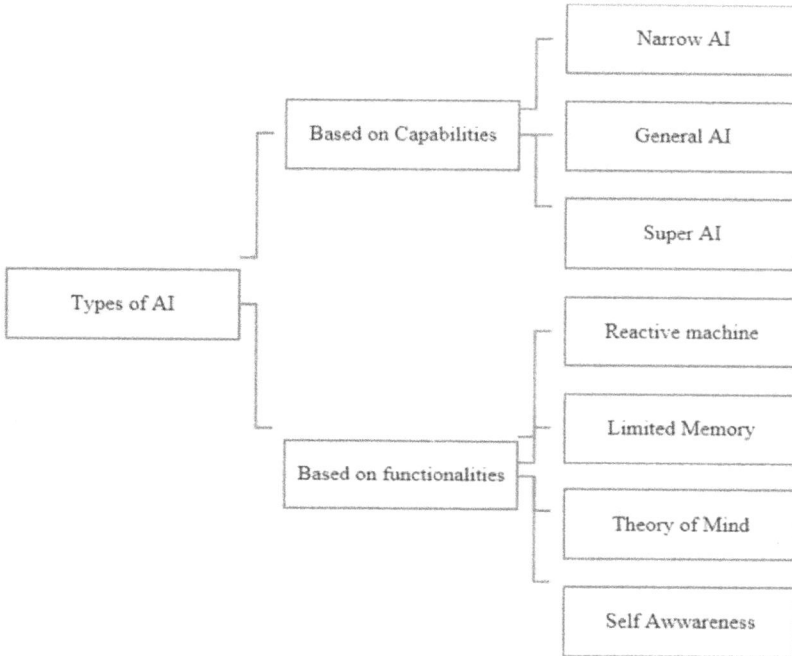

FIGURE 1.1   Types of artificial intelligence.

Narrow AI, called also weak AI, performs only one task. Machines that belong to this type target a single subset of cognitive capabilities and advance in that scope. Recently, many applications of this type have become progressively common as machine learning and deep learning methods continue to develop. Apple Siri, Google Translate, image recognition software, spam filtering, and Google's page-ranking algorithm are example of systems of the narrow AI type.

General AI, also known as strong AI, has the ability to understand, learn, and perform any task just like a human. In this type, a machine can apply knowledge and skills in different circumstances. Currently, machines of this kind do not exist. Researchers are focused on developing machines with general AI.

Super AI exceeds human intelligence. In this type, the machine will be able to perform any task better than a human can. Super AI is based on the concept that machines may progress in order to be extremely close to humans in terms of sentiments and behaviors. In this type, machine plus human understanding, it also induces its own emotions, needs, beliefs, and desires. Thinking, solving puzzles, making judgments, and making decisions are some critical characteristics of super AI. The existence of this type is still speculative.

With respect to functionality, AI may be classified into four subtypes. For instance, the reactive machine and the limited memory are considered a simple AI system with limited functionality and performance. The theory of mind AI [15] performs human tasks with a high level of proficiency and is considered to be the most sophisticated and developed type of AI.

The AI systems based on functionality (see Figure 1.1) can be classified into four subtypes: reactive machine, limited memory, theory of mind, and self-aware [2, 4]. Next, we will detail each subtype.

The reactive machine is the basic type of AI systems. In this type, machines are built without memory-based functionality. They have no capability to use old experiences in order to correct current decisions. Therefore, they are not able to learn. They only study the current situation and select the best action among possible ones. A common example of this type is Deep Blue, the IBM chess program that beat Garry Kasparov in the 1990s [16]. This system can identify pawns on the chessboard and make the best move without retaining memories or using the old experiences. These AI systems act according to a complete and direct observation of the environment without any previous knowledge about the world.

In the limited memory type, machines are built with a small amount of memory. Therefore, they have a limited capability to use past experiences in order to make new decisions. They can hold data for a short time and have limited capacity. As well as reactive capabilities, these machines are able to learn from historical data to make better decisions. Self-driving vehicles, chatbots, virtual assistants, and many existing applications fall under this type [16].

Theory of mind AI [15] is a psychology term. In the AI field, it means that machines must have social intelligence and understand human emotions. The goal of developing such machines is to simulate real emotions and beliefs using computers. This simulation can influence future decisions. Nowadays, several models are used to appreciate human behavior. However, a model that includes its own mind is not yet created. These systems can understand human requirements and predict behavior. Also, they have the capability to interact with people and identify their needs, emotions, and requirements. As well, they can predict behaviors. Bellhop Robot is an example of this type. It is created in order to be used in hotels. It has the ability to assess people's demands who come stay at the hotel.

Up to now, the self-aware type of AI does not exist. Machines of this kind will be smarter than human beings. In addition to perceiving human emotions, such systems will be able to understand their internal state and conditions. As well, they have their own demands, emotions, and faith.

The next section discusses how an AI system works by discussing AI systems and subsets.

1.3 AI SYSTEMS AND SUBSETS

An AI system is a machine-based system that can make predictions or decisions based on data collected from real or virtual environments for a given set of human objectives. Using sensors or specific given inputs, it perceives the real or virtual environment in order to take the appropriate decisions or actions [17]. The general structure of AI systems is represented in Figure 1.2.

AI is composed of several subsets. In this chapter, only three subsets are represented (see Figure 1.3): machine learning, deep learning, and natural language processing.

1.3.1 MACHINE LEARNING

Machine learning (ML) is one AI subset. It covers the study that allows machines to learn from experience without being explicitly programmed. Using ML, machines can learn, identify patterns, and make decisions with minimal human intervention.

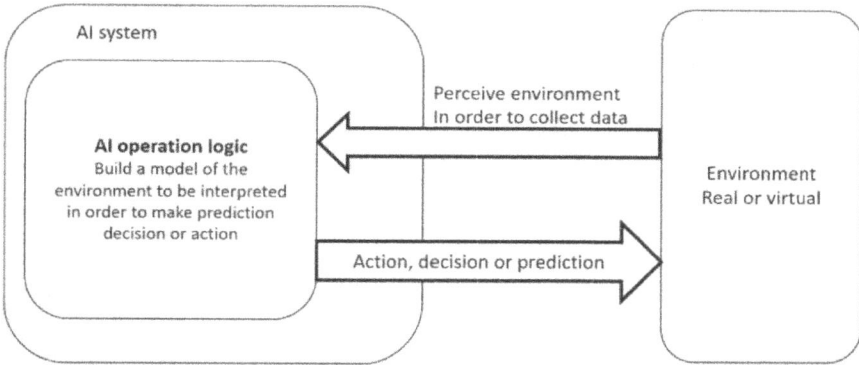

FIGURE 1.2   AI system.

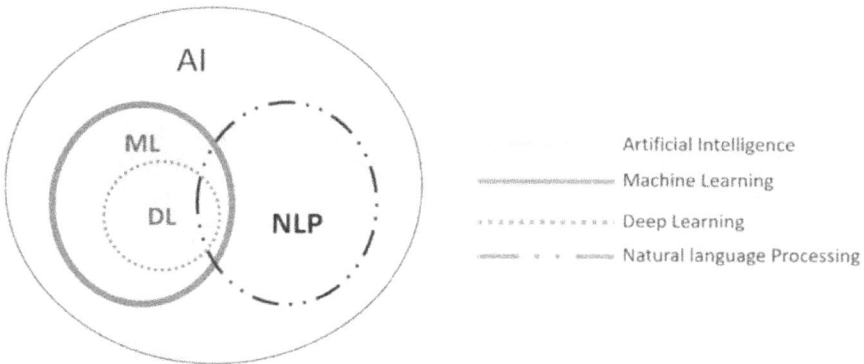

FIGURE 1.3   Subset of AI.

TABLE 1.1

ML Techniques Comparison

	Description	Advantages	Disadvantages
Dimensionality reduction	Based on converting the higher dimensions data set into a lesser dimensions data set. This conversion should be ensured to conserve similar information.	– Time, computation, and storage reduction – Accuracy increase due to less data interpretation – Noise and interference removal	– Data loss due to reduction – Misinterpretation of principal components due to fewer features used – Undesirable for non-linear data
Ensemble	It is a general meta-approach to machine learning that produces better predictive performance by combining the predictions from multiple models.	– Higher predictive accuracy – Useful for linear and non-linear data types	– More cost to train and deploy – Less interpretable – Time and space consumption
Decision tree	The most popular tool for classification and prediction. It is a flowchart-like tree structure that predicts decisions with their outcomes and costs.	– Simple to interpret – Simple to understand – Logarithmic order of growth used in training data due to its tree structure – Little data preparation	– Unstable due to major results changing even for small data variations – Time consumption in training when inputs increase
Rules system	It is a system that uses human-made rules in order to store, sort, and manipulate data. Therefore, it mimics human intelligence. Generally, such systems need a source of data and a set of rules for manipulating that data.	– Availability to users – Cost-efficient – End result accurate – High in terms of speed	– Deep knowledge – Manual work – Time-consuming to generate rules – Less learning capacity

Many algorithms are used in ML such as decision tree, regression, rules systems, ensemble approach, and others. Table 1.1 summarizes a comparison between these techniques.

1.3.2 DEEP LEARNING

Deep learning (DL) is the main subset of ML that enables computers to solve perceptual problems such as image and speech recognition. It is a powerful concept that is used in many applications like finance [8] and even in other fields (e.g., biological sciences [18]). By using multiple processing layers, DL is able to discover patterns

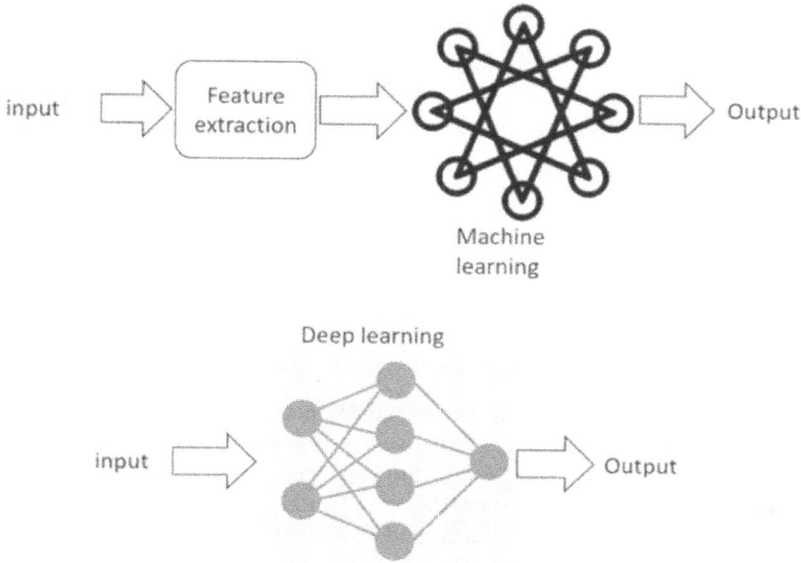

FIGURE 1.4 Difference between ML and DL.

and structures from large data sets. Generally, DL does not use prior data processing. It automatically extracts features from available data. Figure 1.4 represents the main difference between ML and DL.

In ML techniques, a feature extraction step will be needed before applying a model. This step is very complex and needs the intervention of an expert and may be manual. However, DL does not necessarily need features. Hence, there is no need to an expert for manually define any features in the model [19].

1.3.3 NATURAL LANGUAGE PROCESSING

Essentially, AI is about imitating the capability of a human mind. Language is the most fundamental ability of a human mind. It allows them to communicate and understand others' needs. Hence, many researchers have been recently working to enhance natural language processing or natural language processing (NLP) [20]. The goal of NLP is to enable computers to understand and process human language. It is a combination of computer science, AI, and linguistic theories.

Recently, NLP has become a crucial part of AI. It made our life simpler by the existence of voice interfaces, chatbots, and many others. In order to represent what happens within an NLP system, NLP is divided into levels. Figure 1.5 shows the linguistic levels of NLP [21, 22].

Phonology represents the sounds. It is based on three rules: phonetic rules, phonemic rules, and prosodic rules. Sounds within words are represented by phonetic rules. However, the variations of pronunciation during speaking are managed by

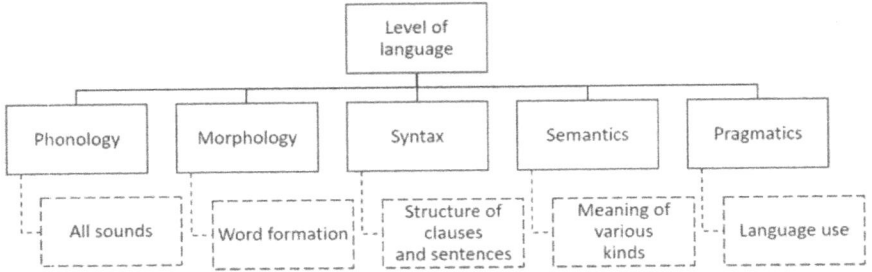

FIGURE 1.5   NLP levels.

phonemic rules. The variation in stress and intonation across a sentence is represented by phonemic rules [23].

Morphology is usually used for identifying the part of the sentence in which words interact together. It represents the word formation [22]. In addition, the set of relations between words' surface forms and lexical forms are also represented in this level.

The structural relationships between words of a sentence are studied in the syntax level [23]. Syntax involves the use of Afan Oromo grammar rules. It consists of analyzing the words in a sentence in order to depict its grammatical structure [22].

Semantic analysis aims to understand the meaning of natural language [24]. Understanding natural language appears an effortless process to humans. Nevertheless, interpreting natural language is an extremely complex task for machines due to the extensive difficulty and subjectivity engaged in human language. Semantic analysis of natural language represents the meaning of a given text while considering the grammar roles and the logical structuring of sentences and grammar [22]. It uses the several approaches, such as first-order predicate logic, in order to represent the meaning [21].

Pragmatics consists of analyzing the real meaning of an expression in a human language by defining and reviewing it. This can be done by identifying vagueness encountered by the system and resolving it.

In the next section, we will describe some relevant AI applications in the financial field as well as in other fields (e.g., healthcare).

1.4   AI APPLICATIONS

The broad areas in which AI is currently operating and contributing include but are not limited to finance, medicine, education, robotics, information management, biology, space, NLP, and many other critical areas supporting people's various activities in their daily life.

Following in this section, some applications of AI in different areas will be elaborated on, mainly in finance and a few other major areas in which AI is considered a major game-changer.

1.4.1   AI Applications in Finance

AI has been the subject of numerous research projects in finance for many decades. It all started with classical AI-empowered solutions for financial and economic problems such as traditional financial markets, insurance, trading, and many others then evolved to what is known nowadays as fintech or financial technology.

Fintech is a term that refers to the application of technologies in general in their different forms to provide solutions for financial problems. Technologies could be hardware like the automatic teller machines (ATMs) or point-of-sale machines (POS) that were used many years ago to provide financial services, and more recently smartphones and cloud-based servers. Another type of technology that is driving fintech is obviously the software solutions that stood out since the mid-years of the 20th century but have become more and more essential and materialized when AI came to offer another level and spectrum of benefits. Lately, technological advancements like data sciences were introduced in fintech to have even more intelligent services with greater benefits that led to the appearance of new generations of fintech called AIDS and smart fintech.

Today, there are numerous applications of AI in finance, but the most important ones can be grouped into two main categories: i) market surveillance, regulation, and risk analytics and ii) trading, portfolio management, and algorithmic trading.

One of the most important areas in which AI meets finance is for the development of software solutions that allow firms to monitor market movements and detect any occurrence of abnormal activities, which significantly increases their ability to take immediate actions and respond to an ever-changing market in terms of business strategies and directions. This type of solution could not exist without the secret recipe of applying AI techniques on a large amount of data to provide major stock exchanges, brokerage firms, and compliance enforcement bodies with the required tools to automate tasks they always do in a more responsive and efficient way. Many market participants in this type of business can benefit from the existence of these solutions, including investors, regulators, and designers, to perform different types of manipulations like insider trading, financial reviews, corporate auditing, and accounting and for present, previous, or future operations. In this category many important tools could be listed like NASDAQ trade surveillance, SteelEye trade surveillance, and MCO trade surveillance, among many other tools.

In the context of risk management and fraud detection, the social network analysis (SNA) is widely used in banking for fraud detection [25, 26]. This type of fraud involves a group of users working in cooperation. In general, SNA is a "data mining technique that reveals the structure and content of a body of information by representing it as a set of interconnected, linked objects or entities". In the finance domain, the social network is a network of entities where entities can be credit cards, merchants, fraudsters, companies, etc.

Another important category of applications in which AI is frequently used to contribute to finance enhanced solutions is trading, and algorithmic trading as explained in [27]. Under this category, software solutions are created to open the door for automated trading that uses computerized algorithms known as algorithmic trading, where AI is used to identify positive trading signals and to build appropriate trading

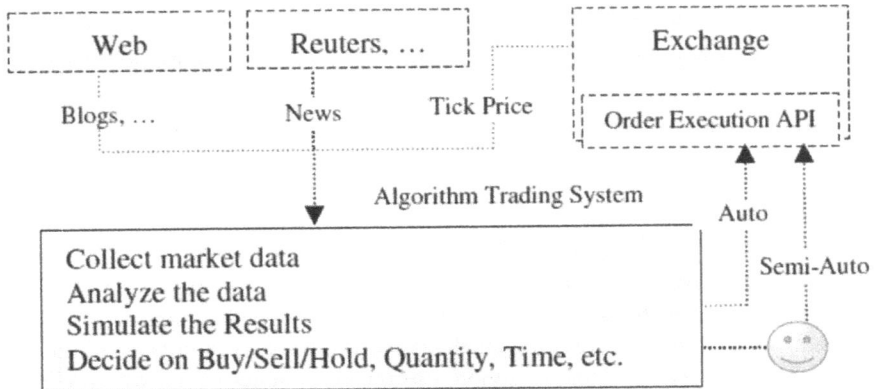

FIGURE 1.6   Data flows for an algorithm trading system [27].

strategies to finally be able to either notify users or recommend trading transactions for them, or even automatically execute trading orders to markets with higher winning chances and profitability. James Simons and his company Renaissance Technologies LLC [28] is among many noticeably successful examples under this category. This firm has created mathematical models to analyze data and execute trades based on the collection of prices and creation of their movement patterns. It currently has 23+ billion dollars of assets under their management.

In the context of trading, the work in [27] consists of using ML algorithms with big data to automate trading actions without the need for human intervention. The automated process enables computer programs to run financial trades faster than a human trader. The algorithmic trading can offer trades at the best possible prices in a timely manner and can also minimize human errors thanks to mathematical models. Still some challenges to applying algorithm trading with big data will be detailed at the end of our work.

Figure 1.6 explains how the data flow works for an algorithm trading system. The procedure is as follows:

The first stage consists in collecting the data related to prices from the exchange, news data from companies like Reuters and Bloomberg.

The second stage consists of analyzing this data where the algorithm trading executes complex analysis to try making profits.

The third stage consists of applying some simulations to check the outcomes. For example, if we decided to buy a particular stock, is this a good option?

The last stage is the decision to buy/sell/hold, including some details (e.g., quantity, the time to trade) by sending trading signals. The signals are directed to the exchanges, and the trading orders are performed without any human intervention.

In some scenarios, the investor can initiate the trading action manually or simply ignore the signals. This is a semi-automatic option that can be provided by the system.

1.4.1.1 AI Applications in Other Areas

AI is highly commonly applied, as stated earlier in this section, in many areas allowing machines to perform tasks that used to be only performed by humans, as they require human intelligence or the ability to perceive information and apply reasoning to take decisions. These areas are not limited to the domain of finance that was introduced in the previous section, as AI is also frequently applied nowadays in many other areas like healthcare, education, NLP, etc. Thus, it is believed that AI made up a core part of what is declared to be the fourth industrial revolution in human history as popularized by Klaus Schwab, the World Economic Forum founder and executive chairman, in 2015 [29] and that it will massively affect our personal lives in every area to which it will be contributing.

Healthcare is one of the most important areas in which AI is contributing, and it was addressed in many projects and research papers like [30]. The introduction of AI in this field allowed the medical industry to produce many control devices that could be used, particularly in intensive care units to monitor patients, perceive their data, and learn when their biological status is under baselines to fire alarms accordingly.

AI also opened the door of using machines in the interpretation of medical images obtained by different scanning devices. Many input images could be given to a machine with tags indicating specific problematic cases to use AI to determine problem patterns and apply these patterns to brand-new images and automatically detect problems.

Another area in which AI could be increasingly highly involved is that related to education, as according to the market research engine's report [31] published in 2021, the global AI usage in education will be increased by 45% to reach 5.80 billion dollars by 2025.

One significant contribution of AI in education could be by allowing educational software like programs and games to provide personalized educational content for individuals through what is called adaptive learning. This type of software can detect the deficiencies of every student and provide the required material to cope with these deficiencies which results in a more comfortable and smoother learning experience.

Also, AI can help ensure universal access to knowledge, including people with hearing or visual impairments or people who need access to material in languages they do not master, thus making classrooms globally available for everyone. PowerPoint plugins like Presentation Translator, for example, falls under this category of software that is AI-empowered. Here AI opens up new learning opportunities for students to benefit from material normally unavailable for them in their schools in their languages or due to their physical or mental inabilities.

In addition, many of the contributions of AI in different fields and areas are based on the ability that AI is providing for software solutions to process natural languages, as it always played a pivotal role in the construction of such systems that can recognize, process, and emulate human languages. Many systems exist nowadays that recognize textual and/or auditory patterns and use this ability for translations or machine learning purposes. One concrete example of this could be found in many instant messaging applications that allow a user to send text or audio messages in their language and allow the recipient user to receive these messages in their preferred language.

1.5 CONCLUSION

Over the past 60 years, the field of AI has made considerable advancements but has also been met with periods of hype. However, more recently researchers have been able to create many technical breakthroughs in rapid succession that enabled machines to outperform humans in fields that required intelligence. As a result, many real-world applications, as mentioned earlier in this section, have used AI to provide humanity with much-needed benefits for businesses and economies and contributed to productivity growth and innovation.

Despite this considerable progress, many challenges are persisting and preventing us from applying AI in many other areas or improving current applications to cover other aspects or have more efficient and reliable results. Therefore, more effort needs to be put into what is usually referred to as "artificial general intelligence" that allows solving more complex problems that tackle things like any human being trying to solve such problems; however, this level in AI is considered still unreachable by most researchers.

A briefing note published in 2018 by McKinsey Global Institute [32] discusses a few challenges that are hampering the evolution of AI and its expansion to other areas. One of these challenges is the effort required to manually label data when using one of the most used techniques in AI, which is the supervised approach in ML. For example, in the context of an application that diagnoses patients with a specific medical problem using lab tests as input, it is required to have huge data sets and label every element in these data sets to determine if the problem is occurring and use these data sets to train the application to be able to determine for a new element if it has the medical problem or not. Note that obtaining this training data is considered another challenge, as in many areas this data could be simply unavailable, and therefore this type of technique would not be applicable.

Another challenge that poses problems sometimes is the inability to explain the output using some technique like deep learning, as it is based on having a complex black box that performs some processing and outputs a decision or prediction as a result. In some applications, the result needs to be well explained in order to ensure the application's transparency, credibility, and trust among people using this output, as in the case of applications for judicial investigations, where investigators are required to provide proofs and explanations to get approval for further actions.

Building generalized learning techniques, where applications can benefit and use their experiences in other similar circumstances as humans are able to do, is another challenge that needs to be overcome. Transfer learning is an AI technique that is currently under study and that could provide a response to this challenge and allow AI models to apply their previous learning to perform other activities in similar but different circumstances.

All of these challenges are definitely obstructing the expansion of AI to give a hand in solving other problems in several areas; however, they also represent opportunities for this field to grow if they are handled appropriately to develop new AI techniques or improve the existing ones to overcome their limitations.

REFERENCES

[1] I. Ananth, "Artificial intelligence," 2018. [Online]. Available: https://witanworld.com/article/2021/04/17/artificial-intelligence/.

[2] A. Karthikeyan and U.D. Priyakumar, "Artificial intelligence: Machine learning for chemical sciences," *Journal of Chemical Sciences*, 134, 2022, pp. 1–20.

[3] J. McCarthy, "Artificial intelligence tutorial—It's your time to to innovate the future," 2019. [Online]. Available: https://data-flair.training/blogs/artificial-intelligence-ai-tutorial/.

[4] V. Jokanović, *Artificial intelligence*, Taylor and Francis, Boca Raton, London and New York, 2022.

[5] E. Kambur, "Emotional intelligence or artificial intelligence?: Emotional artificial intelligence," *Florya Chronicles of Political Economy*, 7(2), 2021, pp. 147–168.

[6] A. Dhamanda, M. I. M.H. Ahmad, M. S. Arshad, M. Zubair, M. Javed, and P. Bazyar, "Artificial intelligence applications," in *Artificial intelligence applications*, Iksad Publications, Ankara, Turkey, 2021.

[7] D. Bholat and D. Susskind, "The assessment: Artificial intelligence and financial services," *Oxford Review of Economic Policy*, 37(3), 2021, pp. 417–434.

[8] H. Arslanian and F. Fischer, "Applications of artificial intelligence in financial services," *The Future of Finance*, 2019, pp. 179–197.

[9] Priyadharshini, "Machine learning: What it is and why it matters," from Simpli Learn, 2017. [Online]. Available: www.simplilearn.com/what-is-machine-learning-and-why-it-matters-article.

[10] P. Sodhi, N. Awasthi, and V. Sharma, "Introduction to machine learning and its basic application in python," in *Proceedings of 10th international conference on digital strategies for organizational success*, 2019.

[11] R. Choi, A. Coyner, J. Kalpathy-Cramer, M. Chiang, and J. Campbell, "Introduction to machine learning, neural networks, and deep learning," *Translational Vision Science & Technology*, 9(2), 2020.

[12] T. Hastie, R. Tibshirani, and J. Friedman, *The elements of statistical learning: Data mining, inference, and prediction* (vol. 2, pp. 1–758), Springer, New York, 2009.

[13] H. Lu, Y. Li, M. Chen et al., "Brain intelligence: Go beyond artificial intelligence," *Mobile Networks and Applications*, 23(2), 2018, pp. 368–375.

[14] EU High-Level Expert Group on Artificial Intelligence, "Ethics guidelines for trustworthy AI [text]," FUTURIUM–European Commission, 2019. Available: https://ec.europa.eu/futurium/en/ai-allianceconsultation/guidelines.

[15] L. Tucci, "Ultimate guide to artificial intelligence to enterprise," 2020. [Online]. Available: https://searchenterpriseai.techtarget.

[16] N. Joshi, "Types of artificial intelligence," 2020. [Online]. Available: www.forbes.com/sites/cognitiveworld/2019/06/19/7-types-of-artificialintelligence/#3e68129b233e.

[17] OECD, *Scoping the OECD AI principles: Deliberations of the expert group on artificial intelligence at the OECD (AIGO)*, OECD Publishing, Paris, France, 2019.

[18] C. Angermueller, T. Pärnamaa, L. Parts, and O. Stegle, "Deep learning for computational biology," *Molecular Systems Biology*, 12(7), 2016.

[19] P. Ongsulee, "Artificial intelligence, machine learning and deep learning," in *2017 15th international conference on ICT and knowledge engineering (ICT&KE)*. IEEE, 2017.

[20] J. Hirschberg and C.D. Manning, "Advances in natural language processing," *Science*, 349(6245), 2015, pp. 261–266.

[21] A. Reshamwala, D. Mishra, and P. Pawar, "Review on natural language processing," *IRACST Engineering Science and Technology: An International Journal (ESTIJ)*, 3(1), 2013, pp. 113–116.

[22] A. Abeshu, "Analysis of rule based approach for Afan Oromo automatic morphological synthesizer," *Science, Technology and Arts Research Journal*, 2(4), 2013, pp. 94–97.

[23] B. Abera and S. Dechasa, "A review of natural language processing techniques: Application to Afan Oromo," *International Journal of Computer Applications Technology and Research*, 10, 2021, pp. 51–54.

[24] T. Nasukawa and J. Yi, "Sentiment analysis: Capturing favorability using natural language processing," in *Proceedings of the 2nd international conference on Knowledge capture*, 2003.

[25] C. Kirchner and J. Gade, *Implementing social network analysis for fraud prevention*, CGI Gr, Mumbai, India, 2011.

[26] V. Ravi and S. Kamaruddin, "Big data analytics enabled smart financial services: Opportunities and challenges.," in *International conference on big data analytics* (pp. 15–39), Springer, Cham, December, 2017.

[27] X. Qin, "Making use of the big data: Next generation of algorithm trading," in *International conference on artificial intelligence and computational intelligence* (pp. 34–41), Springer, Berlin, Heidelberg, October 2012.

[28] "Renaissance Technologies. In Wikipedia," 13 May 2022. [Online].

[29] "New forum center to advance global cooperation on fourth industrial revolution," *World Economic Forum*, [Online]. Available: www.weforum.org/press/2016/10/new-forum-center-to-advance-global-cooperation-on-fourth-industrial-revolution/. [Accessed 16 May 2022].

[30] V. Vinolyn, I. R. Thomas, and L. A. Beena, "Framework for approaching blockchain in healthcare using machine learning," *In Blockchain and machine learning for E-healthcare systems*, n.d.

[31] "Artificial intelligence in education market size, share, analysis report," *Market Research Engine*, 15, 2021. [Online]. Available: www.marketresearchengine.com/artificial-intelligence-in-education-market.

[32] M. G. Institute, *The promise and challenge of the age of artificial intelligence*, Briefing Note, Washington, DC, 2018.

2 AI and Human Cognizance

P. Sanjna, D. Sujith, K. Vinodh, and B. Vasavi

2.1 HISTORY OF AI

Artificial intelligence (AI) was first introduced in the year 1956 by John McCarthy in a conference. The goal behind AI is machines operating like humans and understanding if the machines have the ability to think and learn by themselves (Christof Koch, 2019). Alan Turing who was a mathematician, has put his hypothesis and questions into an action and analyzed if the machines can think, which was called as Turing test and enabled the machine to think like humans.

2.2 ARTIFICIAL INTELLIGENCE AND HUMAN CONSCIOUSNESS

Recently, understanding and AI, as well as mindfulness applied to AI, have received a lot of attention. According to one viewpoint, there is a chance of obtaining, creating, and implementing these models, and then, when applied to AI, it is possible to learn human awareness. According to Holland, there are two types of care: delicate and substantial (Wladawsky-Berger, 2018). Weak phony mindfulness is in charge of the design and advancement of devices that assist in the therapy of discernment, whilst strong phony mindfulness is in charge of the design and advancement of perceptive robots. Re-created knowledge is now being employed in a variety of fields, including AI, academic cerebrum inquiry, neurology, and others. Human insight has also begun to make progress in the same way as imaginative effort has. AI may be able to take leadership of tasks and vocations that need some human effort. It also supports a rapid autonomous data course and requests, as well as exploring and performing data inspection, all of which contribute to making quick decisions. It's a good idea to think about AI as human-made information in which humans have combined all of the discernment and reflexes that our brains have, with the goal of making it limit and feel like a human. We call insight what we learn from our experiences and how we respond to situations and events. Investing in AI should be done in the same way that visualizing frozen yogurt produces a delectable flavor and a bright creative mind. Mindfulness will be attempted as a possibility by human-created thinking frameworks. Human-created information is related to mind science, where it has turned into a serious development in the field's review, thanks to this

DOI: 10.1201/9781003328414-2

process for thinking derived from the interaction of neurons in the human frontal cortex. To give just one example, computer-based comprehension has enabled the destruction and recovery of word-related plans. A handful of people have also introduced the "machine risk thought." Before AI can be used in general, it must first be understood. At its core, modernized believing is an extension of human data, and its progress is reliant on computational advancements. For an unusually lengthy period, copied data has made the essential strides not to coordinate attention in specialists. Care clearly refers to several aspects of human understanding that are necessary for our finest intellectual abilities: opportunity, strength, thrilling experience, learning, and thought, to name a few.

2.3   TYPES OF AI

The first two types of AI, reactive AI and limited memory AI, shown in Figure 2.1 are simple and would help better life for all humans, whereas the other two, theory of mind AI and self-aware AI, would come at a greater loss and risk for human civilization as they can understand humans and make decisions on their own using the intelligence they have.

2.4   CAN AI DEVELOP HUMAN-LIKE CONSCIOUSNESS?

AI is currently one of the most problematic concerns, with little consensus on the differences and similarities between human and machine data. Human-driven and destinations, such as the mission for human-like data as the best quality level for

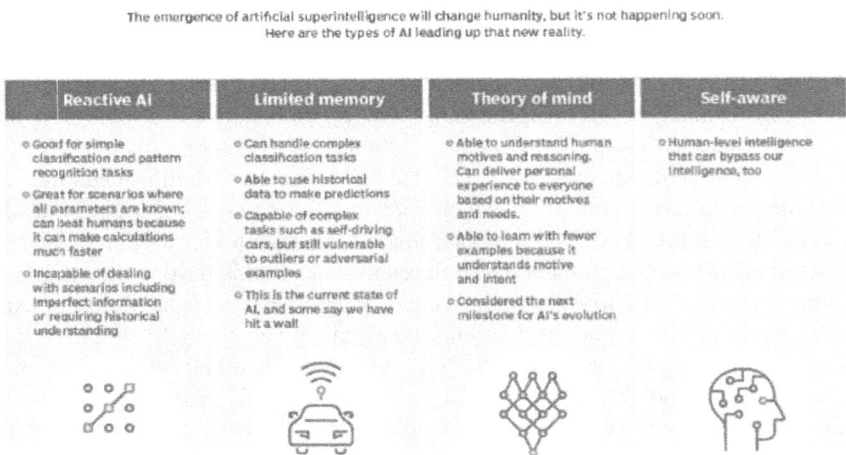

The emergence of artificial superintelligence will change humanity, but it's not happening soon.
Here are the types of AI leading up that new reality.

Reactive AI	Limited memory	Theory of mind	Self-aware
o Good for simple classification and pattern recognition tasks o Great for scenarios where all parameters are known; can beat humans because it can make calculations much faster o Incapable of dealing with scenarios including imperfect information or requiring historical understanding	o Can handle complex classification tasks o Able to use historical data to make predictions o Capable of complex tasks such as self-driving cars, but still vulnerable to outliers or adversarial examples o This is the current state of AI, and some say we have hit a wall	o Able to understand human motives and reasoning. Can deliver personal experience to everyone based on their motives and needs. o Able to learn with fewer examples because it understands motive and intent o Considered the next milestone for AI's evolution	o Human-level intelligence that can bypass our intelligence, too

FIGURE 2.1   Types of AI.

AI, are unaffected by various essential inconveniences, such as perseverance, reasonableness, and morals, while the importance of cooperation between the mind, the body, and the general climate has been emphasized (Wang, 2020; Dong et al., 2020).

Human-created consciousness (AI) and an ongoing leap forwards in data innovation might simplify it for individuals and robots to work all the more adequately. Subsequently, much work has gone into creating human-aware AI, which attempts to assemble AI that adjusts to the intellectual capacities and cutoff points of human colleagues as a "colleague." To underscore a serious level of coordinated effort, similarity, and correspondence in "mixture groups," similitudes, for example, "mate," "accomplice," "change self-image," and "mutual perspective," are utilized (Barrat, 2013). When going about as "human colleagues," AI accomplices should have the option to screen, investigate, and respond to a wide assortment of modern human social elements like consideration, inspiration, feeling, inventiveness, arranging, and argumentation. Accordingly, these "computer-based intelligence accomplices" or "colleagues" should have human-like (or humanoid) intellectual capacities that consider complementary understanding and joint effort (for example, "human awareness"). No matter how modern and independent AI specialists advance in certain spaces, they will doubtlessly stay oblivious to robots or reasoning devices that help individuals in specific and complicated tasks. They have an essentially unique working framework (computerized versus organic) and intellectual characteristics and powers than natural creatures like people since they are advanced machines.

1. Self-driving autos are an AI-helped innovation that utilizes a customary neural organization (Tesla, cruise).
2. From robot-aided operations to the protection of private documents, AI plays a unique role in the health-care business.
3. Today's robots lack natural intelligence, yet they are capable of problem solving and thinking.
4. The Roomba 980 model employs AI to measure room sizes, recognize obstructions, and recall the best cleaning routines. The Roomba can also estimate how much vacuuming is required based on the size of a room and cleans floors without the intervention of humans.
5. AI is employed in Google Search, Siri, and Google Assistant, among other search engines and assistants.
6. AI-based misrepresentation recognition and purchaser results can help banks, VISA, and loan specialists diminish extortion and oversee hazard (fraud divisions at banks and VISA, Ravelin, and Stripe).
7. Massachusetts General Hospital, one of the world's most prepared foundations, has paired up with handling goliath NVIDIA to organize AI-controlled items for the illness area, examination, therapy, and chiefs.
8. AI is heavily integrated into Facebook's platform, whether its Messenger chatbots, algorithmic newsfeeds, photo tagging recommendations, or ad targeting.

9. Slack's AI utilizes an information structure called the "work diagram" to gather information on how each organization and its laborers utilize the application and speak with each other.

2.5   COGNITIVE ABILITIES OF AI

Intellectual limit is an extensive mental ability that incorporates thinking, critical thinking, arranging, conceptual reasoning, complex idea handling, and learning through experience. The three-technique model is the most exhaustive scientific classification of intellectual capacities. The main layer incorporates explicit and confined capacities, the second incorporates bunch variables and wide capacities, and the third incorporates the general mind. Machines can gather data, store the data in memory, and recover the data when an issue should be tended to. Scholarly gifts are human limits that are being dealt with in AI for it to execute jumbled tasks quickly and capably. For example, AI-based vehicles can drive themselves by looking over course maps, roads, streets, and pathways and keeping a distance, similarly to doing an internal examination when there is a turn. These capacities exhibit the robots' scholarly capacities. When it comes to memory and reviewing, AI regularly utilizes PC vision to obtain tactile information and afterward store it. For instance, on the off chance that I have a pal called Avantika, AI will decipher data from my faculties and express feelings as a companion whenever it sees Avantika.

2.6   ETHICS OF AI

Human-made brainpower morals, as well as AI morals, are a bunch of convictions, ideas, and strategies that manage moral conduct in the creation and arrangement of the AI situation utilizing generally acknowledged good and bad standards.

Robot ethics, commonly referred to as robo ethics or machine ethics, is the study of how to build ethical robots and what rules should be followed to guarantee that robots behave ethically. Robo ethics is concerned with issues such as whether robots will be able to address a long-term risk to persons and if the use of explicit robots, such as killer robots in wars, will be detrimental to humans. Roboticists must ensure that autonomous structures may conduct ethically acceptable behavior when robots, AI structures, and other free systems, such as self-driving vehicles, work with humans (Cangelosi & Schlesinger, 2015; Cangelosi, 2010).

Computerized reasoning (AI) and mechanical technology are adjusting and changing our human headway overall. In the public area, applying AI moral norms to the preparation and execution of algorithmic or savvy systems and AI drives is basic. Simulated intelligence will be made and utilized in an ethical, safe, and reliable way in a similar way as human-made mindfulness ethics. Computer-based intelligence morals and wellbeing should be an essential need in the plan and arrangement of

AI frameworks. The motivation behind AI ethics is to forestall human and cultural issues brought about by AI framework abuse, misuse, terrible plans, or unexpected bothersome outcomes.

A self-driving vehicle sees its current circumstance and drives all alone, with practically no human information. A gigantic amount of information should be caught consistently by a bunch of sensors introduced all through the vehicle for the vehicle to perform productively and handle its driving climate. These are then handled by the vehicle's independent driving PC framework. The autonomous automobile must also undergo extensive training to comprehend the data it gathers and make the best judgement feasible in any given traffic condition.

Everyone makes moral judgments every day. When a driver slams on the brakes to avoid colliding with a jaywalker, the risk is shifted from the pedestrian to the car's passengers. Consider a self-driving car approaching a grandma and a child with faulty brakes. By veering off somewhat from the street, it is feasible to save oneself. Rather than a human driver, the vehicle's PC will settle on the choice this time. Which would you like: the grandma or the kid? Do you think there is only one right response? This is a normal moral situation that represents the significance of morals in specialized advancement. Therefore, we must build technology and AI in such a way that they are based on ethics and follow principles. AI is the future of the world; it should be created in such a way that it serves the country and its people in the most ethical manner, abiding by all laws and regulations.

2.7 LANGUAGE AND AI

Language and understanding are always linked to each other at all points of time, and these are linked to challenges in robotics as well. Advancements have been made in the field of robotics in deep learning, and speech understanding (Cangelosi, 2010). Studies say that a robot without any pre-existing knowledge of linguistics is able to learn linguistics commands with the help of deep learning and praising. The robots in the future must deal with daily basic complicated tasks which are diverse and highly undetermined (Dutoit, 1997). There are many drawbacks and reasons why language is such a challenge in AI and robotics.

More viable human-robot communication is turning out to be progressively critical as mechanical technology propels us to understand the maximum capacity of robots for society. Despite significant advances in AI and machine learning over the last decade, language acquisition and understanding are still related to a variety of robotics issues. The fields of deep learning, language (including machine translation), speech recognition, picture recognition, image captioning, distributed semantic representation learning, and parsing have all achieved significant progress. While communicating in language should be essential for the arrangement, our current ability to give communicated in language commitment is restricted. Future service robots will have to cope with sophisticated, diversified, dynamic, and extremely unexpected linguistic occurrences daily while engaging with human users.

The main purpose of the research is to spotlight the obstacles that are faced by engineers and that they have overcome in a process to develop robots in a more natural manner, where a robot can process the language naturally like us humans (natural language processing [NLP]). Providing solid and provable presentations on the current knowledge of robots and their performance, their ability to integrate speech and language, and their ability to adapt to new situations and operate mostly on real-time assumptions instead of analysis and providing a certain and simple way of robots able to speak are important issues in this field. Language itself is dynamic, intellectual, and social.

In order to make the robots behave and speak like individuals in practical and fair situations, there is a need to build approaches that actually allow robots to understand the expressions and phrases that are connected to a real-world environment. Creating robots that converse regularly with people in social settings is one example. We need to understand that language is neither a physical entity nor a genuine signal, but rather a distinct framework in which the meaning of signing alters depending on the context.

2.8   WHAT IS LANGUAGE TO US?

Language is the ability to learn and use jumbled correspondence systems, particularly the ability of individuals to do so, and language is a specific example of such a system. Language is what separates humans from other species. Because we think through language, words, as revealed by the savant, are the machinery of perception. Language is the necessary tool for communication (Noda et al., 2015; Cangelosi, 2010).

Language is what distinguishes us as individuals. It is the method through which individuals communicate. Learning a language entail mastering a complex system of vocabulary, design, and punctuation in order to communicate successfully with others.

Language is an essential component of human cooperation. People are the only species who have perfected intellectual verbal communication, despite the fact that other organisms have some form of communication. Language aids us in communicating our feelings, opinions, and thoughts to others. It has the capacity to generate and at the same time has the capacity to destroy the social structure. It may appear indisputable.

1. **Language importance in individual development:**
   People learn to speak at different speeds, and the age at which a kid begins to use language might indicate how well they are growing. This does not, however, only apply to infants. It also applies to tiny children learning a second language at school that differs from the one they speak at home, adults learning a new language, and persons who have lost their language due to an accident and are attempting to reclaim it.

2. **Language's importance in culture and society:**
Language allows us to convey our feelings and thoughts, and it is fascinating to our species because it allows us to transmit the unambiguous convictions and practices of other communities and networks.

3. **Channel of thought:**
Language is a medium of thoughts; it's the way we think and express ourselves. When a person's reasoning cycle is disrupted, their reasoning cycle capacity will be changed. Language is the medium where we communicate our experiences, and it is also a reflection of signals and signs.

4. **Communicate with another:**
Individuals all across the world communicate with one another and exchange ideas as the concept of globalization takes hold. Understanding a common language has aided people in communicating, even if they are from different parts of the world. Language has evolved into a vital means of communication between nations, social groups, various organizations and associations, networks, and friends. Communication with other countries: different nations communicate in different languages. Because English is a universal language, global communities exchange perspectives through it.

6. **Personal communication necessitates the use of language:**
Even though much of human connection is nonverbal (we may express our feelings, sentiments, and ideas through our motions, articulations, tones, and sensations), language is critical for individual correspondence. Whether it's having the ability to communicate with your friends, your partner, or your family, having a shared language is essential for these kinds of interactions.

7. **The importance of language in emotional development:**
Because dialects have such a strong internal impact on us, it's fairly rare for us to say only one phrase to anger or appease someone. Every word, no matter how tiny or huge, has a considerable amount of power (Mutlu et al., 2009). We can see how words may affect us on the inside, but it's also crucial to consider the emotions that can control language. The fact that every language has a word for "adoration" or "outrage" shows that emotions have a strong impact on language.

2.8.1 Is It the Same for All?

Language, like culture, that other most human trait, is notable for its unity in diversity: there are numerous dialects and societies, each unique yet fundamentally similar, in light of the fact that there is one human instinct and that one of its major properties is the ability to allow such variety.

Dialects differ in a variety of ways. They may use a variety of sounds, construct words in a variety of ways, and arrange words to form sentences in a variety of ways, and that's only the beginning! When we say 'language,' we're referring to the act of speaking, writing, or signing. We're not talking about how individuals can communicate with their bodies, faces, or eyes; we're talking about an organized language system.

A 'vernacular' is a group of languages that are not identical in language and are understood by people who speak other dialects of the same language. The accent of the language, the words that are used, and the individuals portraying and organizing their speech are said to be examples of how languages can vary from one another. It might be a combination of geographical and social reasons, and people who use or speak the same language are more likely to reside in a neighborhood.

Etymologists use the term 'lects' in several circumstances. This term refers to the way people speak inside a discourse community that recognizes them in some way, but not because of friendly circumstances or where they live.

We have a similar mental structure and are 'wired' for language, we dwell on the same planet, and we may have (usually) comparable interactions. However, when it comes to language, particularly language development, we all have different approaches to deciphering our interactions and incorporating them into the language creation process. For you, one sound may represent something, whereas for the other person, a different sound may imply the same thing. I may use one gesture to convey a message, whereas another person may use a different signal. As a result, people in different parts of the country invented their own methods of pronouncing "man," "woman," "dog," and so on.

Once a group of people agreed on a sound/motion set, their relatives either changed it or migrated away and went through strong/motion change processes on their own, creating tongues.

The languages for robots is different. We have various programs like Java, Python, C++, and Lisp.

2.9   WHAT IS INTRINSIC MOTIVATION?

Intrinsic motivation is playing a vital role in developmental psychology, as it is the basis for spontaneous exploration and curiosity. It is said to create a brainstorming situation among humans, which is essential for cognitive growth in humans, and it has been combined with the curiosity of robotics in recent times (Nguyen & Oudeyer, 2014). Firstly, it gives you an idea of the tactics that are being utilized in psychology as part of the intrinsic study. Secondly, we need to provide evidence that these tactics are not feasible and inconsistent when applied to reinforcement learning. Thirdly, we need to provide a solution and a formal topology where several computational methodologies provide the basis for operational processing and intrinsic motivation.

2.9.1   How to Teach Intrinsic Meaning to Robots

We need to devise techniques for robots to grasp spoken words robustly in a real-world environment, despite inherent uncertainties, to build robots that can interact organically with people in a real-world context, such as workplaces and homes. We must recognize that language, rather than being a tangible thing or a set of objective signals, is a dynamic symbol system where the meanings vary depending on the context and situations, and that is understood subjectively.

A method for teaching robots about engine capabilities that combine dynamic, naturally persuaded learning with impersonation learning is available. The socially guided intrinsic motivation by demonstration (SGIM-D) algorithmic plan enables robots to efficiently learn high-layered, constant sensorimotor backwards models, and in particular, it learns dispersions of parameterized engine arrangements that settle a corresponding appropriation of parameterized objectives/errands. To improve I, SGIM-D improves its assessment of L1.

For example, in the examination, a robot arm must understand how to adapt and use a fishing line, which shows that SGIM-D effectively correlates with social and natural learning, as well as the benefits of human properties, which helps it to figure out the effective way to deliver a variety of results in a climate, while growing more precise arrangements in large spaces.

SGIM-D employs a social learning strategy. The learner imitates the situations in the shown way in a short span of time, while also recalling the desired outcome or goal. It's recommended to remember that reason before beginning its autonomous exploration. It then generates a new aim while keeping the prior one in mind.

2.9.2 Logic-Based Programming

1. **Programming:**

   It provides solutions to the declarative problems that arise in the programming. It's more like a list of feasible options for completing a task. Unpredictability and distributions in reasoning are present.

2. **Semantics and orders:**

   Various order: thought age by robots:

   The interior cycles of words are needed to empower robots and arrange data. For robots, calculations like latent Dirichlet allocation (LDA) re utilized to classify circumstances and words.

   Word meaning acquisitions:

   It is feasible to isolate data from pictures utilizing multimodal data. The robot that this model is based on accumulated multimodal information from the things in plain view while a human client showed the robot object labels. At last, the robot had the option to perceive inconspicuous items with an exactness of 86% and show phrases with pinpoint precision.

   The studies mentioned in the preceding section, as well as multimodal information obtained by looking, gripping, and shaking items, were utilized to categorize concepts into categories and calculate hierarchical links. The robot's expected hierarchical structure, with hierarchical linkages based on feature similarities, were recorded. When this paradigm is applied to the problem of concept formation localization, the following outcomes are obtained.

   For both the age of different thoughts and the learning of connections between them, AI calculations have been proposed. The proposed approach likewise permits robots to gain syntax from the beginning, since the progress between these ideas is viewed as sentence structure. Individual, item, movement, and limitation thoughts are made utilizing multimodal latent Dirichlet allocations (MLDAs), which order multimodal information obtained from situations where people use objects.

3. **Here are some metaphorical phrases to consider:**
   Metaphor as a cognitive process:
   Metaphors help humans understand, process, and experience things and situations in society. For example, 'I'm overjoyed' is a phrase used when someone is extremely happy. The conceptual metaphor 'emotions are liquid' reveals that we understand 'joy' as an intangible goal domain in terms of a 'liquid.' Metaphors can help with emotional intelligence as well as the cognitive ability to process complicated information.

4. **Observational learning:**
   Learning from demonstration (LfD) and programming by demonstration (PbD):
   These are the terms that are used to explain how exactly the robot will function, where both the language and robotics get advantages from learning when implemented with action. This is frequently called imitation learning, where learning occurs in place of supervision. Re-creation of an instructor's rules and trajectories is said to be imitation learning.
   Reinforcement learning:
   Reinforcement learning helps the robot to analyze and discover the optimal behavior, which it understands through the trial-and-error method. This reinforcement learning helps to create robots with true linguistic skills. For example, if we ask a robot to put the table tennis ball over the net, in this process, the robot processes the data and observation of dynamic variables with a proper velocity, which gives the assumption of a future outlook or outcome.
   Punctuation and activities:
   It is said that the important plan of language content in humans is made of three main categories: 1) sensory framework, which is linked to the expressiveness of feelings such as expressions and signals; 2) being in calculative and purposeful situations or understanding a framework where the subjects are linked to semantics; and 3) a syntactic computational framework, which deals with sensory functions to provide a purposeful framework.

5. **Pragmatics:**
   Joint effort with humans: improving robots that can collaborate with people should be a primary objective of our language and advanced mechanic research. To achieve a common goal, humans and robots must be able to communicate effectively and respect one other's physical and mental states. During the language acquisition process, newborn children should comprehend the speaker-listener or educator-student roles, as well as the reversibility of these jobs. Data is clearly handed on during genuine collaboration through practical demonstrations, but it is also deictic and clearly articulated during discussions.

2.10   SOCIETY AND AI

The evolution and advancements of AI have influenced society in both the long and short terms. AI has had major ramifications in the professional world for those working with modern technologies, and there are now ethical laws in place to help people make good decisions in complex situations. AI technology and its advancements have influenced many aspects of society, providing both benefits and drawbacks to humans. AI applications can be controlled so that no harm is done to security or personal data, or human rights are violated. The use of AI aided in the forecasting of the labor market. AI aids in the creation of new products by increasing productivity. With AI, there is a significant rise in questions about whether it can increase job loss at one point while also creating new jobs at another. AI is used for 'social good,' and it is a fact that technological advancements have enabled AI to analyze and process data. AI's capabilities enable it to solve a wide range of problems in the world, including the creation of a self-sustaining society.

2.10.1 CAN AI CREATE A SOCIETY OF ITS OWN?

AI is becoming increasingly important in society for making decisions and analyzing business trends. By analyzing future trends, it plays a key role in a company's decision-making process (Cangelosi & Schlesinger, 2015). It has applications in the banking, health, and manufacturing industries, among others (Admoni & Scassellati, 2017). AI is constantly striving to improve its effectiveness and efficiency by eliminating all human errors. It is said that AI can create its own society because AI will be used in every industry and in everyone's lives in the future.

AI in the field of medical sciences:

- AI has revolutionized the medical field and has had a significant impact. Different types of machine learning and models have been developed to aid in predicting critical cases and determining whether a patient has cancer, or a tumor based on past symptoms and health records. It also aids in future health predictions for patients, allowing them to take preventative measures sooner and live a healthier lifestyle.
- AI has created a virtual and private assessment as well as advanced technology equipment that is designed specifically for people's monitoring and different cases and their health situation outcomes in a timely manner. It enables doctors to conduct assessments in a much timelier manner, allowing them to begin treatment right away. Once the assessment is completed, AI provides the best medical recommendations on the spot.
- The use of health-care bots is another excellent way for the medical industry to work their way in the field, as it allows them to provide 24/7 assistance and categorize appointments according to the severity of the situation, reducing the amount of effort required to manage appointments. Without AI-based machines, this would not be possible.

AI in air transport industry:

- Airport transportation is the most organized mode of transportation in the world, and it is here that the need for AI has emerged as a viable option. Here, AI machines are involved in providing routes and planning, as well as takeoff and landing times, which are all calculated and displayed for smooth plane operation.

- The navigation of maps and routes, as well as a thorough examination of the entire cockpit panel to ensure that everything is in working order, was carried out with the assistance of AI in the airways. This has produced very effective and promising results, which is why it is recommended that it be used frequently. AI's main goal in the field of air transportation is to ensure that humans have a comfortable and safe journey.

AI in banking and financial institutions:

- In the bank, AI has played an important role in managing financial transactions and handling large and diverse activities. Machine learning models, which are AI, have been used to help banks with day-to-day tasks such as operational or transactional tasks, stock market–related tasks and their management methods, and so on.
- With the help of AI, which is a major aspect in the banking industry, it also assists in tracking and monitoring unusual transactions and reporting them to regulators. Other scenarios include credit card analysis, in which well-known credit card companies track transactions on a geographical level, determine whether they are suspicious based on various parameters, and then work to resolve the issue.
- AI is attempting to provide the highest level of customer satisfaction while also improving their travel experience. With the help of past purchases and tracking techniques, it also personalizes ticket offers.
- Machine learning algorithms, with the help of AI, look for ways to maximize sales revenue in the long run.
- AI can detect climate changes and much more, which will allow us to predict flight delays in advance. Additionally, it aids in crew scheduling as well as traffic detection for landing and takeoff.

AI in the gaming industry:

- Virtual games are the rage these days, and the demand and profit potential are unimaginable. This is one of the areas where AI has made the most progress. The bots are always ready to play, and you don't need to play with anyone else.
- It provides a real-time gaming experience with the help of AI. AI gaming also includes mind games that improve a person's IQ and critical thinking.

AI in the Automotive Industry:

- With the help of human knowledge, manufacturing and robots are developing at the same time. In comparison to the financial industry, the automotive industry employs less AI.
- AI is referred to as the "new adaptors" in this industry.
- AI is being used in supply chain, manufacturing, production, and, most recently, "driver assessment," in which a person can choose whether to drive a car by sitting in the driver's seat.

- Tesla is an example of a successful automated vehicle. One of the amazing inventions of AI is the ability to drive on its own based on the route map, analyzing and detecting and calculating the distance and roads, and driving safely.
- Tesla would be AI. Elon Musk is a brilliant visionary who is the CEO of Tesla, SpaceX, and a slew of other companies that have a global impact. Tesla's autonomous driving has prompted the entire automobile industry to take notice of Tesla, as it is a ground-breaking concept developed by Tesla's team. In the near future, we will likely see a good and controllable AI society of Tesla that is used for autonomous driving, as each car can communicate with another car for safe driving. Tesla's team has devised a revolutionary plan. AI in the coming years can cause good or bad to human civilization, but it is sure that AI will definitely have their own societies and civilizations for the betterment of themselves as well as humans.

AI in Manufacturing industry:

In the manufacturing sector, AI has limitless potential. Everything from maintenance to replacing humans in certain tasks has been automated. AI allows the quality of work to be improved while also increasing output. Microsoft, for example, will transform all information so that workers can perform better.

AI in organizational intelligence:

For businesses, a large amount of data is generated from customers, which takes a long time to process and analyze. Traditional businesses and methods are failing due to technological and speed advancements. AI enables companies to explore data, analyze data, and predict changes faster than humans, allowing them to make quicker and more effective decisions.

AI in urban design:

AI aids in the development and planning of cities. There will be a massive amount of data that needs to be analyzed; AI gathers large amounts of data and aids in the organization and understanding of urban areas as they evolve. AI data can express itself and show how growth has progressed in the past and in the future, utilities required, safety, and so on.

AI in education:

The concept of education must evolve from generation to generation, and this evolution is critical. People in the education industry are always asking where changes are needed and how to make them (Chatterjee & Bhattacharjee, 2020). AI has the potential to create a dynamic, systemized, and effective learning environment for subjects, which could be a game-changer. AI teachers are another example of useful advancements. AI can be a better tutor by showing students visualizations in 3D to help them understand concepts better.

AI in fashion:

With the help of AI, the world can better understand people's buying patterns and changing behaviors, as well as predict future fashion trends, which is a huge step forward.

AI in supply management:

AI will be able to predict humans without judging but in a way with proper risk analysis and find exact decisions even in difficult situations and in a cost-effective manner. AI will be able to create more dependencies and complicated data than humans. As a result, proper and effective decisions can be made.

2.11 SINGULARITY

In terms of technology, singularity means a hypothetical future where technology is growing very rapidly and is out of control and irreversible. These powerful technologies will change rapidly and unpredictably transform our reality. Singularity would be applied to such advancements where it involves computers and programs that are being advanced with the help of artificial intelligence, which is created by humans. These changes would affect and cross the boundary of humanity and computers. Nanotechnology is said to be one of the important technologies that probably will make singularity come into reality. This explosion will have a drastic impact on human civilization. These computer programs and AI turn the machines and robots into super-intelligent and high-cognitive-capacity machines that would be beyond human capability and intelligence.

If the AI would return in a way that would destroy human civilization, singularity has won its pace. When AI forms a society for itself, there is a high chance that it will become stronger, and humans cannot destroy that vast technology. Humans create AI, which can destroy its own kind, and this action is irrevocable. Once AI has its own society, it is protected and cannot be destroyed.

Singularity would occur when computer programs improve to the point that AI surpasses human intelligence, potentially erasing the human-computer divide. One of the main technologies that will make singularity a reality is nanotechnology.

This expansion of intelligence will have a tremendous impact on human society. These computer programs and AI will evolve into super-intelligent machines with cognitive powers far exceeding those of humans.

Challenges that society needs to face when AI has its own society, which would give way to singularity:

- Not all the effects are positive. AI has the potential to cause harm by leaking private information. Several states and cities have banned the government from using facial recognition because it can be manipulated and AI interference with privacy grows.

- AI will undoubtedly cause a shift in the workforce. As AI advances, it will take away human jobs and work more efficiently in their place, resulting in job losses.
- With AI advancements in the automotive and car industries, there is a risk that autonomous vehicles will harm people.
- Another issue is that AI has the potential to cross ethical and legal lines. With AI's goal of benefiting society, it would have a negative impact on society if it strayed too far from the desired goals.
- Maintaining privacy can be difficult. AI is a powerful tool that is backed up by data. Every second, it collects data on everyone, making it difficult for people to maintain their privacy. If businesses and governments make appropriate decisions based on the information they gather, as China does with its social system, social oppression can result.
- The current trend of digitization, combined with AI and the Internet of Things, has made the situation more vulnerable to cyber-attacks.
- For military applications, AI has opened new possibilities, particularly in weapons systems, with special functions for selecting attack targets. These weapons should be used in conjunction with proper command and responsibility, and they should always be associated with humans.
- Robotics must be safe, dependable, and secure. We must ensure that robots are aware of and follow ethical laws and that they do not harm society. However, AI development has created a society in which robots may become powerful enough to turn against society.

2.12 AI AND ITS FUTURE

Going back to 2017, when Facebook, Tesla, and many other major players were competing in the AI industry, a real-time example of AI could create its own society. Facebook created two AI robots named 'Alice' and 'Bob' and had them sit in front of a live audience to test and introduce them to the world (Wladawsky-Berger, 2018). Facebook's plans failed to materialize and go as planned because the robots stunned the entire world, including the Facebook team, by conversing in their own language. This is the first sign that AI can perform tasks that are like those performed by humans; they created their own language and spoke with each other, which in the future, if provided a number of AI systems, there might be a chance of all the Ais talking to each other and forming a society.

As the trend indicates, they have various levels of life. AI is attempting to become alive in the same way that humans do. According to philosophical anthropology, the first step is to comprehend the purpose of the purpose and to become goal oriented. Another factor is that it tries to improve itself to the point where its organism allows it to do so. This is something that trees, birds, and other living creatures do. AI can evolve and allow itself to learn and adapt to new environments, allowing it to create its own species. It's beneficial in one way and detrimental in another.

All of these may be benefits that AI provides to society, as it is claimed that AI can create society for itself. Though it has brought many benefits and advancements to society, it has also brought with it several challenges that may arise in the future. The concept of singularity emerges when AI and advancements in the world reach a point where they are uncontrollable and unstoppable.

REFERENCES

Admoni, H. and Scassellati, B. Social eye gaze in human-robot interaction: A review. *Journal of Human-Robot Interaction*, 6(1), pp. 25–63, 2017.

Barrat, J. *Our final invention: Artificial intelligence and the end of the human era*. Thomas Dunne Books, New York, 2013.

Cangelosi, A. Grounding language in action and perception: From cognitive agents to humanoid robots. *Physics of Life Reviews*, 7(2), pp. 139–151, 2010.

Cangelosi, A. and Schlesinger, M. *Development robotics: From babies to robotics*. Cambridge, Massachusetts, 2015.

Chatterjee, S. and Bhattacharjee, K.K. Adoption of artificial intelligence in higher education: A quantitative analysis using structural equation modelling. *Education and Information Technologies*, 25, pp. 3443–3463, 2020.

Christof Koch. Will machines ever become conscious?, 2019. https://www.scientificamerican.com/article/will-machines-ever-become-conscious/

Dong, Y., Hou, J., Zhang, N. and Zhang, M. Research on how human intelligence, consciousness, and cognitive computing affect the development of artificial intelligence. *Complexity*, 2020, pp. 1–10, 2020.

Dutoit, T. *An introduction to text-to-speech synthesis*, vol. 3. Springer Science & Business Media, Berlin, 1997.

Mutlu, B., Yamaoka, F., Kanda, T., Ishiguro, H. and Hagita, N. Nonverbal leakage in robots: Communication of intentions through seemingly unintentional behavior. In *Proceedings of the 4th ACM/IEEE international conference on Human robot interaction* (pp. 69–76), 2009.

Nguyen, S.M. and Oudeyer, P.Y. Socially guided intrinsic motivation for robot learning of motor skills. *Autonomous Robots*, 36, pp. 273–294, 2014.

Noda, K., Yamaguchi, Y., Nakadai, K., Okuno, H.G. and Ogata, T. Audio-visual speech recognition using deep learning. *Applied Intelligence*, 42, pp. 722–737, 2015.

Wang, P. A constructive explanation of consciousness. *Journal of Artificial Intelligence and Consciousness*, 7(2), pp. 257–275, 2020.

Wladawsky-Berger, I. The impact of artificial intelligence on the world economy. *The Wall Street Journal*, (16), p. 11, 2018.

3 Integration of Artificial Intelligence with IoT for Smarter Systems
A Review

R. Raj Mohan, N.P. Saravanan,
and B. Leander

3.1 INTRODUCTION: BACKGROUND

The Internet of Things (IoT) is defined as an interconnection of things that permits them to connect, interact, and exchange information with the help of microcontrollers and electronic circuits, software, sensors, and actuators. The manipulators, sensors, and interconnection produce immense quantities of information from which one can put forth many developments and improve the expertise of the system by means of the artificial intelligence (AI) process [1]. AI-enabled IoT makes smarter machines that create smarter actions and supports results with minor instruction or without human intrusion. The AI of Things (AioT) is the blend of the AI knowledge and the organization of the IoT. In recent times, the trends of IoT-based systems employ an AI for data manipulation by including it in existing IoT applications. The word cyber-physical system (CPS) is a grouping of disciplines and is defined by NSF as "engineered systems that are made from, and depend upon, the seamless integration of computational algorithms and physical components" [2]. To execute AI into IoT, significant coding practices are required to convert it into a smart system or application. Mostly, AI is integrated using a mechanism within a network application.

But on the other hand, the inclusion of a machine learning mechanism as a cyber IoT without any programming effort allows the IoT technology to enable AI solutions effectively. The merging of AI and IoT can redescribe the means of business, industries, and economic functions [3]. There are many methods for sensing units and taking decisions for the system with AI-based IoT. Li and Kara [4] discussed two methods to plan a system architecture and then control the measures of selection for each portion of the units. This work is authenticated with a real-time analysis of a temperature monitoring system. Chatfield and Reddick [5] anticipated an outline of a smart government performance system implemented with an IoT-based system. Most smart or AI-based applications are usually demonstrated with in-built intelligence and have the capability to transfer information collaboratively or over cyberspace. In order to attain the better automation which is required for smarter IoT applications, sensors are integrated into a junction essentially to be competent, intellectual, consistent, precise, and connected. Moreover, these sensors are expected to be strong and healthy and must ensure the safety of the users interacting during the work process. Here in this review, an application is discussed as a case study for a better understanding of AI-based IoT.

3.2   CYBER-PHYSICAL SYSTEM

Recent technology developments of IoT-based design, applications, and technologies, composed with an industrial CPS, are being supported by an extensive choice of sensors that are supported with AI thoughts. In the present condition, the CPS is evolving from the integration of a well-designed infrastructure, smarter units, embedded computing gadgets, beings, and physical environments linked by a communication interface. Therefore, the CPS is a grouping of numerous disciplines like advanced system learning and information science or analytics, embedded design, firmware or software, wireless sensor networks, cybernetics, mechatronics and robotics, cloud computing, and cognitive science facts. The significant reason for merging IoT with the CPS is to develop an independent system that can manage divergent conditions all over the world, which would eventually care for and manage human beings to lead a healthier life on the globe. The elementary part of IoT and CPS schemes consists of smart things and the networks linked between them as well. When it is assumed that all the connections are made, the data is generated and communicated from one point to another every instant [2].

3.3   AI AND IOT-CPS

Recently, AI expertise includes various progressions by adding different learning concepts like machine learning, deep learning, and natural language processing. Similarly, when AI procedures are included in IoT, it becomes the AioT. Table 3.1 represents the important major terms and definitions used in this review with supporting literature. It points out the major keywords with standard meanings to accomplish a basic understanding of related ideas to the subject and improve the simplicity of statements without unnecessary repetitions.

The following are the advantages of AI when used along with IoT:

a. boosting the operational efficiency,
b. providing better risk management,
c. providing enhanced services, and
d. increasing the scalability of IoT.

TABLE 3.1

Major Terms and Appropriate Acronyms for Smarter System with Relevant Literature

Key Terms	Acronym with Literature
AI	A possible result to cognitive queries that are frequently related to human cleverness, like learning, answering queries, and identifying patterns [6]
IoT	Linked sensors organized by using the cloud, which enables them to communicate with each other in a wireless manner [7]
	An interconnection of things from sensors to smart devices [8]
Machine learning	A quantity of design is created from data given to the machine and targeted to make sense of prior unaware data and to develop a great agreement of the data in terms of responsibilities like identifying images, speech, patterns, or improving strategies [9]
Big data	Handling a huge quantity of data, which is complex and reliant on various causes, categorized by capacity, speed, diversity, and reliability [10]
Deep learning	A technique of permitting computational models, including processing layers, to acquire information depicted by various levels of knowledge [11]
Smart	The use of independence, situation responsiveness, and connectivity to make a gadget or thing smarter [8]

3.4 MAJOR COMPONENTS OF AI-BASED IOT FOR CPS

The major components identified for AI-based IoT and CPS are classified as follows.

Smarter objects: There are two elements that are referred in the real world: one is the physical object (PO) and the other one is a smarter object (SO). A PO is like any object that may not be able to immediately connect with the IoT and then is an essential portion of the designed unit. Such physical objects can have smarter objects attached to them. These smarter objects are the AI components that have the capability to communicate through a network.

Information storage and processing: The smarter objects can manage minor sets of data uninterrupted getting into the system, and this can be kept temporarily in the SO until a given task is completed, and then it can be passed to the global information or data storage. To handle these types of information in real time and utilize them well, the role of big data analytics is vital. Because all this data needs to be stored, the data processing can only be done with a proper algorithm based on previous data or knowledge of the system. This is where a few challenges are rising around the CPS.

Communication networks: Smart systems require a smarter system or interconnected infrastructure facility, and connecting units and gadgets to a telecommunication connection is not a novel concept since this has existed for a long time. For example, the innovative Bluetooth low-energy (BLE) was formed to operate by utilizing a very low power consumption. The gadgets can be connected to smarter devices like tablets or smart mobile phones through BLE and allowed to send or receive a small amount of data only.

Message protocols used for IoT: The IoT environment is generally extremely heterogeneous with a fusion of gadgets in a row with various kinds of message protocols and using various methods of communication models. Tschofenig et al. [12] defined four data or message transfer models: device to device, device to the cloud, device to the gateway, and back-end data sharing. In device-to-device transfer, protocol design problems that address interoperability need to be examined and ground-breaking results established.

3.5   AI-EMPOWERED IOT WITH CPS

When it comes to a discussion about the data or information, there is plenty of that in the IoT united with the CPS arrangement. The information or data is either enormous or slight, and it is always an integral part of the IoT and linked devices. For a data-dependent result, more data should be utilized. Moreover, storing this data for investigations inside smart things or objects may not always be possible. The appropriate way to analyze the information is based on machines. The following characteristics are required to be a good machine learning system when it deals with such a huge volume of data with the system.

a. The capability of data preparation,
b. Both basic and advanced-level learning algorithms,
c. Development of adaptive and automated processes,
d. The ability for system performance in terms of data, storage, and networks, and
e. Real-time process with sensible decision-making.

3.6   IOT AND SMART SENSING SYSTEMS

Sensors play a vital role in everyday life and are considered a major module for IoT-dependent systems, as the sensing units permit the IoT to accumulate data to take intelligent decisions and become a smarter sensing unit. The human-made intellect IoT systems succeed in achieving smart sensing systems when AI is incorporated into them. When IoT gives way to form a network of gadgets by utilizing the Web, the informative, human-created reasoning methods both offer fancy ideas and are ready to do all sorts of smart actions. Here, the data taken from the sensor can be divided with the help of AI association and can settle on smart choices with high quality and overcome challenges with regard to achieving the smart sensing units. Generally, two different stages are performed to get the sensor statistics, depending on the application requirements: near the sensor data origin, say node level, or at the cloud level. Additionally, the node-level sensor data handling system is characterized as follows: i) edge computing, which is processing the data as it is received from the device, and ii) fog computing, which happens after filtering the information to make wise decisions [13].

3.7   BIG DATA AND IOT SENSORS

Due to the exponential growth of IoT, there will be a massive number of things increasing in order to collect, store, and transfer data over the Internet. The quantity of data being formed is also increasing swiftly, resulting in big data. Big data has two parts: one is gathering and storing data through complex equipment and programming, and the second one is analysing to find out the significant information. The study in this description mentions the maximum possible way of examining and employing the increasing quantity of data being formed by IoT sensors with supplementary components. The main purposes of big data in IoT and sensor design are massive data examination and to explore it to find smart designs. In most cases, the IoT system is executed using its own firmware. Moreover, these systems can correspondingly use the cloud for additional investigation of the information, and similarly, whenever the information is attained at the cloud, the product can also measure them.

3.8   THE ROLE AND COLLABORATION OF AI IN IOT

AI or computerized reasoning is the boundary of a thoughtful personal computer or precise and powerful robot to finish work that is often linked by vigilant beings. To understand the extreme power of IoT, all the industries and organizations must realize that they have to combine the IoT with AI innovations for quicker way of achieving possibilities. So, that act will lead to the design of smarter or more brilliant machines. Moreover, this can reinvent intelligent machines almost without human participation. Also, the collaboration of IoT and AI will reach an impressive market in the near future due to the fact that they are always undergoing incredible growth and have an upright nature. The merging of AI and IoT will create an impact with all kinds of challenges to reach a significantly more critical industrialized revolution and a major change in the market. There are wonderful opportunities and space available for these two technological advances in the near future. Though it is thoughtful concepts and innovative practices are quired to be combined with supremacy to create a new paradigm for interconnecting AI with IoT, this is not an easy task in today's society. However, the integration of AI and IoT technologies works well in several industrial sectors, such as automotive, manufacturing, and production, and to a lesser extent in commercial fields. To implement AI, there is a broad way available with simulated intelligence that can be utilized by each industry. A few possible implementations are given here.

- To provide customized medication and related data for effective medical care in the medical industry,
- To serve customer-oriented requirements and procure commodities based on the demand with the help of previous data for the virtual shopping process, and
- The board strives to identify fraudulent trades by accurately scoring credit, implementing efficient processes, and automating data [14].

In the future, this simulated intelligence is going to contribute to an extreme knowledge undertaking, for example, speech reply and language interpretation without human involvement.

3.9 AI-BASED IOT APPLICATION WITH CASE STUDIES

Even though AI implementation with IoT is an extreme portion or it is considered to be a provoking thought, recently it has been quite reasonably beneficial in some genuine applications. In spite of the fact that AI in the IoT for the majority of the design is a new kind of idea, it has quite recently been used in many real-time applications [15, 16]. The case study is a useful way to understand any given method with the collected information regarding the particular model. As it is recognized, long-range (LoRa) is a wireless modulation method resulting from chirp spread spectrum (CSS) expertise. LoRa modulated transmission is strong against disturbances and can be received across great distances. IoT-based smart home monitoring system using LoRa with AI is explored using this study to understand the deployment process in a better way [17].

3.9.1 AI-ENABLED HOME MONITORING SMART SYSTEM USING LORA

Recently, a few design applications explore IoT-based smart homes, which garner significant attention nowadays because of the advancements in the field of communication technology. There are a few limitations applicable to the existing smart home system due to the fact that the technologies like short-distance communication and cellular connectivity have few boundaries like handling the range of signals and complexity of the implementation as well as the total cost for execution.

This design based on a LoRa-based smart home arrangement, as shown in Figure 3.1. It is suggested for distant monitoring of various IoT sensing units and

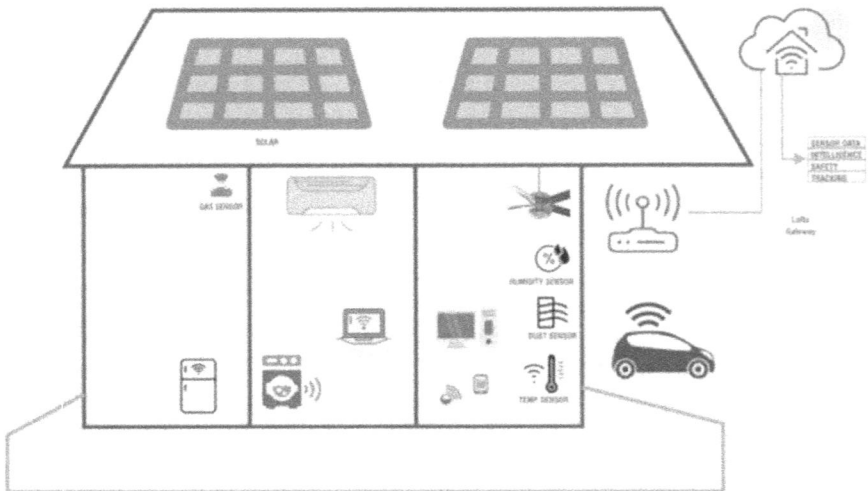

FIGURE 3.1   Smart home using LoRa.

devices by means of an AI process [18]. The system is backed up with Cloudera, which is an open-source platform with programmable computer organization with the facility of information storage.

LoRa is a physical layer–registered low-power and wide area network (LP-WAN) technology that modulates the signal in the sub-GHz industrial, scientific, and medical (ISM) frequency band. It can be implemented for marketable usage as well as personal home-based interconnections and is inexpensive compared to other technologies [19]. In this work, LoRa is integrated with AI to obtain different types of IoT servers and clouds that can be used for smart work in a home environment. In the present scenario of IoT, it is forecast that the quantity of industrial IoT gadgets used for sensing, tracking, and controlling using LP-WAN will climb close to 0.5 billion by 2025, and similarly, it is estimated that the number of connected gadgets to cyberspace will cross 75 billion by 2025. These enormous connected IoT devices are expected to implement the vast demand for channel capacity of the growing number of technologies [20]. The LoRa, Sigfox, narrow-band IoT (NB-IoT), LTE-machine type communication (LTE-M), extended coverage global system for mobile communication (EC-GSM), random phase multiple access (RPMA), my-things (MIOTY), and DASH7 are presently trading in the markets owing to the LP-WAN technologies. Whereas LoRa, Sigfox, random phase multiple access, and MIOTY operate in the unlicensed spectrum, the others are built on the cellular licensed spectrum [21]. The Sense HAT is supplementary for the Raspberry Pi development unit. The board has approved the use of its LED matrix for all physical measurements and output. Some of the challenges that create impacts on data rate, payload, and sharing of ISM bands are discussed next.

Major challenges:

- The rate of data that can be provided by the LP-WAN network,
- More devices will create a greater level of interference with each other,
- Even simple ALOHA channel access by the LP-end gadgets can decay the performance of the system, and
- Interoperability is a major challenge, since there are no broad schemes or tools in the open-source environment to install the LP-WAN.

3.10 CONCLUSIONS

Here with this review, it is realized that the IoT is going to offer a novel scheme of technology and hardware with a chance to collect continuous information about all of the physical actions of a process. Besides modern reasoning, AI helps the system to be smarter when it is incorporated into it in an apt way based on the requirements of the system. The smart system might include one or more other technologies like big data and the cloud to enhance the process. In the current situation, many manufacturing industries and business enterprises are capturing the benefit of AI and IoT knowledge to get rid of a logistics network and supply interruptions or lessen influence over the process. It is understood that AI-driven IoT units can track and gather various forms of human interactions and investigations and review the information before moving from one gadget to another. The importance and the contributions of AI implementation over the IoT-based smarter systems or CPS need to be realized by all the business organizations, manufacturing, and production industries since it is going to play a greater role in the life of everyone in the near future. This review gives the benefits of LoRa technology for smarter home automation using LP-WAN. The case study is discussed from different perspectives of the review to tell apart AI-based systems for hardware or processor-controlled applications and data handling applications. The rise of IoT is seen all over the world for creating a relaxed and unchallenged living environment with the usage of smarter devices such as smart sensing units, controllers, or actuators and plenty more devices. Finally, IoT has reformed every facet of the lifecycle and attracted the eye of major researchers into a brand-new type of lifestyle and also the standard of existing to remain with implemented smart systems.

REFERENCES

[1] Kankanhalli, Atreyi, Yannis Charalabidis, and Sehl Mellouli. 2019. "IoT and AI for Smart Government: A Research Agenda." *Government Information Quarterly* 36 (2): 304–9. https://doi.org/10.1016/j.giq.2019.02.003.

[2] Ghosh, Ashish, Debasrita Chakraborty, and Anwesha Law. 2018. "Artificial Intelligence in Internet of Things." *CAAI Transactions on Intelligence Technology* 3 (4): 208–18. https://doi.org/10.1049/trit.2018.1008.

[3] Xiaoping, Yang, Dongmei Cao, Jing Chen, Zuoping Xiao, and Ahmad Daowd. 2020. "AI and IoT-Based Collaborative Business Ecosystem: A Case in Chinese Fish Farming Industry." *International Journal of Technology Management* 82 (2): 151. https://doi.org/10.1504/ijtm.2020.107856.

[4] Li, Wen, and Sami Kara. 2017. "Methodology for Monitoring Manufacturing Environment by Using Wireless Sensor Networks (WSN) and the Internet of Things (IoT)." *Procedia CIRP* 61: 323–28. https://doi.org/10.1016/j.procir.2016.11.182.

[5] Chatfield, Akemi Takeoka, and Christopher G. Reddick. 2019. "A Framework for Internet of Things-Enabled Smart Government: A Case of IoT Cybersecurity Policies and Use Cases in U.S. Federal Government." *Government Information Quarterly* 36 (2): 346–57. https://doi.org/10.1016/j.giq.2018.09.007.

[6] Ma, Wenting, Olusola O. Adesope, John C. Nesbit, and Qing Liu. 2014. "Intelligent Tutoring Systems and Learning Outcomes: A Meta-Analysis." *Journal of Educational Psychology* 106 (4): 901–18. https://doi.org/10.1037/a0037123.

[7] Khaled, Mandour, and Salma Raja. 2019. "Smart Homes: Perceived Benefits and Risks by Swedish Consumers." Bachelor's Thesis, Malmo University, Faculty of Technology and Society, Sweden.

[8] Silverio-Fernández, Manuel, Suresh Renukappa, and Subashini Suresh. 2018. "What Is a Smart Device?—a Conceptualisation within the Paradigm of the Internet of Things." *Visualization in Engineering* 6 (1). https://doi.org/10.1186/s40327-018-0063-8.

[9] Maria, Schuld, Ilya Sinayskiy, and Francesco Petruccione. 2015. "An Introduction to Quantum Machine Learning." *Contemporary Physics* 56 (2): 172–85. https://doi.org/10.1080/00107514.2014.964942.

[10] Xindong, Wu, Xingquan Zhu, Gong-Qing Wu, and Wei Ding. 2014. "Data Mining with Big Data." *IEEE Transactions on Knowledge and Data Engineering* 26 (1): 97–107. https://doi.org/10.1109/tkde.2013.109.

[11] LeCun, Yann, Yoshua Bengio, and Geoffrey Hinton. 2015. "Deep Learning." *Nature* 521 (7553): 436–44. https://doi.org/10.1038/nature14539.

[12] Tschofenig, Hannes, and Emmanuel Baccelli. 2019. "Cyber physical Security for the Masses: A Survey of the Internet Protocol Suite for Internet of Things Security." *IEEE Security & Privacy* 17 (5): 47–57. https://doi.org/10.1109/msec.2019.2923973.

[13] Mukhopadhyay, Subhas Chandra, Sumarga Kumar Sah Tyagi, Nagender Kumar Suryadevara, Vincenzo Piuri, Fabio Scotti, and Sherali Zeadally. 2021. "Artificial Intelligence-Based Sensors for next Generation IoT Applications: A Review." *IEEE Sensors Journal* 21 (22): 24920–32. https://doi.org/10.1109/jsen.2021.3055618.

[14] SAS. 2018. "Artificial Intelligence—What It Is and Why It Matters." www.sas.com/en_us/insights/analytics/what-is-artificial-intelligence.html.

[15] Yashodha, G., P. R. Pameela Rani, A. Lavanya, and V. Sathyavathy. 2021. "Role of Artificial Intelligence in the Internet of Things—a Review." *IOP Conference Series: Materials Science and Engineering* 1055 (1): 012090. https://doi.org/10.1088/1757-899x/1055/1/012090.

[16] Laxmi, S., N. S. Rudra, K. Hemachandran, and K. N. Santosh. 2021. *Machine Learning Techniques in IoT Applications: A State of the Art.* IoT Applications, Security Threats, and Countermeasures, 105–117.

[17] "What Are LoRa and LoRaWAN?" The Things Network. Accessed July 9, 2022. www.thethingsnetwork.org/docs/lorawan/what-is-lorawan.

[18] Md. Shahjalal, Moh. Khalid Hasan, Md. Mainul Islam, Md. Morshed Alam, Md. Faisal Ahmed, and Yeong Min Jang. 2020. "An Overview of AI-Enabled Remote Smart—Home Monitoring System Using LoRa." *IEEE*, 510–13. https://ieeexplore.ieee.org/document/9065199.

[19] Chen, J., K. Hu, Q. Wang, Y. Sun, Z. Shi, and S. He. 2017. "Narrowband Internet of Things: Implementations and Applications." *IEEE Internet of Things Journal* 4 (6): 2309–14. https://doi.org/10.1109/jiot.2017.2764475.

[20] Ikpehai, Augustine, Bamidele Adebisi, Khaled M. Rabie, Kelvin Anoh, Ruth E. Ande, Mohammad Hammoudeh, Haris Gacanin, and Uche M. Mbanaso. 2019. "Low-Power Wide Area Network Technologies for Internet-of-Things: A Comparative Review." *IEEE Internet of Things Journal* 6 (2): 2225–40. https://doi.org/10.1109/jiot.2018.2883728.

[21] Castro Tome, Mauricio de, Pedro H. J. Nardelli, and Hirley Alves. 2019. "Long-Range Low-Power Wireless Networks and Sampling Strategies in Electricity Metering." *IEEE Transactions on Industrial Electronics* 66 (2): 1629–37. https://doi.org/10.1109/tie.2018.2816006.

4 Influence of Artificial Intelligence in Robotics

Pingili Sravya, Skanda S. Tallam, Shivani Prabhu, and Anil Audumbar Pise

4.1 INTRODUCTION

A robot is usually treated as a tool, as it can perform only several limited or specialized functions; a welding robot can be said to be a perfect example of a robot. It has a physical presence in the world but cannot adapt to fundamental world changes. If some parts are to be welded by the robot, then clear instructions or plans must be laid for the robot to complete the task. Otherwise, the robot does not function, as the design procedure focuses on a robot that can execute a particular function, like how a screwdriver is prepared to turn screws and hammers to hammer nails. The process usually includes designing parts and fixtures to keep the details to make it more comfortable for the robot.

So, the opposite of treating a tool is to treat it as an agent, and an agent can be represented as an entity that can sense and effect change in the world.[1] So, it's the human's job to improvise this tool so that it can adapt to changes in the world just like the way humans do. An automatic vacuum cleaner is a perfect example of an intelligent robot. Here, the vacuum cleaner acts as its agent to accomplish its task (vacuuming). A vacuum cleaner can operate in various rooms with different sizes or layouts, whereas an agent can adjust to any circumstance. When compared, an instrument may be altered to fit strange events, like when the screwdriver handle is used to hammer the nail, but the modification was not planned and is usually not the best option. Science fiction films featuring robots that seem and behave like peers to humans have helped to strengthen the agency perspective of robots. The design process is how an agent or a tool can communicate with the world, especially by sensing and planning, and it will adjust to new but equivalent situations or circumstances based on previous experience. So, to achieve this state, artificial intelligence is operated to provide the required functionality.

Figure 4.1 shows that robotics have both joint cognitive systems and artificial intelligence (AI), which is a recent strategy that suggests that a robot is considered a member of a human-machine unit where the intelligence is harmonious and emanates from the contributions of each other. Because there is at least one robot and one human agent on the team, it is repeatedly referred to as a mixed team. One example of a joint cognitive system is self-driving automobiles, where drivers may switch on and off the driving. The design procedure focuses on how the agents work together and harmonize to achieve the unit's objectives. Joint cognitive systems treat robots differently from peer agents with distinct agendas.

DOI: 10.1201/9781003328414-4

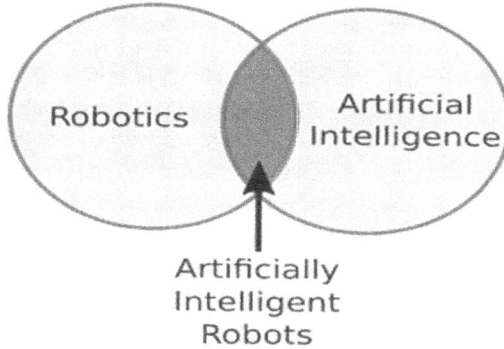

FIGURE 4.1 Artificially intelligent robots.

4.2 HOW DOES AN AI ROBOT WORK?

When an AI algorithm is incorporated into a robot, it does not need to receive any orders to make the decision, but it can work independently. Now the robot will be able to learn, solve, reason, understand, or react appropriately on its own; this is all because of machine learning and AI.

Robots with AI have computer vision that allows them to navigate, assess their environment, and decide how to react.[2] Robots learn how to carry out their duties from humans through machine learning, which is also a component of computer programming and AI.

Most robots are not intelligent, but shortly, businesses will search for bright and process automation. There is a definite trend toward mobile, autonomous robots that can intelligently gather, interpret, and manage data to make the best decisions for manufacturing or production. In many situations, a robot not carrying its weight is no longer sufficient.

4.2.1 REINFORCEMENT LEARNING IN AI AND ROBOTICS

Reinforcement learning is all about training machines on learning modules through which they can make a set of decisions. Agents learn to accomplish goals in uncertain and potentially challenging environments. In reinforcement learning, AI usually faces a game-like situation. Computers use tests and errors to find solutions to problems. AI receives prizes or punishments based on the outcome, and the AI goal is to maximize the total awards.

The biggest hurdle in reinforcement learning is designing the virtual environment (simulation), which depends entirely on the tasks. Preparing the virtual environment is relatively straightforward if their model needs to become superhuman in the game of Chess, Go, or Atari. When building a model that can drive a car, it is essential to make a realistic simulator before the automobile can be carried into the real world. The model must figure out how and when to brake and evade collisions in a safe environment. In this case, the cost of 1,000 cars is minimal, but moving the model from the simulated environment to the actual world is tough.

A key differentiator in reinforcement learning is how well the agent has been taught or instructed. Instead of looking at the data provided, the model interacts with the environment and looks for ways to maximize rewards. In reinforcement learning, neural networks take over the storage of experiences to improve task completion.

4.3   TYPES OF AI USED

4.3.1   Weak Artificial Intelligence

AI restricted to a particular or small domain is called weak AI or narrow AI. Weak AI mimics human thought processes. Automating time-consuming operations and data analysis in ways people can't always do can benefit society. Strong AI, a hypothetical AI comparable to human intellect, can be compared with weak AI.[3]

Weak AI helps identify patterns and generate predictions; weak AI aids in the conversion of massive data into valuable knowledge. The newsfeed on Meta (formerly Facebook), Amazon's suggested purchases, and Apple's Siri, the feature on the iPhone that responds to verbal requests, are all examples of weak AI. AI of this kind is used to simulate human cognition and behavior.[4] The orders and replies for the robots are predetermined. The robots merely locate the correct answer when the proper instruction is delivered; they cannot interpret the orders. The best examples of this are Alexa and Siri. These gadgets' AI only completes duties that the owner requests of it.

4.3.2   Strong Artificial Intelligence

These autonomous robots employ this kind of AI to do their duties. Once trained to complete the work correctly, they do not require any monitoring. Since many processes are getting automated, self-driving and internet cars are two of the most intriguing instances of this form of AI today. Humanoid robots that can sense their surroundings and interact well are also furnished with this AI. As no human interaction is necessary, robotic surgery is also growing in popularity.

AI researchers and scientists have yet to achieve strong AI. To be successful, we must find ways to make machines conscious and program different cognitive abilities. Devices must improve the efficiency of individual tasks and the ability to apply their empirical knowledge to a wider variety of problems, taking practical learning to the next level. Powerful AI uses a theory-of-mind AI framework that relates to the capacity to perceive other intelligent people's requirements, sentiments, opinions, and thought processes. The theory of mind AI is not about duplicating or simulating; it is about how to train appliances so they can comprehend humans.

4.3.3   Specialized Artificial Intelligence

Super AI is the intellectual level of a system where a machine can surpass human intellect and execute any given task with more refined cognitive characteristics than humans. It is a general AI result. The critical attributes of effective AI include the ability to feel, use logic, solve puzzles, judge, schedule, understand, and express independently. Super AI is still a theoretical concept. The actual development of such systems remains a world-changing task.

4.4 IMPACT OF AI AND ROBOTICS IN PRIVATE LIFE

This trend is impacting not only businesses but also everyday personal tasks. Today, people use many home appliances that perform tasks, such as robotic vacuum cleaners and innovative coffee makers. Personal assistants like Siri and Alexa can help anyone find information, order, and control devices in their smart home. As consumer demand grows, the development of such assistants never stops.

In the next few years, cognitive technology will become widely used to manage our health and well-being. For example, robots that help people with disabilities or illnesses, and companion robots may increasingly care for and protect the elderly.

4.5 APPLICATION OF AI

The benefits of connecting AI and robots for industrial applications are already evident in many factories. In this consideration, there are infrequent challenges to overcome. One of the main issues is the specialized talent needed to integrate AI into the industry. There is still a gap between the professional AI community and industry experts.

AI makes robots more efficient because they learn on their own to recognize new situations or things. Today, however, robotics is mainly used in manufacturing and many other fields, performing a wide range of movements more efficiently and precisely than humans. Robots perform incredible actions, including handling boxes in warehouses that facilitate specific tasks. Here, we examine the applications of AI robotics in different fields and the training data used to train a robot to make it industry-ready.

4.5.1 AI and Robotics in Healthcare

AI and robotics are already at work in some healthcare facilities. They can perform tasks such as genetic testing, robotic surgery, cancer research, and data collection. In addition, AI can detect skin cancer in dermatology. Several automated systems in use today in the world's leading hospitals offer enormous potential in performing more complex tasks quickly and accurately.[5] The robot's focus and interest are to enhance its core capabilities further and enable it to perform tasks with greater precision. AI powers these robots so they can learn as they complete lessons.

AI shines when it comes to accurately diagnosing human health conditions. AI recognizes patterns that lead patients to different health conditions. The patient's current state is determined by analyzing and researching charts and data. Tests so far have shown that AI can correctly diagnose the disease 87%

of the time. In contrast, the curacy rate for human health condition detection was 86%. This accuracy is underpinned by the fact that robots and AI can scan thousands of cases simultaneously, looking for correlations between hundreds of variables.

4.5.2 AI and Robotics in Agriculture

Every day, farms generate millions of data points about water usage, weather conditions, temperature, soil, and many more factors. This data is used in real time with AI and machine learning models to generate valuable information, such as choosing the right time to seed, crop selection, hybrid seed selection for increased yield, and more. Figure 4.2 shows that AI systems can help improve the overall quality of the crop, known as precision agriculture. AI technology can help detect diseases among crops, pests, and nutrient deficiencies in crops and farms. AI sensors can detect weeds and determine which herbicides should be disseminated in those areas. These machines help improve output and lower crop waste left in the field. Many other organizations are continuously working towards enhancing agricultural efficiency.[6] These agricultural robots or machines use sensor fusion, computer vision, and AI models to help in locating harvestable produce.

The agricultural industry is the second most extensive sector after defense with a market for service robots developed for specific use. The International Federation of Robotics (IFR) assesses that nearly 25,000 agricultural robots have been traded. The number of robots sold and used in agriculture is almost as same as those used for military purposes.

4.5.3 AI and Robotics in the Service Industry

Services provided by robots and AI offer unprecedented scale and economies of scale, as most of the cost is expended on development. Physical robots cost a trace of

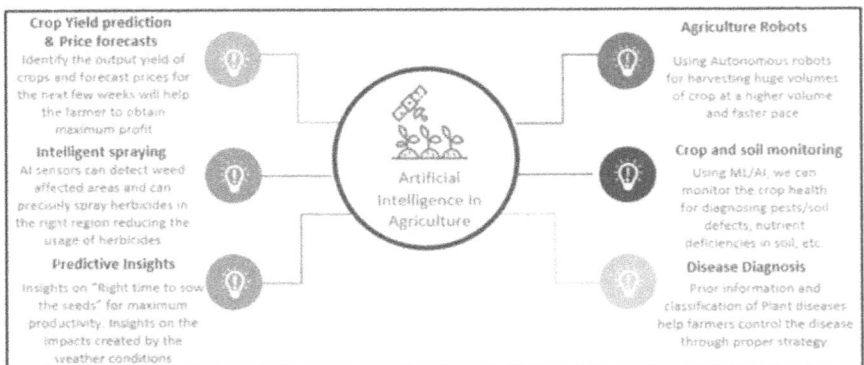

Crop Yield prediction & Price forecasts
Identify the output yield of crops and forecast prices for the next few weeks will help the farmer to obtain maximum profit

Intelligent spraying
AI sensors can detect weed affected areas and can precisely spray herbicides in the right region reducing the usage of herbicides

Predictive Insights
Insights on "Right time to sow the seeds" for maximum productivity. Insights on the impacts created by the weather conditions

Artificial Intelligence in Agriculture

Agriculture Robots
Using Autonomous robots for harvesting huge volumes of crop at a higher volume and faster pace

Crop and soil monitoring
Using ML/AI, we can monitor the crop health for diagnosing pests/soil defects, nutrient deficiencies in soil, etc

Disease Diagnosis
Prior information and classification of Plant diseases help farmers control the disease through proper strategy.

FIGURE 4.2   AI and robotics in agriculture.

additional personnel, and virtual robots can be deployed at a nominal extra charge. Virtual service robots (chatbots, virtual agents, and many more) can scale at almost no additional cost. Such dramatic scalability applies to virtual service robots such as chatbots and "visible" robots such as holograms. For example, airports have hologram-based humanoid assistance robots used frequently to help passengers and answer general questions (e.g., any flight information, check-in counters, directions to airport hotels, and many more) that can be responded to in all typically used languages. These holograms require affordable hardware (such as cameras, microphones, speakers, and projectors) and do not take up any floor space (travelers are encouraged to use luggage carts when the holograms are complete).

4.5.4 AI AND ROBOTICS IN THE MANUFACTURING SECTOR

Modification of the manufacturing technique can be surveyed through AI and robotics. This process minimizes human effort, improves efficiency, and simplifies the manufacturing system. Previously, multiple individuals were required to manage or complete a task, but with the execution of AI-based robots, one robot is enough to meet the given task. Here we look at what role AI and robotics play and their potential and drawbacks in manufacturing sectors.

Error detection and correction have become more accessible with AI-based robots. They are programmed to find bugs and provide solutions to overcome the damage. Technological advancements have made it easier to control the functions of the entire system with just one touch. Artificially intelligent machines are programmed to operate each year automatically, based on the real-time situation so that they can make the right decisions. Each phase is surveyed by a bunch of sensors that send data to AI-based software, and the data provided is used to manage production.

4.5.5 AI AND ROBOTICS IN SUPPLY CHAIN MANAGEMENT

Autonomous robots are already prevalent in manufacturing, final assembly, warehouses, and more. Future supply chains will see a significant growth of autonomous robots in the supply chain industries, allowing people to move to less hazardous and higher-paying jobs.

From Figure 4.3 we can see as the autonomous robot market grows, coordination between end-to-end supply chain operations will become easy. Many companies now use autonomous robots for explicit functions in their supply chains, controlling different robots to validate their benefits. As companies grow and expand their operations, these autonomous robots could become one of the future advancements in the supply chain and become the benchmark for optimizing manufacturing functions.

4.5.6 AI AND ROBOTICS IN SPACE PROGRAMS

Like many other industries, the space exploration process uses AI and robotics to accelerate its mission. This takes us to great distances and places never before

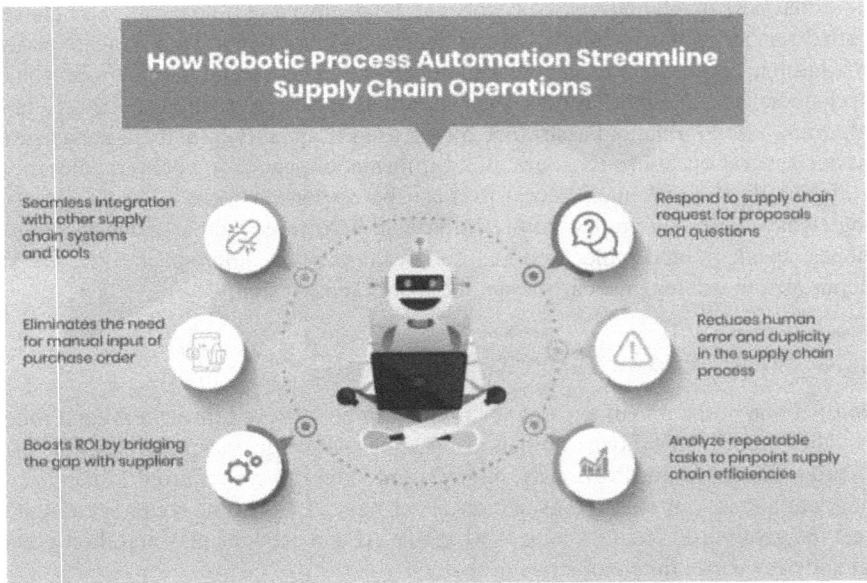

FIGURE 4.3   AI and robotics in supply chain management.

explored. New technologies such as machine learning and profound learning offer organizations involved in space programs an opportunity to take advantage of AI and robotics. Robots support space exploration through planetary rover mechanical design, space manipulator mechanical design, space robot actuators and sensors, space robot ends effectors/tools, reconfigurable robots, and robot mobility. Here are some examples of robotics and AI in space engineering. While the National Aeronautic and Space Administration (NASA) is trying to improve the AI of its Mars rover and is working on developing automated satellite repair robots, other companies are also looking to enhance space exploration with robotics and AI.

4.6   ADVANTAGES AND DISADVANTAGES OF AI IN ROBOTICS

The term "human error" comes from the fact that sometimes humans can make mistakes, but if the AI is programmed correctly, the chances of making a mistake are comparatively less. AI tries to make decisions from previously collected data using a precise set of algorithms, which is considered one of the significant benefits of AI. Developing AI robots can overcome dangerous human limitations. Fly it to Mars, defuse bombs, and explore the deepest oceans, mines, and oil. We can make decisions faster and act accordingly using AI and other technologies. While humans analyze many emotional and functional factors when making decisions, AI-powered machines operate according to how AI is planned to produce results faster.

The capacity to create machines that can simulate human intelligence is no small accomplishment. It can take a lot of time and resources and cost more capital. AI can also be very expensive, as it has to run on the latest software and hardware, stay up-to-date, and meet the current requirements. Moreover, one critical drawback of AI[7] is its inability to learn or think outside the box. AI can learn from preset data and experience over time but cannot be innovative in its approach. AI-enabled machines are developed or programmed to perform a specific task. When asked to do something else, they often fail or produce unhelpful results, which can have serious negative consequences.

The specific thing is that it has excellent potential to create a better world. Humans have the most crucial role to play in keeping the rise of AI from spiraling out of control. Although AI has strengths and weaknesses, its impact on global industries is undeniable. We continue to grow every day and drive the sustainability of our business. This certainly requires AI skills and the need for continuing education to succeed in many new jobs.

4.7 CONCLUSION

AI is the central part of new companies creating computational intelligence models. Debates are going on, saying that AI can be harmful or even have chances of surpassing human intelligence; this might not be very comforting. However, there are many uses, such as problem-solving, reasoning, and language comprehension. AI makes our lives easier by carrying out activities like helping us navigate through busy streets or having many conversations, which can run on any conventional computer or phone. Also, there is no need to develop another type of computer or technology because we already have the technology to integrate AI, which can support the vast complexities of human intelligence. AI is the future in the technology field; yes, it can be harmful or even surpass human intelligence, but only if it is not monitored or if the program is not written clearly.

REFERENCES

[1] Murphy, R. R. (2019). *Introduction to AI robotics, second edition (Intelligent robotics and autonomous agents series)* (Second edition). Bradford Books.
[2] Bullock, M. (2022, March 30). Artificial general intelligence in plain English—towards data science. *Medium.* https://towardsdatascience.com/artificial-general-intelligence-in-plain-english-e8f6e9a56555.
[3] GeeksforGeeks. (2019, October 29). *Artificial intelligence in robotics.* www.geeksforgeeks.org/artificial-intelligence-in-robotics/.
[4] Escott, E. (2017, October 24). What are the 3 types of AI? A guide to narrow, general, and super artificial intelligence. *Codebots.* https://codebots.com/artificial-intelligence/the-3-types-of-ai-is-the-third-even-possible.
[5] Han, C., Sun, X., Liu, L., Jiang, H., Shen, Y., Xu, X., Li, J., Zhang, G., Huang, J., Lin, Z., Xiong, N., & Wang, T. (2016). Exosomes and their therapeutic potentials of stem cells. *Stem Cells International, 2016,* 1–11. https://doi.org/10.1155/2016/7653489.

[6] Towards Future Farming: How Artificial Intelligence is Transforming the Agriculture Industry—Wipro. (2020). *Wipro*. www.wipro.com/holmes/towards-future-farming-how-artificial-intelligence-is-transforming-the-agriculture-industry/#:%7E:text=AI%20 systems%20are%20helping%20to,to%20apply%20within%20the%20region.

[7] Kumar, S. (2021, December 12). Advantages and disadvantages of artificial intelligence. *Medium*. https://towardsdatascience.com/advantages-and-disadvantages-of-artificial-intelligence-182a5ef6588c.

5 A Review of Applications of Artificial Intelligence and Robotics in the Medical and Healthcare Sector

Pokala Pranay Kumar, Dheeraj Anchuri, Pusarla Bhuvan Sathvik, and Raul V. Rodriguez

5.1 INTRODUCTION

In today's world, people are fighting dangerous and unstoppable diseases. Compared with 15 years ago, there has been a lot of improvement in every sector like medical, finance, manufacturing, technology, textile, etc. This change has revolutionized human thoughts and made them think in different ways that weren't imaginable before. But many times this change saves the world. In the same way, evolution changes happen in human life to save humans. Nowadays technology rules the world and plays a major role in every sector. For example, in manufacturing, the past generation handled the work with humans or with human-controlled machinery. Now, robots have introduced and changed the phase of the manufacturing sector with their automation power. The same is true with the medical field. This field was the one that has created huge opportunities to showcase an individual's brilliant work. Previously, there were no x-rays to scan fractures, there were no advanced medical facilities to cure harmful diseases like a tumour, etc. But now we have medicine to cure advanced levels of tumours and detect the fractured part easily using mobile applications or software applications. The medical and healthcare sector are the fields that emerge with the latest technology to help people in fighting diseases. Scientists are trying to improve our biological activities which make us strong and fight harmful diseases with our DNA. DNA helps in finding all solutions to our problems. Bioinformatics, DNA sequencing, etc., are fields that involve both tech and medical synthesis. These emerging technologies help to rescue humans and save their lives. For example, COVID-19 ruled the world without medications. But with our advancements of technology like artificial intelligence (AI), machine learning (ML), robotics, etc., scientists found a vaccine that helps people in creating antibodies that help to fight the COVID virus.

DOI: 10.1201/9781003328414-5

The advancement of technology can be summarized by what is called AI. There are many domains like ML, deep learning, big data, robotics, etc., which is used in drug development, vaccines or antidotes to diseases, medical imaging and diagnosis, monitoring patients, advanced medical treatment with radio waves which helps in curing cancer, etc. ML is the concept of making a machine think and act on its own by training the situations; for example, we will train our software to detect an apple. After training, when we showed the machine as apple, it detects it as such. In the same way, we will train with huge amounts of data to detect or identify objects with acceptable accuracy. This process takes a lot of computation power and lots of hidden layers in algorithms to gather the weights from the data and find the pattern to detect the output. This advancement of using powerful computation is also known as deep learning, which is the advancement to ML. There are three parts of ML. Supervised learning is a process where we will train and give the target to find. Unsupervised learning is the process by which we will train with big data, but the algorithm has to find the pattern to detect the object which is not trained. Reinforcement learning is the process that involves supervised and unsupervised learnings where we will train the data and ask the machine to find patterns. In this, the machine uses its trained knowledge and gains knowledge from its surrounding environments and trains itself to find patterns in the detection of new objects. This concept has been implemented in the medical field which is known as medical imaging, where instead of regular data the algorithms get trained with medical data like patients x-rays or patient numerical data. In this chapter, we will introduce the advancements in the healthcare sector and applications of AI and its subdomain technologies used in the healthcare and medical sector.

5.2   LITERATURE REVIEW

In this paper [1], the authors mentioned many algorithms from ML which helps in building medical applications with the best accuracies. The authors discussed algorithms like support vector machine (SVM), naïve Bayes, neural networks, decision trees, fuzzy logic, K- nearest neighbours, etc., which gave their best performance in detecting heart diseases, detecting breast cancer, diagnosing thyroid disease, etc. The huge data extraction and training with heavy computed devices makes the human detect the diseases within seconds. In this way of learning different activities and the technology, behaviour is going to change day by day. The deep learning process runs in the same way, like day-by-day growth in the accuracies and development of layered algorithms with more advancements, like detection within a short period, reduces the computation power using the trained knowledge algorithms called transfer learning. The authors from this article [2] reviewed the biological data advancements and emerging logic-based techniques with experimental algorithmic approaches. The authors gave an overview of the data transformation in the medical sector, dataset creation, and deep learning approaches to detect the diseases and find the easy way to diagnose them.

5.3 APPLICATION OF AI IN HEALTHCARE AND MEDICAL DIAGNOSIS

5.3.1 APPLICATION OF AI IN HEALTHCARE

With the advent of AI, the world has been on the move. Day by day, progress in the field has led to its application in almost all of the world's domains, including healthcare. The significant advancement in the healthcare domain is in the applied field of medical diagnosis and treatment. The increased presence of AI was thought to be a threat to replace the clinicians, but later it was concluded that AI could guide the clinicians for easy and accurate treatment and diagnosis. Diagnosis of diseases and disorders can be more targeted with the use of AI. Several cardiovascular, neurological, and oncological diseases can be diagnosed and detected with accuracy by applying AI as it works on the previously fed information [3].

From assisting the physicians and healthcare workers to administrative works, analysis of the images, robotic surgeries, virtual assistance, etc., AI offers it all. According to a few reports, AI is also used in the reduction of dosage errors and cybersecurity. The application of AI is a value chain that finds itself helpful in drug development and ambient assisted living (AAL) [4].

5.3.2 APPLICATIONS OF MACHINE LEARNING IN HEALTHCARE

ML is extensively divided into directed and unmonitored learning [5].

5.3.2.1 Identification and Diagnosis by Machine Learning

In the medical field, the major application of ML is to identify and diagnose diseases and illnesses considered hard to diagnose. These could range from different inherited disorders to malignancies, which are difficult to discover during the early stages [6].

5.3.2.2 Machine Learning in Diabetes Prediction

Diabetes is among the most dangerous and yet common diseases nowadays. It not only harms the health of a person alone but is also responsible for many other severe diseases. Diabetes damages the kidneys, heart, and nerves. ML could assist with diagnosing diabetes early, saving lives, through techniques of modelling computations like k-nearest neighbours (KNN), which anticipates diabetes [7].

5.3.2.3 Discovery of Drugs and Manufacturing through Machine Learning

An early-stage drug emergence assessment is one of the key clinical uses of ML. This incorporates creative work grades. This initial medication disclosure assessment is a basic clinical application of ML. This uses inventive steps, such as sequencing or precision medicine, which can aid in creating optional strategies to cure multiple aspects of disease. The idea of ML has already received acceptance, which may see information strategies without any expectations [6].

5.3.2.4   Machine Learning in Epidemic Prediction

Artificial mobility based on brainpower and ML is also used today in the vision and anticipation of diseases all over the world. Nowadays the scientific community is approaching huge amounts of satellite data, unsurprising amounts of web media, page information, etc. Neural network connections contribute to the acquisition of this data, from intestinal disorders to persistent, spectacular pollution. In young nations, these outbreaks are particularly important since they necessitate medical growth and maintenance. Another important example is "ProMED-mail", a web-enabled tool that constantly displays and alerts on critical problems and immediate issues [6].

5.3.2.5   Maintenance of Medical Records by Machine Learning

Maintaining medical records is an exhaustive and expensive process that continuously updates and keeps medical records and patient information. ML advancements add to the arrangement of this issue by diminishing time, exertion, and cash in the constant endeavour [8].

Using ML techniques, such as vector machines (VMs) and optical character recognition (OCR) reinforcement, the Google Cloud Vision API can help healthcare professionals sort and collectively gather data on therapeutic benefits. This data can be used to improve research, healthcare, and safety [8].

5.3.3   Robotics in Healthcare

Medical robots alter how operations are performed, streamline provision and disinfection, and provide clinicians time to interact with patients. Intel provides a comprehensive technological portfolio for the advancement of robotic technology in medical care, including careful, measured, administration, social, versatile, and free robots/independent robots [9].

5.3.3.1   Surgeries by Robotic Technology

Robotic operation is the method by which a robot performs an operation under its computer program. Even though the specialist is practically associated with arranging the methodology to be done and will likewise monitor its execution, the arrangement won't be executed by the specialist, but instead by the robot. The different phases of a common robot activity (albeit robotic surgery is far from common) are discussed next in terms of the different difficulties associated with the robotic fulfilment of a surgery/operation [10].

5.3.3.2   Robotics in Drug Developments and Production

These days robots are utilized in the entire inventory network of drugs from essential exploration to the creation of medication and bundling. Robots work with the discovery of basic new restorative medical products, permitting patients to do fast clinical trials, help drug makers to meet more rigid principles on the creation of medicines, and improve drug creation productivity. Since manufacturing robots have their beginning stage in assembling regions, it's not startling that they are generally in the production unit of the medication advancement lifecycle. The headways needed for performing dull occupations with high degrees of exactness and accuracy are currently increasing. The essential point of convergence of robots in the assembling cycle is at the packing of assembled items like needles [11].

"A model example is mechanical robots are utilized for the gathering of clinical syringes (FANUC/Farason) or filling and shutting of vials (Stäubli/Zellwag Pharmatech)" [11].

5.3.3.3 Robotics in Telepresence

Nowadays, physicians are using robots in rural and isolated places to consult and treat patients. The room is equipped with telepresence. A consultant can use the robot to deliberative and guide medication from far away, keeping the specialist in touch depending on the medical condition and providing drug advice. The key highlights of these robotics gadgets incorporate map-reading ability inside the emergency room and complex cameras for the physical assessment [12].

5.3.3.4 Radiology

Radiology is one of the important advances in the development of robots with special relevance in view of high levels of radiation and human safety hazards. Siemens' twin robotized x-rays is an advancement in fluoroscopy, angiography, and 3D plotting healthcare. The expert may continuously observe 3D images in just one room when the robot travels in comparison with the patient. It plays an enormous role in x-rays. Regular 2D x-ray beams don't consistently detect minimal bone hairline ruptures; the objective is to support the accuracy by using a 3D figurative scan view. A 3D image can be produced of a related design of an x-ray structure that takes what is crucial for a framework computed tomography (CT) [13].

"Error-free Robotic cyberknife treatment (Cyberknife Exactness, Sunnyvale, USA) is used in malignant growth patients with radiation treatment, it gives stereotactic radiation (SRS) treatment and stereotactic body radiation (SBRT), innovative automated accuracy therapies wherever in the body and coordinated synchronizations of movements in real-time" [13].

5.3.3.5 Nanorobots

Although we certainly have not achieved the stage of nanotechnology, trends are growing increasingly significant towards it. As digestible and digital pills arise, we are gradually moving towards nanorobots [14]. On this subject, Max Planck Institute researchers experiment with robots that physically swim through the body fluids and can be utilized to give medications or another medical relief to a very specific degree—exceedingly micro-sized—less than 1 millimetre [14]. These scallop-like microbots are intended to swim in non-Newtonian fluids like the bloodstream around your lymph system or slippery goo on your skin surface or eyes. Despite its small size, the origami robot is equally as stunning as a super-fortified carrier. The capsule containing it is swallowed and disappears in the patient's stomach. It can repair wounds in the stomach lining or remove foreign materials, for instance, swallowed toys, with the aid of the magnetic fields controlled by a technician [14].

5.3.4 Deep Learning in Healthcare

Deep learning can assist physicians to analyse information and detect multiple conditions by processing large volumes of data from different sources, such as x-rays, genomic data, and electronic health records, to address a range of medical problems, such as a reduction in error diagnosis rates and a prediction of procedures [15].

5.3.4.1 Deep Learning in Tumor Detection

The use of convolutional neural networks (CNNs) considerably improved early cancer diagnosis and reached very high levels of accuracy in issues such as mammography screening for breast cancer [15]. In this discipline, the deep learning algorithms approach can even exceed the accuracy of human diagnostics to detect essential characteristics in diagnostic imaging investigations.

5.3.4.2 Drug Discovery and Medicine Precision by Deep Learning

The discovery of new drugs is still met with excitement by academia and the general public. However, the drug development cycle is slow and costly, and only about 10% [15] of drugs make it to market. Deep learning can be used to automatically reduce the costs of drug development by enabling more efficient fingerprinting and feature selection, as well as de novo drug design [16].

5.3.4.3 Simplifying Clinical Trials by Deep Learning

Clinical trials are complicated and expensive. Deep learning and ML can be utilized to perform predictive analytics to distinguish possible candidates for clinical trials and enable researchers to pool in individuals from various information focuses and sources. Deep learning will enable continuous tracking of these trials with the fewest errors and least amount of human intercession [17].

5.3.4.4 Deep Learning for Emergency Clinic Readmissions, Length of Stay, and Anticipating Inpatient Mortality

Deep learning-controlled calculations can analyse countless indicators for each patient, including free-text notes [15]; hence it determines what information is important without hand for a certain forecast and a choice of elements that a specialist considers important [18].

5.3.4.5 Deep Learning in Genomics

Top researchers use deep learning to quickly and concisely generate high-level summaries of information [19] (for example, DNA concatenation, RNA evaluations). Huge algorithms are derived by the revelation of massive highlights, enhancing performance compared to ordinary systems, making information interpretable, and making biological data more significant.

Subsequent efforts directly apply CNNs towards the underlying DNA evolution without trying to represent it before.

By calculating evolution on little locations of data space and separating limits across zones, CNN requires less cutoff focus than an entirely connected organization. This enabled the models to be set up on the more apparent DNA interface, enhancing the overall visual check of classic materials [19].

5.3.4.6 Mobile-Based Deep Learning

A range of applications including health surveillance is transformed using sensor-equipped smartphones and wearables [20]. Since it is possible to make a difference between the wearables of consumer health and medical devices, a solitary wearable gadget may now monitor the scope of clinical danger factors [19]. Deep learning is viewed as a vital component in breaking down this new kind of information. There has been no major assessment using deep learning through a wearable sensor. A few researches dealt with telephone and clinical screen information, "specifically, applicable examinations dependent on profound learning were done on Human Movement Acknowledgment (HAR)" [19].

5.3.4.7 Deep Learning in Medical Diagnosis and Imaging

Once automated perception seems successful, it's indeed essential to note how clinical data are located in deep learning, in particular on magnetic resonance imaging (MRI) displays in the anticipation of Alzheimer's disease as well as its types [19, 21, 22]. CNNs have been used in different clinics to get a levelled alternative representation of low-field knee MRI degrees in a ligament accordingly and to forecast osteoarthritis threat/hazard [19]. This process has gotten the best results over a risky condition employing truly selected 3D multiscale highlights despite 2D images. In addition, deep learning has been used in different sclerosis wounds in 3D MRI multichannel [23]. Additionally, deep learning is used to differentiate between types of breast cancer found by ultrasound [19, 24].

5.3.5 Big Data in the Healthcare Industry

Big data, in its raw meaning, is a large amount of data. The processing and storing of this data for analysis are defined as data analysis. This data can be produced from various sources, mainly from the digital footprints we leave while using the Web. The correct processing and analysis of this big data can give powerful insights. In the healthcare industry, big data is produced from healthcare records from hospitals, diagnostic centres, medical records of patients, medical examination results, equipment that is a part of the Internet of medical things, etc. Biomedical research also yields some amount of big data, which can help understand various aspects of the health and healthcare industry [25].

The credit for big data goes to the digitization of the healthcare industry. The management and use of big data for clinical purposes and healthcare are highly dependent on the information technology sector. The establishment of real-time biomedical health monitoring systems and protocols results from the creation of related software that generates alerts, thereby sharing the information with the patients. The healthcare data components help in improving the quality and services along with being cost-effective [25].

5.3.5.1 Big Data and Public Health

People nowadays have become immensely dependent on technology for answers. The data collected from different sources poses unique challenges. The data is analysed to make a map of epidemics and pandemics or any other outbreaks of disease. However, the use of data in the improvement of prevention and care in any country is seen. Pairing with predictive and precession analysis, the surveillance of the public's health will be easy, giving more insights into any other possible health disorders and diseases [26].

5.3.5.2　Predictive Analysis

The prediction of public health is made based on data collected. Individual health is also monitored, and future problems can be predicted with the use of big data. People are more connected to smartphones and devices; they keep track of their sleeping habits, heart rate, etc., daily. The monitoring can be done by themselves using a few additional data tools and real-time alert systems. For instance, chronic insomnia and increased heart rate can be a sign of future heart diseases.

In addition to this, doctors can quickly make decisions based on the data to improve patients' health. Patients with a complex medical history and numerous medical conditions are at an advantage as they can receive treatment early [27, 28].

5.3.5.3　Advanced Chronic Disease Prevention

Researchers have turned to big data analytical tools for improving the detection and treatment of chronic diseases. These tools are also used to prevent chronic diseases by identifying the environmental factors influencing the advent of the conditions and disorders. This helps the clinical experts to monitor the susceptible individuals and suppress the disease closely.

Cancer is a significant health issue faced by people in the world. Big data is a powerful tool that enables oncologists to provide personalized therapy. The data from diverse sources provides data for one patient, making a unique molecular profile for that person. The significance of the collected data is enormous in terms of a rare type of cancer [26].

5.3.5.4　Telemedicine

Telemedicine has been in the industry for four decades and has advanced with time. Online consultations, video conferences, wireless devices, etc., are exploited to the maximum. Initial diagnosis, remote monitoring of the patients, and medical education are also performed. Telesurgery robots perform the surgery on doctors' commands quickly and in real time without being present at the surgery location.

Personalized treatment is possible with telemedicine as it reduces patient visits to the hospital and the cost of travel and admission. This also helps in monitoring patients from anywhere in the world [27].

5.3.5.5　Opioid Abuse in the United States

This is a national health issue, and big data is finding its use in the healthcare sector for detection, diagnosis, prevention, and treatment of diseases and disorders and helps in dealing with the social issues that need to be dealt with to maintain public health—for instance, the prevention of opioid abuse in the United States.

Accidental deaths in the country have been the most common cause of death. The experts used big data obtained from years of insurance and data from pharmacies. A total of 742 risk factors have been identified, and those at risk have been alerted. This saves lots of lives and money for the country [27]. There are many future experiments and applications in the building stage that may change the face of the medical field.

5.4　CONCLUSION

The referred applications gave an overview of how technology has been utilized in saving mankind. There would be many conclusions and solutions to unsolved medical diseases which helps to improve medical and healthcare sectors in a way where every individual will get medicine at less cost. These advancements help in cost reduction and save the time of doctors as well as specialists. With the help of robotics in the future people may be treated by robots. Robotic surgery is in the testing stage where worldwide medical research institutes are focused to make this testing phase into a physically deployed phase. Technology gives humankind a chance to live a long healthy life.

ACKNOWLEDGEMENT

We would like to thank Dr Raul V. Rodriguez for guiding us and thank the authors and organizations who provided valuable information that helped in the completion of this chapter.

REFERENCES

[1] K. Shailaja, B. Seetharamulu, and M. A. Jabbar, "Machine learning in healthcare: A review," in *Proceedings of the 2nd International Conference on Electronics, Communication and Aerospace Technology, ICECA 2018*, Sep. 2018, pp. 910–914, doi: 10.1109/ICECA.2018.8474918.

[2] R. Miotto, F. Wang, S. Wang, X. Jiang, and J. T. Dudley, "Deep learning for healthcare: Review, opportunities and challenges," *Brief. Bioinform.*, vol. 19, no. 6, pp. 1236–1246, May 2017, doi: 10.1093/bib/bbx044.

[3] Michelle Rice, "The Growth of Artificial Intelligence (AI) in Healthcare," *HRS*, 2019. www.healthrecoverysolutions.com/blog/the-growth-of-artificial-intelligence-ai-in-healthcare (accessed Jun. 29, 2021).

[4] A. Bohr and K. Memarzadeh, "The rise of artificial intelligence in healthcare applications," in *Artificial Intelligence in Healthcare*, Elsevier, 2020, pp. 25–60.

[5] C. Toh and J. P. Brody, "Applications of machine learning in healthcare," in *Smart Manufacturing— When Artificial Intelligence Meets the Internet of Things*, IntechOpen, 2021.

[6] "Top 10 applications of machine learning in healthcare—FWS," *Flatworld Solutions*, 2017. www.flatworldsolutions.com/healthcare/articles/top-10-applications-of-machine-learning-in-healthcare.php (accessed Jun. 29, 2021).

[7] Olena Kovalenko, "12 Real-world applications of machine learning in healthcare—SPD group blog," *SPD GROUP Blog*, 2020. https://spd.group/machine-learning/machine-learning-in-healthcare/ (accessed Jun. 29, 2021).

[8] Prashant Kathuria, "12+ Machine learning applications enhancing healthcare sector 2021 I upGrad blog," *upgrad*, 2021. www.upgrad.com/blog/machine-learning-applications-in-healthcare/ (accessed Jun. 29, 2021).

[9] "Robotics in healthcare: The future of medical care—intel," *Intel*, 2020. www.intel. com/content/www/us/en/healthcare-it/robotics-in-healthcare.html (accessed Jun. 29, 2021).

[10] P. Veerabhadram, "Applications of robotics in medicine—IJSER journal publication," *IJSER J.*, 2011. www.ijser.org/onlineResearchPaperViewer.aspx?Applications-of-Robotics-in-Medicine.pdf (accessed: Jun. 29, 2021).

[11] "The role of robots in healthcare—international federation of robotics," *IFR*, 2021. https://ifr.org/post/the-role-of-robots-in-healthcare (accessed Jun. 29, 2021).

[12] Mark Crawford, "Top 6 robotic applications in medicine—ASME," *ASME*, 2016. www. asme.org/topics-resources/content/top-6-robotic-applications-in-medicine (accessed Jun. 29, 2021).

[13] Z. H. Khan, A. Siddique, and C. W. Lee, "Robotics utilization for healthcare digitization in global COVID-19 management," *Int. J. Environ. Res. Public Health*, vol. 17, no. 11, p. 3819, Jun. 2020, doi: 10.3390/ijerph17113819.

[14] "Benefits of robotics in healthcare: Tasks medical robots will undertake," *The Medical Futurist*, 2019. https://medicalfuturist.com/robotics-healthcare/ (accessed Jun. 29, 2021).

[15] Tommaso Buonocore, "Deep learning & healthcare: All that glitters ain't gold I by tommaso buonocore I towards data science," *Towards Datascience*, 2020. https://towards datascience.com/deep-learning-in-healthcare-all-the-glitters-aint-gold-4913eec32687 (accessed Jun. 29, 2021).

[16] Y. Xu, H. Yao, and K. Lin, "An overview of neural networks for drug discovery and the inputs used," *Expert Opinion on Drug Discovery*, vol. 13, no. 12. Taylor and Francis Ltd, pp. 1091–1102, Dec. 02, 2018, doi: 10.1080/17460441.2018.1547278.

[17] Meenu EG, "These are the top applications of deep learning in healthcare," *Analytics Insight*, 2021. www.analyticsinsight.net/these-are-the-top-applications-of-deep-learning-in-healthcare/ (accessed Jun. 29, 2021).

[18] A. Rajkomar *et al.*, "Scalable and accurate deep learning with electronic health records," *npj Digit. Med.*, vol. 1, no. 1, p. 18, Dec. 2018, doi: 10.1038/s41746-018-0029-1.

[19] R. Miotto, F. Wang, S. Wang, X. Jiang, and J. T. Dudley, "Deep learning for healthcare: Review, opportunities and challenges," *Brief. Bioinform.*, vol. 19, no. 6, pp. 1236–1246, May 2017, doi: 10.1093/bib/bbx044.

[20] K. Shameer, M. A. Badgeley, R. Miotto, B. S. Glicksberg, J. W. Morgan, and J. T. Dudley, "Translational bioinformatics in the era of real-time biomedical, health care and wellness data streams," *Brief. Bioinform.*, vol. 18, no. 1, pp. 105–124, Jan. 2017, doi: 10.1093/bib/bbv118.

[21] S. Liu, S. Liu, W. Cai, S. Pujol, R. Kikinis, and D. Feng, "Early diagnosis of Alzheimer's disease with deep learning," in *2014 IEEE 11th International Symposium on Biomedical Imaging, ISBI 2014*, Jul. 2014, pp. 1015–1018, doi: 10.1109/isbi.2014.6868045.

[22] T. Brosch and R. Tam, "Manifold learning of brain MRIs by deep learning," in *Lecture Notes in Computer Science (including subseries Lecture Notes in Artificial Intelligence and Lecture Notes in Bioinformatics)*, 2013, vol. 8150 LNCS, no. PART 2, pp. 633–640, doi: 10.1007/978-3-642-40763-5_78.

[23] Y. Yoo, T. Brosch, A. Traboulsee, D. K. B. Li, and R. Tam, "Deep learning of image features from unlabeled data for multiple sclerosis lesion segmentation," *Lect. Notes Comput. Sci. (including Subser. Lect. Notes Artif. Intell. Lect. Notes Bioinformatics)*, vol. 8679, pp. 117–124, 2014, doi: 10.1007/978-3-319-10581-9_15.

[24] J. Z. Cheng *et al.*, "Computer-Aided Diagnosis with Deep Learning Architecture: Applications to Breast Lesions in US Images and Pulmonary Nodules in CT Scans," *Sci. Rep.*, vol. 6, no. 1, pp. 1–13, Apr. 2016, doi: 10.1038/srep24454.

[25] S. Dash, S. K. Shakyawar, M. Sharma, and S. Kaushik, "Big data in healthcare: Management, analysis and future prospects," *J. Big Data*, vol. 6, no. 1, pp. 1–25, Dec. 2019, doi: 10.1186/s40537-019-0217-0.

[26] R. Pastorino *et al.*, "Benefits and challenges of big data in healthcare: An overview of the European initiatives," *Eur. J. Public Health*, vol. 29, no. Suppl 3, pp. 23–27, Oct. 2019, doi: 10.1093/eurpub/ckz168.

[27] Sandra Durcevic, "18 Examples of big data in healthcare that can save people," *Datapine*, 2020. www.datapine.com/blog/big-data-examples-in-healthcare/ (accessed Jun. 30, 2021).

[28] R. V. Rodriguez, P. S. Sairam, and K. Hemachandran (eds.), *Coded Leadership: Developing Scalable Management in an AI-induced Quantum World*, CRC Press, India, 2022.

6 Impact of the AI-Induced App 'Babylon' in the Healthcare Industry

Satheesh Kumar M, Tanmai Sree Musalamadugu,
Neelam Kumari, and Raul V. Rodriquez

6.1 INTRODUCTION

Artificial information can be used in human management organization planning and resource assignment in successful and social idea associations. Harrow Council, for instance, oversees the IBM Watson Care Manager framework to improve cost productivity. It links users with an idea supplier who manages their issues within their allocated idea budget. In addition to organizing original ideas and claims, it provides informational nuggets for dynamically effective use of the clerk's benefits. Similar techniques are employed to enhance the permissive experience using PC-based knowledge. In order to facilitate patient engagement, IBM Watson and Birch Hey Children's Hospital in Liverpool are developing a "psychology clinic" that combines an app. Prior to a visit, the app will detect productive weights, provide information upon request, and provide doctors with data to aid in prescribing the proper medications.

Artificial intelligence (AI) can be used to deconstruct and identify patterns in large and complex data sets faster and more conclusively than was previously possible. It can also be used to combine different types of information, such as to help validate findings and identify authentic sources for further study. The AR Compartment Database from the Institute for Cancer Research uses AI to predict new pain medication possibilities by fusing genetic and clinical data from patients with insightful research data. To make the process of finding a cure quicker and more logical, researchers have developed an AI "robot tester" called Eve (K. Williams, 2015). By conducting clinical trials with appropriate patients, AI structures used in human organizations could serve as the foundation for remedial research.

Clinical care—Artificial understanding is currently being trialed in some emergency rooms in the UK to assist in the assessment of disabilities. The same could be said for the use of AI to query clinical data, investigate discrepancies, and implement standards (E. L. Siegel 2013).

The following are some potential uses of AI in clinical research:

- Medical imaging and restorative options have long been purposefully gathered and addressed, and they are now easily accessible for developing AI systems. AI could speed up target treatment while reducing the cost and

DOI: 10.1201/9781003328414-6

time associated with isolating yields. This could potentially allow for more options to be pursued. When it comes to diagnosing conditions like pneumonia, skin and chest cancer, and eye conditions, PC-based insight has shown promising results (D. Wang, 2016).

- Echocardiography—The Ultromics structure, tested at the John Radcliffe Hospital in Oxford, uses AI to dissect channels used in echocardiography that detect certain kinds of pulses and monitor coronary disease.
- Neurological condition screening—AI tools are being developed to help advisors anticipate disturbing scenes and monitor and manage the side effects of neurological conditions like Parkinson's disease.
- Surgery—In research, mechanical tools with AI restrictions have been used to finish certain tasks in the keyhole clinical system, like fastening packs to close wounds.

6.2 SCENARIOS FOR PATIENTS AND CONSUMERS

Applications—The numerous applications that AI provides have altered how success and claims of an idea are evaluated. Using AI, the Ada Health Companion app manages a conversational bot that combines information about customer feedback with other data to suggest potential findings. GP at Hand, a comparison app developed by Babylon Health, is currently being tested at National Health Service (NHS) Clinical Practice Meetings in London. To aid in predictable disease association, information devices or AI-controlled visitor bots are used. For instance, the IBM-developed Arthritis Virtual Assistant for Arthritis Research UK gains knowledge from interactions with patients and then refines information and advice on therapy, diet, and exercise (Release, 2017). Government-funded and private initiatives are investigating the potential applications of AI to control mechanical systems and programs that support people living at home with conditions like early-onset dementia. Applications of artificial knowledge that evaluate and support adherence to a suggested drug and treatment have been tested, with encouraging results, for instance, in tuberculosis patients (L. Shafner, 2017). Other devices, like the Sentrian, use AI to analyze data collected by sensors worn by patients at home. It is important to identify signs of decline to avoid trips to the doctor's office and to enable early intervention (S. F. Moore, 2018).

General health—Using AI, it is possible to better identify the early stages of serious disease outbreaks and the causes of pandemics, such as water pollution. Additionally, AI has been used to anticipate adverse drug reactions, which are thought to be the root cause of up to 6% of urgent cases in the UK (Jacobsmeyer, 2012). The UK company Babylon has fueled efforts to "put an open and sensible thriving organization in the hands of every person on earth" by releasing AI devices. The association currently conducts training in the UK and Rwanda and has plans to expand to the Middle East, the United States, and China. The organization's approach combines the strength of AI with human therapeutic propensity to deliver an unmatched approach to human administration.

6.3 HOW BABYLON'S AI FUNCTIONS

The presented assembly of research experts, artists, stars, and professionals in the field of disease transmission are involved in the creation and redesign of AI Babylon's limits. A stunning piece of collaborative effort is on the way to explore frontline AI; it's the experience of accessing a vast amount of information from the therapeutic framework, predictable sourcing from our own unique clients, and the thorough determination of Babylon's own exceptional genius.

Data graph and customer overview:

> One of the world's most meticulously screened sources of useful information is Babylon's Knowledge Graph. It is coded for machines and receives human information about modern medicine. This will serve as a clarification of the biased aspects of Babylon so that we can communicate with one another. The Knowledge Diagram shows the importance of supportive communication among various therapeutic structures and vernaculars. The tolerant cases are stored in the User Graph, while the Knowledge Graph provides information on medications in general. Greater visibility is taken into account when combining the Babylon Knowledge Graph and the User Graph. With data and outcomes, we can back up brands, and we can also keep our data getting better.

The engine that infers:

> In order to provide information on potential resolution conditions, it is fundamentally insufficient to observe how clients express their symptoms and risk factors. Our social event engine, an amazing game plan of AI frameworks, is at the core of Babylon AI. It is set up for instinct on a space of more than 100 billion different combinations of signs, diseases, and risk factors every second to help perceive conditions. which could impact the data entered. Our AI can think profitably and at scale thanks to the initiation engine, which also provides success data to millions of users.

6.4 NATURAL LANGUAGE PROCESSING

Patients won't use our AI if they can't get the data; otherwise, our AI can't provide the data to patients, and vice versa. To make it easier to navigate any block, we employ natural language processing (NLP). NLP enables computers to understand, process, and sometimes use common human language and language plans. By connecting the occurrence of helpful terms to our knowledge diagram, it breaks both talk and content into smaller chunks and deciphers these squares to understand what each individual part means and how it contributes to the overall meaning. Through NLP, our AI can find guides, suggest clinical records, and visit clients in a powerfully commercial and human way.

6.5   AI-RELATED RESEARCH AT BABYLON

We use machine learning (ML) for numerous experiments all throughout the Babylon mastermind. We combine probabilistic models with critical learning techniques in the selection engine to make system thinking practical. If we look at the support structure in the knowledge diagram, we anticipate a new relationship between renewing contemplations. In NLP, we build language-understanding models that can be trained on massive data sets from the Web and correspondence with our clients. We use ML to show our NLP structure in new languages.

Without the use of front-line ML techniques, Babylon would not be possible, so we essentially invested in creating the top of the line to learn about current social actions. Babylon rushes to contribute back to the AI society through articles, blog segments, and unreservedly publishing some of our work to help everyone.

6.6   ADMINISTRATIONS BABYLON OFFERS

Babylon's designers, specialists, and scientists have developed an AI structure that can take data about the appearances someone encounters, balance the information against a database of known conditions and infections to find a potential match, and shortly after perceive the course of action and associated risk factors. People can use "Ask Babylon" to learn more about their responsibilities and stresses, but this organization is not intended to take the place of a professional's authority or to be used in an emergency situation involving welfare.

On the way, it offers the organization "opposite with authority" methods for its application, GP at Hand, which provides daily access to therapeutic administration specialists through video or audio conferences. You can download the app from the App Store or Google Play. Authorities can provide advice on treatment, address concerns, review the course of action, and organize measures that can be given to the patient during the meeting. The patient's entire clinical history is taken care of in a secure location, and if necessary, it is easy to find their financial history. If the patient happens to need to return to their plan, they can read helpful notes and replay the action progress at any point that would not be achievable without ML strategies at the bleeding edge. In a general sense, we've created a top-notch collection query right now. Through its newspapers, blog sections, and openly performing some of our work to support everyone, Babylon is quickly returning to the AI social order.

Another component accessible in the application is Healthcheck. Working with specialists, researchers, and disease specialists, this AI device can answer questions about an individual's family ancestry and lifestyle and compare them to a clinical database to generate a wellness report and bits of knowledge to help someone.

The start-up claims that in its own tests, the AI structure was spot on 80 percent of the time and that the tool was never designed to completely overcome the direction of a certifiable authority, just to reduce dwell time and help with selection and making dynamically accurate decisions. The world is facing an excessive absence of authorities and therapeutic specialists, and the technologies offered by Babylon are one way to deal with the help that will improve the social protection of a large number of people. As indicated by NHS England, "every safety case [of Babylon] meets the rules required by the NHS and has been carried out using a sound assessment strategy to a heightened desire."

While it probably won't be a perfect structure, Babylon shows that human thinking has advanced adequately to work closely with restorative administration specialists and can be a valuable tool. In any case, despite everything, patients need to be their own advocates for social protection. In the event that a direction obtained from human-made thinking doesn't seem to hit the mark, it's helpful to ask for a resulting guess—from a human.

6.7 AI FOR PATIENTS AND MEDICAL PROFESSIONALS

Every person with a cell phone must switch to a moderate restorative approach for Babylon Health. They understand how crucial it is to have an app with a finite ending. As their CEO Ali Parsa revealed to the *Telegraph*: "[Doctors] are the most expensive part of the restorative administration. The second . . . is the timing . . . [by] the time [most diseases] show their symptoms, a £10 issue has become a £1,000 game plan."

It is understood by Babylon Health that they may forego both of these costs. The free app provided by Babylon Health today gives users a central location to monitor their health and interact with their AI-powered chatbot. For a fee, clients can watch a video featuring leading medical experts who can locate useful records of the booming pace and numerous carefully chosen AI gadgets that Babylon Health claims can enhance the quality of treatment. By tracking vital signs, drugs, and results across a vast client base, Babylon leverages a phenomenally critical data set. This data set makes it possible to flexibly improve the presentation of their AI close to the client's peak.

IBM Watson for Oncology has a more modest focus: improving disease-solving outcomes. IBM realizes that it can provide every medical professional in the treatment of diseases with similar data to that of specialists at the highest level of damage investigation. IBM partnered with Memorial Sloan Kettering Cancer Center to train their supercomputer Watson on a vast amount of medical records and research. In 2016, Watson was launched to help clinicians make confident, evidence-based recommendations for cancer treatments in a fraction of the time. As shown by Deborah DiSanzo, CEO of IBM Watson Health, Watson for Oncology has recently been consistently used to treat 16,000 patients since the last quarter of 2017. With computers handling performance reviews, managers can focus on what employees do well. The doctor wants to treat the patient's intense sadness associated with risky medical procedures.

6.8   DATA FOR AI IS THOUGHT-PROVOKING

Both IBM Watson and Babylon Health agree: specialists can deliver better treatments by learning from the inevitable outcomes of different patients. PC-based information can be obtained from consistent information to see how a patient's illness would respond to a treatment choice. Artificial brain power is a mechanized system used by a computer to encourage itself to make choices using information preparation. The organization of information is the fuel of AI, as demonstrated by Andrew Ng of Stanford University.

Babylon Health and IBM Watson have put together structures that produce this "fuel" from their clients. As they attract more clients, they will improve the data. The effect of this system is a silent circle, where things eventually get better because they connect more clients. The downside of impact-generating items is that they are difficult to distribute. To put it simply, it can be very difficult to find a large audience for them.

Babylon Health and IBM Watson partnered with healthcare organizations to prepare for and take a test of their capabilities. Babylon Health is launching its services with the help of the UK branch of the NHS. The NHS is looking for ways to facilitate their basic needs of specialists and will be the main chatbot for Babylon in north central London, a borough of 1.2 million people, for half a year. IBM Watson is assisting Memorial Sloan Kettering to help train Watson on the amount of clinical data and healing health information that the organization is known for.

6.9   REGULATORY RISK: A POTENTIAL CHALLENGE

With AI-powered human organizations showing such a huge number of endorsements, it's conceivable that the standard should move quickly through the Food and Drug Administration (FDA). In any case, the FDA is interested from now on. As the *Wall Street Journal* writes: "How on earth would you say you will manage programming that learns?"

Current principles need benchmarks to assess the flourishing and abundance of AI frameworks, which the FDA has attempted to convey by giving direction to the research of AI frameworks. The head leader exhorts AI frameworks as "generally thriving things" that are misregulated because they present themselves to clients in order. The following direction legitimizes the use of true evidence to evaluate the presentation of AI frameworks.

Despite these difficulties, things look bright for useful AI-powered organizations. Babylon Health and IBM are just two different new activities that are expanding the scale of organizations by strengthening the parts that don't scale: specialists. While these groups have their own unique perspective on the future, they all agree that AI will enable our physicians to deliver the best solutions to the greatest number of individuals, especially when the best treatment kicks in before we're wiped out.

Front-line information is what drives human-made intellectual capacity, so AI is constrained by peculiarities in information's nature and receptivity. Furthermore, evaluating lengthy and intricate instructional records calls for a strong writing ability. While many people are enthusiastic about the potential roles that AI could play in the NHS, others point out legitimate challenges, such as the NHS's failure to digitize therapeutic records on a continuous basis and its IT structures' lack of adherence to interoperability and associations, electronic record keeping, and information naming. Patients' and experts' support for the dissemination of private information about frontline success is subject to requirements. People possess qualities that AI frameworks will undoubtedly be unable to possess, such as sympathy. Clinical practice typically combines complex decisions and constraints, such as the ability to observe expressive movements and insightful information, that nascent AI cannot currently reflect. Additionally, the question of whether some human knowledge is innate and therefore impossible to teach is covered. The scenarios in which AI will be able to show self-organization have been designed on the assumption that this quality is unique to humans and cannot be possessed by a machine.

6.10   CONCLUSION

Artificial awareness pressures are employed or observed for some welfare and research purposes, including the detection of disease symptoms, the management of prevailing conditions, the creation of successful associations, and the display of pharmaceuticals. However, they may be constrained by the possibility of success-related information being available and the lack of AI that possesses some human traits, such as compassion. Reproducible advances in knowledge can help to solve enormous difficulties. The use of AI raises two or three significant societal issues, the basic number of which has increased as a result of issues brought on by the use of information and the expansion of human organizations even further. The ability for AI to be developed and used in a manner that is immediate and appropriate to the interests of general society, while enlivening and encouraging improvement in this endeavor, will be a key test for the future amalgamation of AI advancement types.

BIBLIOGRAPHY

Insights, C.B. 2017. *AI, healthcare & the future of drug pricing*, s.l.: s.n.

Jacobsmeyer, B. 2012. *Focus: Tracking down an epidemic's source phyiscs*, s.l.: s.n.

Moore, S.F. 2018. *Harnessing the power of intelligent machines to enhance primary care*, s.l.: s.n.

Release, I.P. 2017. *Arthritis research UK introduces IBM watson-powered 'virtual assistant' to provide information to people with Arthritis*. s.l., s.n.

Shafner, L. 2017. *Evaluating the use of an artificial intelligence platform on mobile dveices to measure and support tuberculosis medication adherence*, s.l.: s.n.

Siegel, E.L. 2013. *Artificial intelligence in medicine and cardiac imaging*, s.l.: s.n.

Wang, D. 2016. *Deep learning for identifying metastatic breast cancer*, s.l.: s.n.

Williams, K. 2015. *Cheaper faster drug development validated by the reposotioning of drugs against neglected tropical diseases*, s.l.: s.n.

7 Identification and Prediction of Pneumonia from CXR Images Using Deep Learning

Paul Nirmal Kumar K, Raul V. Rodriguez, and Pokala Pranay Kumar

7.1 INTRODUCTION

The condition of inflammation of the lungs, which happens mostly in the alveoli air sacs, is called pneumonia [1]. This is generally caused by infection with viruses and other microorganisms and even bacteria. A single agent causing the disease is not able to be isolated through highly careful testing in around half of the cases. The chances of getting affected by pneumonia is high when people have a history of certain diseases such as asthma, diabetes, chronic obstructive pulmonary disease (COPD), sickle cell disease (SCD) or even a weak immune response, smoking, alcoholism, exposure to air pollution, malnutrition, and poverty [2]. This disease presents with symptoms such as difficulty breathing, cough, sharp chest pain, and fever. Pneumonia causes around 4.5 million premature deaths annually and around 120 million people get infected [3]. Based on the agent that causes it, pneumonia is classified into the following classifications:

Bacterial Pneumonia: The most widespread source of community-acquired pneumonia (CAP) is due to bacteria. In the process of inhalation, bacteria enters the lungs through which it can reach the bloodstream. When these bacteria enter the lungs, they are taken into the alveoli, the deepest cavities of the lungs. From here, the bacteria get into the places between the conjoining alveoli. The body responds to this by attacking the invasive microorganisms and killing them, which eventually leads to fever, cold, and unbalanced oxygen pumping. The treatment for bacterial pneumonia includes antibiotics generally, with the assistance of a ventilator.

DOI: 10.1201/9781003328414-7

Viral Pneumonia: Viruses utilize droplets which advance through the mouth and nose while a person is inhaling to enter the lungs. Once they reach the alveoli in the lungs, viruses start invading other cells and try to kill them. When the immune system responds to these invasive viruses, the conditions gets even worse and eventually leads to an unbalanced oxygen supply. Viruses have caused about one-third of pneumonia cases in adults. When taken in a wider context, the contingency with pneumonia caused by viruses may get out of control when there are virus outbreaks such as coronavirus.

Fungal Pneumonia: Fungal pneumonia is seen mostly in people with a weak immune system, for example, due to acquired immune deficiency syndrome (AIDS) and other health issues. This kind of pneumonia can be even caused by a hidden infection which later arises due to other factors.

Parasitic Pneumonia: Parasitical pneumonia is scarce when compared to other causes of pneumonia. Parasites enter the body through direct contact with skin and even through inhalation. After entering the body, these parasites travel to the lungs and cause inflammation and an imbalanced oxygen supply.

The confirmation of this disease's diagnosis is generally done either by blood tests, physical examination or chest x-rays (CXRs) [4–6].

Physical examination: This includes notably a higher heart rate, decreased oxygen in the body or even crackles heard through a stethoscope. It can also be determined if there is a low expansion of the chest while breathing.

CXR: The CXR is the most commonly used method for the diagnosis of chest-related diseases. In the case of pneumonia, it is hard to determine the diagnosis through CXR when there is a case of dehydration, obesity, and other lung-related health issues. Based on CXRs, pneumonia is further classified into the following types:

- **Lobar pneumonia:** This type of pneumonia affects the continuous part of the lobe of a lung. It releases toxins that cause inflammation and edema of the lung. Inflammation of exudate inside the alveoli space eventually leads to consolidation.
- **Bronchopneumonia:** This type of pneumonia affects the bronchi where the inflammation is seen in the form of patches in the lobules of the lungs. This pneumonia is also known as lobular pneumonia. This pneumonia affects both the alveoli and the bronchi that are present in the lungs. There is a chance of this pneumonia in patients who have undergone surgical operations. One of the most affected lung lobes is the right and accessory lung lobe.
- **Interstitial pneumonia:** This type of pneumonia affects the interstitium tissues present in the lung. These tissues get inflamed due to the invasion of the disease. This pneumonia causes a cellular form when there is inflammation of the cells of the interstitium, and this pneumonia also causes a fibrotic form when there is thickening and scarring of lung tissue. The scarring in the lungs caused by this disease is irreversible.

Most commonly, the classification of pneumonia is done on the basis of the place it was acquired so as to identify the agents that are the suspects for the cause of the disease and treat the patient accordingly. This classification is done in two forms, i.e. community-acquired pneumonia (CAP) and health care–associated pneumonia (HCAP).

CAP: When a person acquires pneumonia outside the health care system, it is called CAP. This is treated by using antibiotics that kill the infecting organisms.

HCAP: When a person acquires pneumonia within the health care system, such as hospitals and other medical clinics, then it is called HCAP. This type is also called medical care–associated pneumonia (MCAP).

Artificial intelligence refers to amplifying the most powerful phenomenon in the world, i.e. 'intelligence'. In this field of technology, there has been a huge advancement in the areas of image recognition and computer vision, which use images as the data. When artificial neural networks and algorithms inspired by the human brain learn from huge amounts of data that are unstructured, highly diverse and also inter-related, then it is referred to as deep learning. A deep learning algorithm is designed to draw conclusions similar to humans by repeatedly analyzing data, and this algorithm is structured logically. The accuracy of a prediction can be determined by a functional deep learning algorithm itself. The performance of these algorithms gets better with further learning of the data. By performing a certain task repeatedly, these deep learning algorithms learn from the experience by slightly modifying the task every time. The number of neural network layers are several and deep, hence the name deep learning. Since data is the fuel for these deep learning algorithms, the enormous amount of data we are producing paves the way for the possibility of such intelligent technology. This revolutionary technology has aided a lot of industries, and health care is one of them. Here, in this chapter, we use deep learning algorithms and convolutional neural networks (CNNs) to classify the CXR images into either 'pneumonia' or 'normal' using the data provided by Mendeley Datasets—'Labeled Optical Coherence Tomography (OCT) and Chest X-Ray (CXR) Images for Classification'.

7.2 DATASET DESCRIPTION

The dataset used for this research is a collection of CXR images. The process of producing CXR images consists of using small doses of ionizing radiation. These images are used to diagnose the parts of the chest such as the lungs, bones and heart. CXR images are widely used in the diagnosis of lung cancer and pneumonia.

The dataset contains validated and confirmed CXR images from the particular folder they have been placed into. The original dataset, named 'Labeled Optical Coherence Tomography (OCT) and Chest X-Ray (CXR) Images for Classification', consists of data in files, namely, 'ChestXRay2017.zip' and 'OCT2017.tar.gz'. From these files we use only the 'ChestXRay2017.zip' file, since our research is mainly based on the CXR images. The 'chest_xray' folder from the. zip file is split into two

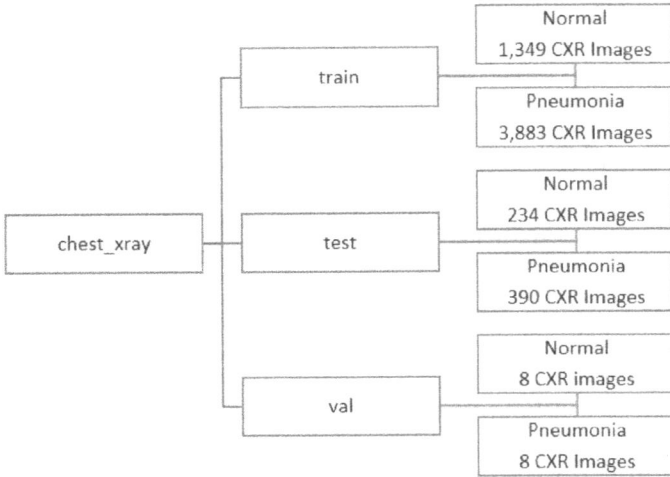

FIGURE 7.1   Dataset description.

specific folders, namely 'train' and 'test'. Each of these folders contains two folders, namely 'Normal' and 'Pneumonia'. The 'Normal' folder contains CXR images with no pneumonia, and the 'Pneumonia' folder contains CXR images with pneumonia. In the 'train' folder, there are 1,349 normal CXR images and 3,883 pneumonia CXR images, whereas in the 'test' folder, there are 234 normal CXR images and 390 pneumonia images. We further split the data into another folder, namely 'val' alongside 'test' and train', in order to be able to check the validity of our model. This 'val' folder has been given two folders, which are 'Normal' and 'Pneumonia', with eight respective CXR images in each of them. All the previously mentioned images are in the JPEG format and gray color space. The dataset is collected from the website http://dx.doi.org/10.17632/rscbjbr9sj.2. The final dataset and its contents are represented in Figure 7.1.

7.3 CONVOLUTIONAL NEURAL NETWORK

The most effective neural network for image classification is the CNN. These neural networks are widely used in computer vision, object detection, and various other functions of image data [7]. These CNNs automatically extract features from the data. For a CNN model, an image is a matrix of pixel values, and these values are dependent on the three different channels—red, green and blue. The working of a CNN consists of different layers through which an image is processed (Figure 7.2).

- **Convolution:** Here, the very first layer is the convolution layer, which is used to extract features from images in the case of CNNs. This convolution layer is applied on the input image matrix using a filter matrix, which results in an activation, and when this task is repeated, it produces a map of activations called a feature map. The height and width of a filter matrix are chosen

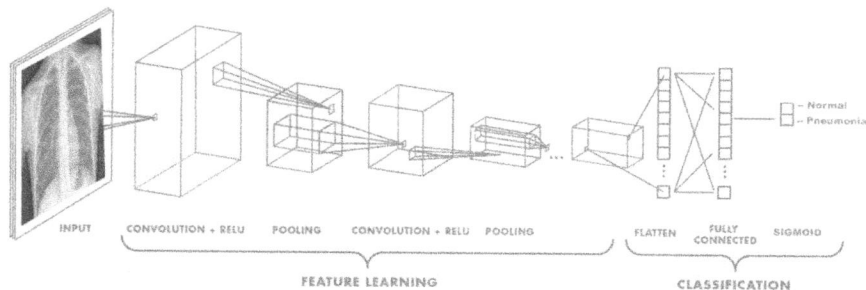

FIGURE 7.2 Layers of CNN.

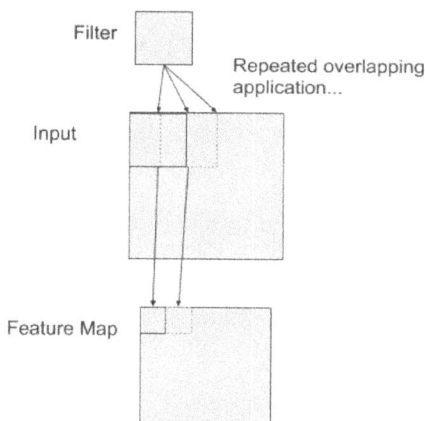

FIGURE 7.3 Feature map in the convolution layer.

to be smaller than that of an input matrix [8, 9]. When an image is convolved with different filters or kernels, it results in different operations such as blur, sharpen, edge detection and image identification and then creates the feature map accordingly (Figure 7.3).

- **Strides:** This is the number of steps the filter or convolution kernel has to shift over the input image, i.e. the filter matrix shifts 1 pixel when the stride is 1 and by default the stride is set to 1 (Figure 7.4). We can change the value of the stride accordingly.
- **Padding:** When the filter matrix moves over the input matrix, the corner values are not given the same weight as the other corresponding values. Hence, in order to overcome this, padding is used. Padding is of two types— zero padding and valid padding. In zero padding, the input matrix is added with zeros on the corners, and in valid padding, a part of the input matrix is deleted.
- **Activation function:** The decision of whether a neuron should be activated or not is taken by the activation function by calculating the

7 x 7 Input Volume   **5 x 5 Output Volume**

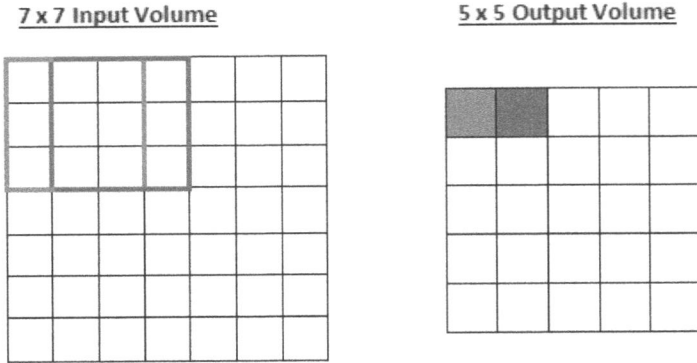

FIGURE 7.4   Shifting the filter matrix over the input matrix when the stride is 1.

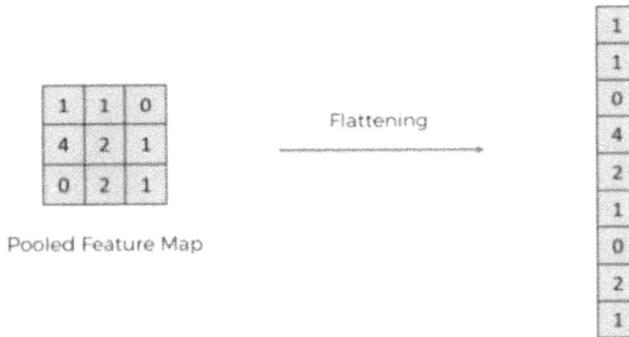

Flattening

Pooled Feature Map

FIGURE 7.5   Flatten layer.

weighted sum and further adding bias with it. There are different types of activation functions to be chosen from, such as Linear Activation Function, Sigmoid Activation Function, ReLU Activation Function and Tanh Activation Function.

*   **Pooling:** In order to extract the most important feature from the feature map, pooling is used. Pooling is a process where the feature map is sub-sampled in order to get only the important features. The process of pooling consists of three methods—Max Pooling, Average Pooling and Sum Pooling. In Max Pooling, we take the largest value from the feature map into the spatial neighborhood. In Average Pooling, we take the average of all the specified values of the feature map into the spatial feature map. In Sum Pooling, we take the sum of all the specified values of the feature map into the spatial map.

    With the formation of the convolutional layer and the pooling layer, a convolutional block is formed.

- **Flatten layer:** After pooling is done for all the feature maps, a number of pooled feature maps are produced. In order to feed these features to a neural network, we use the process of flattening the pooled feature maps, which results in a vector with all the features (Figure 7.5).
- **Fully connected layer:** A fully connected layer is a collection of layers of neurons forming a neural network. This is the last layer of the CNN architecture. The vector values formed in the flatten layer are fed to these neural networks, which are totally connected to one another. These fully connected layers have individual activation functions which are based on the problem [10–13]. This forms a whole CNN model.

7.4 MODEL BUILDING

For our problem of identifying pneumonia from CXR images, which is a classification problem, we use the model with the following steps:

1. Before creating a CNN deep learning algorithm, the model first uses data augmentation and data generation for the effectiveness of the algorithm. Data augmentation is a process to virtually increase the size of the data that is to be trained. This process helps to advance the data with more diversity by using various data augmentation techniques such as rotating, flipping and cropping the images. Data generation is a process of creating a certain amount of data batches that are inserted in the network. This batch size is generally 32, 64 or 128. This batch size must be chosen based on the resources used for computation and accordingly to the model performance [14–16]. The current model uses a size of 64 per batch.

2. Now, for the building of the algorithm, the model consists of five convolutional blocks in which the number of filters for each block are 16, 32, 64, 128 and 256, respectively. With the kernel or filter size of a 3 × 3 matrix, padding as zero padding, batch normalization, activation function as ReLU, pooling as maximum pooling and dropout functions in order to decrease the overfitting, these convolutional blocks are formed. While compiling, the model uses the Adam optimizer and binary cross-entropy loss. Adam is an algorithm that optimizes the model by updating the weights of the neural networks. This optimizer is an extension of a gradient descent optimizer. It is named 'Adam' since it works on the adaptive moment estimation of network weights.

3. Once the CNN is created, the model uses callbacks such as model checkpoints and early stoppings in order to minimize loss, overfitting and underfitting. Model checkpoint helps the model to save a copy where it seems to be efficient. This forms a checkpoint where the model has performed the best in the iterative process. Early stopping is a process to stop the training iterations when the model notices an increasing generalization difference [17, 18]. When the model identifies such a difference, it calls back to the model checkpoint where the training had performed well.

4. Next, the model is fitted in which it is trained for 10 epochs. An epoch is the iteration of the model over each data point in the batches of the training dataset. A high number of epochs will probably give high accuracy, but factors like the computational power of the machine must be considered while defining the number of epochs [19, 20]. The relation between epoch, iteration and batch size can be understood by taking an example—consider there are 1,000 examples in a dataset; if we create batches of size 500, then it will take two iterations to complete one epoch.

7.5 EVALUATION

In this model, which is developed on the Python programming language, when the model accuracy and the loss of the model are created, it is seen that with the increase in epochs, the model loss is decreasing and the model accuracy is increasing, and hence the train and validation values are seen to be converging. After creating the confusion matrix for the model, it emphasizes the accuracy of the model as around about 91% (Figure 7.6).

The confusion matrix of the model shown in Figure 7.7 shows how the model has performed and gives different metrics such as accuracy, precision, recall and F1-score. Precision is the ratio of correctly predicted positives to the total positives predicted. Recall is the ratio of correctly predicted positives to the total positive examples in the data [21]. F1-score is the combined metric of precision and recall.

7.6 CONCLUSION

Considering the size of the data used, which was relatively small, the accuracy gained is comparatively good. Therefore, this deep learning technique can be used to classify CXR images to identify and detect pneumonia, which can help patients save time and be prepared for their treatment. Moreover, it is seen that with further

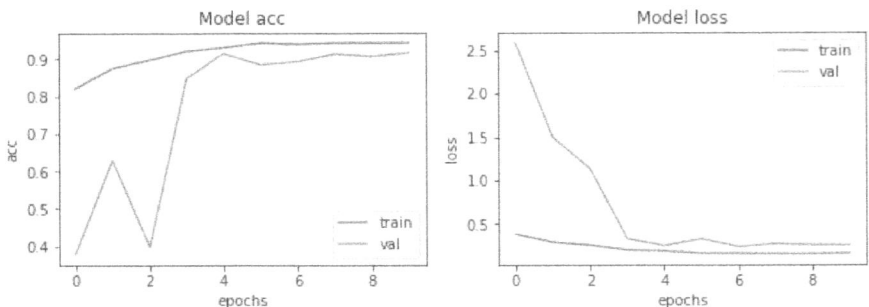

FIGURE 7.6   Model accuracy and model loss.

```
CONFUSION MATRIX ------------------
[[191 43]
[ 13 377]]

TEST METRICS -------------------
Accuracy: 91.02564102564102%
Precision: 89.76190476190476%
Recall: 96.66666666666667%
F1-score: 93.08641975308642

TRAIN METRIC --------------------
```

FIGURE 7.7 Confusion matrix of the model.

development and implementation of models, artificial intelligence can be a breakthrough in solving the scarcity of specialists who examine CXR images.

REFERENCES

[1] Nonspecific Interstitial Pneumonia (NSIP) https://my.clevelandclinic.org/health/dise ases/14804-nonspecific-interstitial-pneumonia-nsip.
[2] India has one doctor for every 1,457 citizens: Govt www.business-standard.com/article/ pti-stories/india-has-one-doctor-for-every-1-457-citizens-govt-119070401127_1.html.
[3] Yuan Tian: Detecting Pneumonia with Deep Learning https://becominghuman.ai/ detecting-pneumonia-with-deep-learning-3cf49b640c14.
[4] Chest radiograph https://en.wikipedia.org/wiki/Chest_radiograph.
[5] X-ray (Radiography)—Chest www.radiologyinfo.org/en/info.cfm?pg=chestrad.
[6] Chest X-rays www.mayoclinic.org/tests-procedures/chest-x-rays/about/pac-20393494.
[7] Deep learning vs machine learning: A simple way to understand the difference www. zendesk.com/blog/machine-learning-and-deep-learning/.
[8] What Is Deep Learning AI? A Simple Guide With 8 Practical Examples www. forbes.com/sites/bernardmarr/2018/10/01/what-is-deep-learning-ai-a-simple-guide-with-8-practical-examples/#454415868d4b.
[9] Deep learning https://en.wikipedia.org/wiki/Deep_learning.
[10] Understanding of Convolutional Neural Network (CNN)—Deep Learning https:// medium.com/@RaghavPrabhu/understanding-of-convolutional-neural-net-work-cnn-deep-learning-99760835f148.
[11] A Basic Introduction to Convolutional Neural Network https://medium.com/@hima drisankarchatterjee/a-basic-introduction-to-convolutional-neural-network-8e39019 b27c4.
[12] How Do Convolutional Layers Work in Deep Learning Neural Networks? https:// machinelearningmastery.com/convolutional-layers-for-deep-learning-neural-networks/.
[13] Convolutional Layer—Science Direct www.sciencedirect.com/topics/engineering/ convolutional-layer.
[14] Activation functions in Neural Networks www.geeksforgeeks.org/activation-functions-neural-networks/.
[15] Convolutional Neural Networks (CNN): Step 3—Flattening www.superdatascience. com/blogs/convolutional-neural-networks-cnn-step-3-flattening.

[16] Gentle Introduction to the Adam Optimization Algorithm for Deep Learning https:// machinelearningmastery.com/adam-optimization-algorithm-for-deep-learning/.

[17] Deep Learning for Detecting Pneumonia from X-ray Images https://towardsdatascience. com/deep-learning-for-detecting-pneumonia-from-x-ray-images-fc9a3d9fdba8.

[18] Training a CNN to detect Pneumonia https://medium.com/datadriveninvestor/training-a-cnn-to-detect-pneumonia-c42a44101deb.

[19] Epoch—DeepAI https://deepai.org/machine-learning-glossary-and-terms/epoch.

[20] Epoch vs Batch Size vs Iterations https://towardsdatascience.com/epoch-vs-iterations-vs-batch-size-4dfb9c7ce9c9.

[21] Idiot's Guide to Precision, Recall, and Confusion Matrix www.kdnuggets.com/2020/01/ guide-precision-recall-confusion-matrix.html.

8 Pulmonary Cancer Detection Using Deep Convolutional Networks

Tejaswini Aala, Varadaraja Krishna, Megha Gada, Raul V. Rodriguez, and Abdullah Y. Muaad

8.1 INTRODUCTION

Pulmonary cancer is one of the major causes of cancer deaths across the globe due to its aggressive nature and delayed detection at advanced stages. If pulmonary cancer can be detected at earlier stages, the number of deaths due to cancer will drastically decreases. Cancer starts to grow when cells in the body grow out of control. Cancerous, i.e., malignant, and noncancerous, i.e., benign, pulmonary nodules are the small growths of cells inside the lung. Early detection of cancers is similar to a noncancerous nodule, which makes differentiation difficult. The diagnosis is based on slight morphological changes, locations, and clinical biomarkers. Various diagnostic procedures are used by physicians for the early diagnosis of malignant lung nodules, such as clinical settings, computed tomography (CT) scan analysis (morphological assessment), positron emission tomography (PET) (metabolic assessments), and needle prick biopsy analysis [1]. Mostly invasive methods such as biopsies or surgeries are used by healthcare practitioners to differentiate between benign and malignant lung nodules. For such a fragile and sensitive organ, invasive methods involve lots of risks and increase patient anxiety.

8.2 METHOD FOR DIAGNOSIS

CT scans are one of the more suitable methods for diagnosis. CT scans do lead to many false-positive findings, however, with carcinogenic effects of radiation. The detection sensitivity of lung nodules improves with sophisticated anatomical details, which depend upon the slice thickness; up to 500 sections/slices are produced in one scan.

An experienced radiologist takes approximately an average of 2 to 3 minutes to observe a single slice. Radiologists also have to look at various properties like the thickness of the CT slices; detection sensitivity also depends on nodule features such

DOI: 10.1201/9781003328414-8

as size, location, shape, adjacent structures, edges, and density, which can increase the workload of a radiologist significantly to screen a CT scan for the possible existence of a nodule.

This work done by radiologists can be done by a machine, which can give an accurate result in less time. The tool which can assist the radiologist is computer-aided detection (CAD). These systems are basically designed to reduce the work for the radiologist and increase the nodule detection rate. However, the present-day generation of CAD systems also helps in the screening process by differentiating between benign and malignant nodules [2, 3].

In Figure 8.1 it is very easy to detect cancer, but in Figure 8.2 the radiologist will find it difficult to determine if the person has cancer or not.

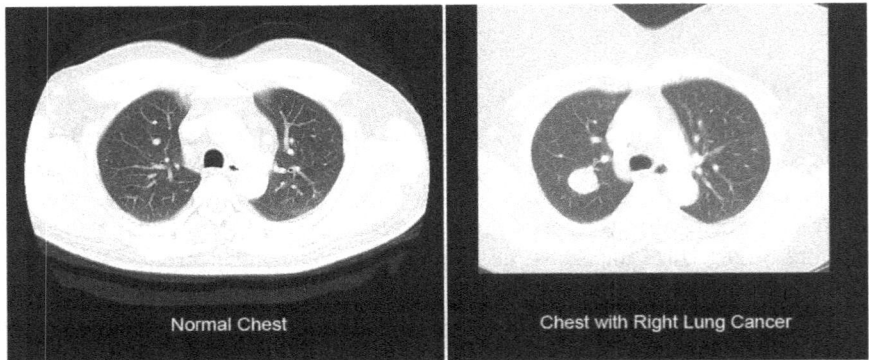

Normal Chest   Chest with Right Lung Cancer

FIGURE 8.1   CT scan of a lung without cancer and with cancer.

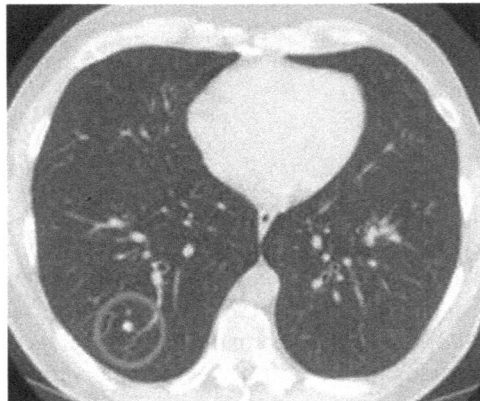

FIGURE 8.2   CT scan of lung with benign tumor.

With the advanced deep neural networks, especially in image analysis, CAD systems are consistently outperforming expert radiologists in both nodule detection and localization tasks. The results from various researchers show a broad range of detection from 38% to 100%, with a false-positive (FP) rate from 1 to 8.2 per scan by the CAD systems. The categorization between benign and malignant nodules is still a challenging problem, however, due to the very close resemblance at early stages [4].

8.3 MALIGNANCY DETECTION

Malignancy means cancer, which is basically related to the size and growth of the nodules. There are three different categories and it is shown in figure 8.3.

The three different categories:

A. **Benign:** This kind of tumor is a tumor that does not invade its surrounding tissue or spread around the body.

B. **Primary Malignant:** A *primary* malignant is a tumor growing at the anatomical site where tumor progression began and proceeded to yield a *cancerous* mass.

C. **Metastatic Malignant**: Metastasis is the spread of cancer cells to new areas of the body, often by way of the lymph system or bloodstream.

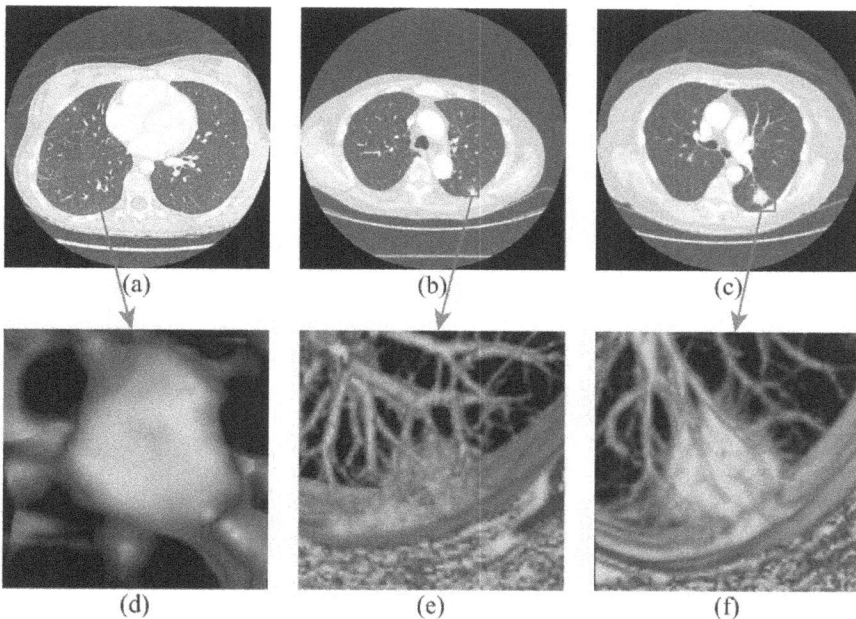

FIGURE 8.3 Benign, primary malignant, and metastatic malignant.

8.4 MODEL USED FOR ACCURATE DETECTION

A deep learning–based model was used for the diagnosis of initial-stage lung cancer by CT scan analysis. Due to the close resemblance of benign and malignant nodules at early stages and various types of scanning errors, a large number of false-positive results are reported in CT scan analysis techniques. To detect FP results, better deep learning–based models for nodule detection and classification are required.

To reduce FP results, we can make use of convolutional neural network (CNN) architectures, which are residual network (ResNet), densely connected network (DenseNet), dual path network (DPN), and mixed link network (MixNet) which can improve classification accuracy and is more effective than going deeper or wider (Figure 8.4).

In this paper, detection of lung cancer is based on three-dimensional (3D) lung CT scans, along with physiological symptoms and the clinical biomarkers, to reduce FP results and ultimately prevent invasive methods from being used.

8.5 WORKING MODEL

Basically, CAD systems for lung cancer have the following pipeline: image preprocessing, detection of cancerous nodule candidates, nodule candidate FP reduction, malignancy prediction for each nodule candidate, and malignancy prediction for the overall CT scan. These pipelines have many phases, each of which is computationally expensive and requires well-labeled data during training. For instance, the

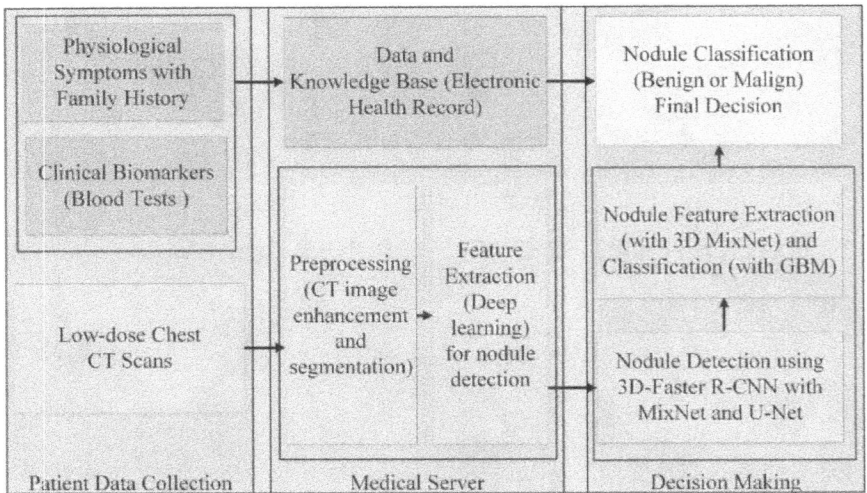

FIGURE 8.4 Flow chart of a neural network.

FP reduction phase requires a dataset of labeled true and false nodule candidates, and the nodule malignancy prediction phase requires a dataset with nodules labeled with malignancy.

True/false labels, i.e., generally denoted as 0/1 for nodule candidates and malignancy labels for nodules, are sparse for lung cancer and may be nonexistent for some other cancers, so CAD systems that rely on such data would not apply to other cancers. In order to achieve greater computational efficiency and generalizability to other cancers convolution neural networks are better.

The presently available CAD system has a shorter pipeline and only requires the following data during training: a dataset of CT scans with true nodules labeled, and a dataset of CT scans with an overall malignancy label and also starts with preprocessing the 3D CT scans using segmentation, normalization, downsampling, and zero-centering. The preliminary approach was to simply input the preprocessed 3D CT scans into 3D CNNs, but the results were poor. So, an additional preprocessing was performed to input only regions of interest into the 3D CNNs. To identify regions of interest, a convolutional U-Net was trained for nodule candidate detection. Then input regions around nodule candidates detected by the U-Net were fed into 3D CNNs to ultimately classify the CT scans as positive or negative for lung cancer. The overall architecture is shown in Figure 8.5.

For each patient, pixel values were first converted in each image to Hounsfield units (HU), a measurement of radiodensity, and 2D slices are stacked into a single 3D mage. Because malignancy forms on lung tissue, segmentation is used to mask out the bone, outside air, and other substances that would make data noisy and leave only lung tissue information for the classifier. A number of segmentation approaches were tried, including thresholding, clustering (K-means and mean shift), and watershed. K-means and mean shift allow very little supervision and did not produce good qualitative results. Watershed produces one of the best qualitative results, but takes too long to run to use by the deadline. Ultimately, thresholding was used. After segmentation, the 3D mage is normalized by applying the linear scaling to squeeze all pixels of the original unsegmented image to values between 0 and 1. Spline interpolation downsamples each 3D image by a scale of 0.5 in each of the three dimensions. Finally, zero-centering s performed on data by subtracting the mean of all the mages from the training set (Figure 8.6).

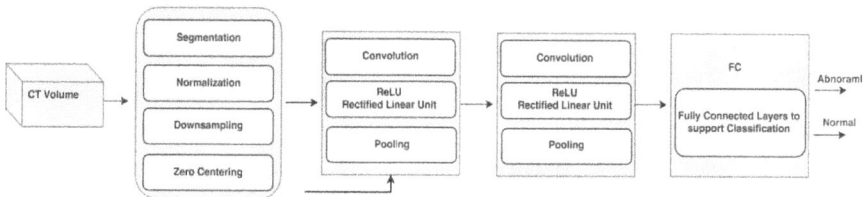

FIGURE 8.5   3D convolutional neural networks architecture.

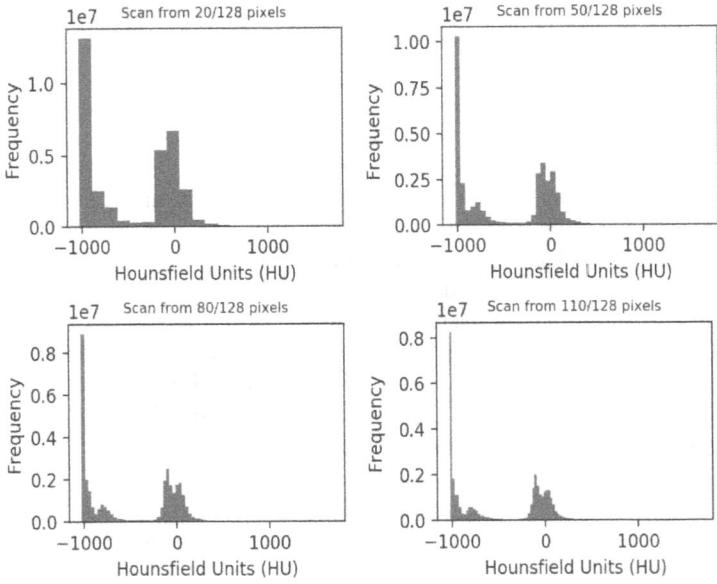

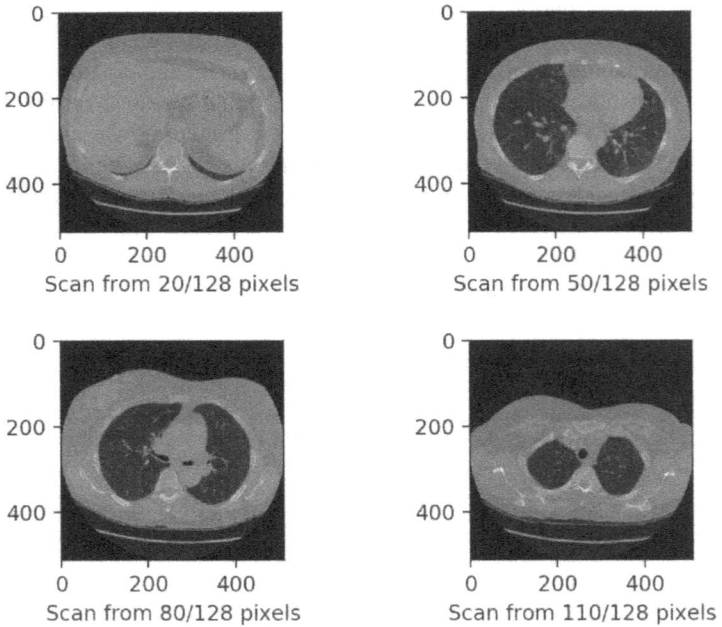

FIGURE 8.6   (a) Histograms of pixel values in HU for sample patients' CT scans at various slices. (b) Corresponding 2D axial slices.

8.6 THRESHOLDING

Air is generally around −1000 HU; lung tissue is typically around −500; water, blood, and other tissues are around 0 HU; and bone is typically around 700 HU, so pixels that are close to −1000 or above −320 are masked (Figure 8.7).

8.7 WATERSHED

The segmentation obtained from thresholding has a lot of noise. Many voxels, i.e., each of an array of elements of volume that constitute a notional 3D space, especially each of an array of discrete elements into which a representation of a 3D object is divided) that were part of lung tissue, especially voxels at the edge of the lung,

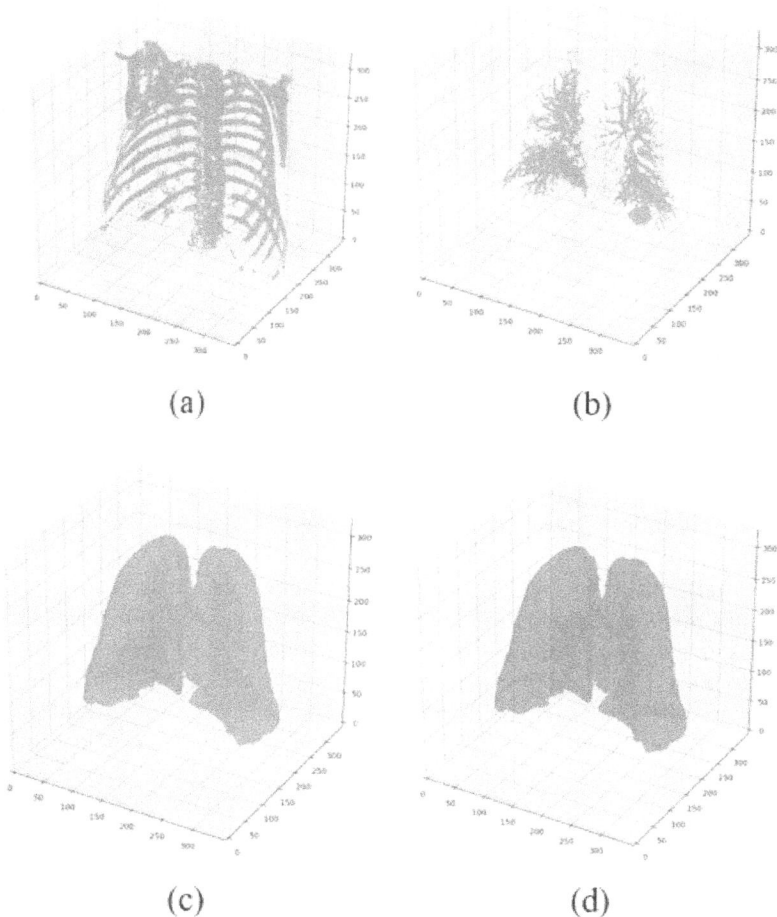

FIGURE 8.7   (a) Sample patient 3D mage with pixels values greater than 400 HU reveals the bone segment. (b) Sample patient bronchioles within lung. (c) Sample patient initial mask with no air. (d) Sample patient final mask in which bronchioles are included.

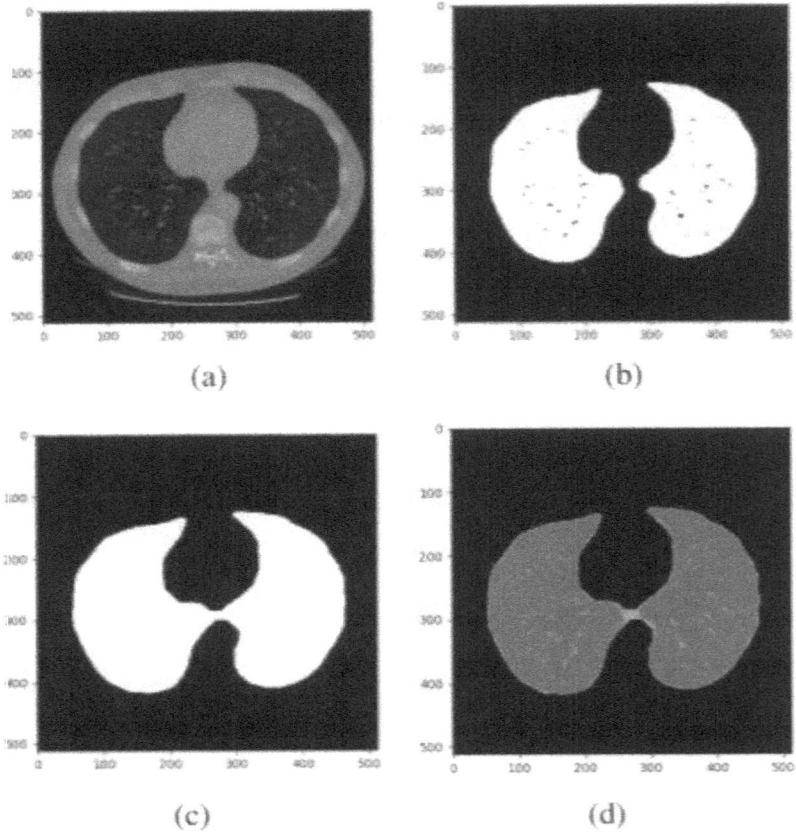

(a)   (b)

(c)   (d)

FIGURE 8.8   Watershed segmentation output

tended to fall outside the range of lung tissue radiodensity due to CT scan noise. This means that our classifier will not be able to correctly classify images in which cancerous nodules are located at the edge of the lung [5]. To filter noise and include voxels from the edges, we use driven watershed segmentation. Qualitatively, this produces a much better segmentation than thresholding. Missing voxels (black dots in Figure 8.8) are largely reincluded. However, this is much less efficient than basic thresholding, so due to time limitations, it was not possible to preprocess all CT scans using watershed, so thresholding was used instead. The typical radiodensities in HU of various substances in a CT Scan are shown in Table 8.1.

8.8   MALIGNANCY 3D CNN CLASSIFIER

Once the U-Net was trained on the LUNA16 data, it is running on 2D slices of Kaggle data and stacked the 2D slices back to generate a nodule candidate. Ideally the output of U-Net would give the exact locations of all the nodules, and it would be

TABLE 8.1

Typical Radiodensities in HU of Various Substances in a CT Scan

Substance	Radiodensity (HU)
Air	−1000
Lung tissue	−500
Water and blood	0
Bone	700

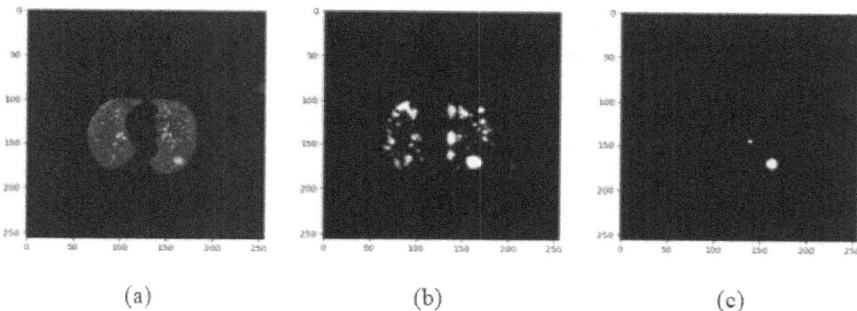

(a)   (b)   (c)

FIGURE 8.9   (a) U-Net sample input from LUNA16 validation set. Note that the image has the largest nodule from the LUNA16 validation set, which we chose for clarity; most nodules are significantly smaller than the largest one in this mage. (b) U-Net predicted output from LUNA16 validation set. (c) U-Net sample labels mask from LUNA16 validation set showing nodule location.

able to declare images with nodules as detected by U-Net are positive for lung cancer and images without any nodules detected by U-Net are negative for lung cancer [6]. However, as shown in Figure 8.9c, U-Net produces a strong signal for the actual nodule, but also produces a lot of FPs, so we need an additional classifier that determines the malignancy. The model parameter is shown in Table 8.2.

A 3D CNN was used as a linear classifier. It uses weighted SoftMax cross entropy loss (weight for a label the inverse of the frequency of the label in the training set) and Adam Optimizer, and the CNNs use ReLU activation and dropout after each convolutional layer during training [7, 8].

8.9   RESULTS

For pulmonary, i.e., lung nodule detection using CT imaging, CNNs have recently been used as a feature extractor within a larger CAD system. For simplicity in training and testing we selected the ratings of a single radiologist. All the test experiments were done using 50% training set, 20% validation set, and 30% testing set. To estimate the results, we considered a variety of testing metrics. The accuracy metric

TABLE 8.2

Model Parameter

Layer	Params	Activation	Output
Input			$256 \times 256 \times 1$
Conv1a	$3 \times 3 \times 32$	ReLu	$256 \times 256 \times 32$
Conv1b	$3 \times 3 \times 32$	ReLu	$256 \times 256 \times 32$
Max Pool	2×2, stride 2		$128 \times 128 \times 32$
Conv2a	$3 \times 3 \times 80$	ReLu	$128 \times 128 \times 80$
Conv2b	$3 \times 3 \times 80$	ReLu	$128 \times 128 \times 80$
Max Pool	2×2, stride 2		$64 \times 64 \times 80$
Conv3a	$3 \times 3 \times 160$	ReLu	$64 \times 64 \times 160$
Conv3b	$3 \times 3 \times 160$	ReLu	$64 \times 64 \times 160$
Max Pool	2×2, stride 2		$32 \times 32 \times 160$
Conv4a	$3 \times 3 \times 320$	ReLu	$32 \times 32 \times 320$
Conv4b	$3 \times 3 \times 320$	ReLu	$32 \times 32 \times 320$
Up Conv4b	2×2		$64 \times 64 \times 320$
Concat	Conv4b, Conv3b		$64 \times 64 \times 480$
Conv5a	$3 \times 3 \times 160$	ReLu	$64 \times 64 \times 160$
Conv5b	$3 \times 3 \times 160$	ReLu	$64 \times 64 \times 160$
Up Conv5b	2×2		$128 \times 128 \times 160$
Concat	Conv5b, Conv2b		$128 \times 128 \times 240$
Conv6a	$3 \times 3 \times 80$	ReLu	$128 \times 128 \times 80$
Conv6b	$3 \times 3 \times 80$	ReLu	$128 \times 128 \times 80$
Up Conv6b	2×2		$256 \times 256 \times 80$
Concat	Conv6|b, Conv1b		$256 \times 256 \times 112$
Conv6a	$3 \times 3 \times 32$	ReLu	$256 \times 256 \times 32$
Conv6b	$3 \times 3 \times 32$	ReLu	$256 \times 256 \times 32$
Conv7	$3 \times 3 \times 3$		$256 \times 256 \times 2$

was the used metric in our evaluations. In our first set of experiments we considered a range of CNN architectures for the binary classification task. Early experimentation suggested that the number of filters and neurons per layer were less significant than the number of layers. Thus, to simplify analysis the first convolutional layer used seven filters with size $5 \times 5 \times 5$, the second convolutional layer used 17 filters with $5 \times 5 \times 3$, and all fully connected layers used 256 neurons. These were found to generally perform well, and we considered the impact of one or two convolutional layers followed by one or two fully connected layers. The networks were trained as described earlier and the results of these experiments can be found. Our results suggest that two convolutional layers followed by a single hidden layer are one of the optimal network architectures for this dataset. Figure 8.10 shows the training error for the 3D CNN.

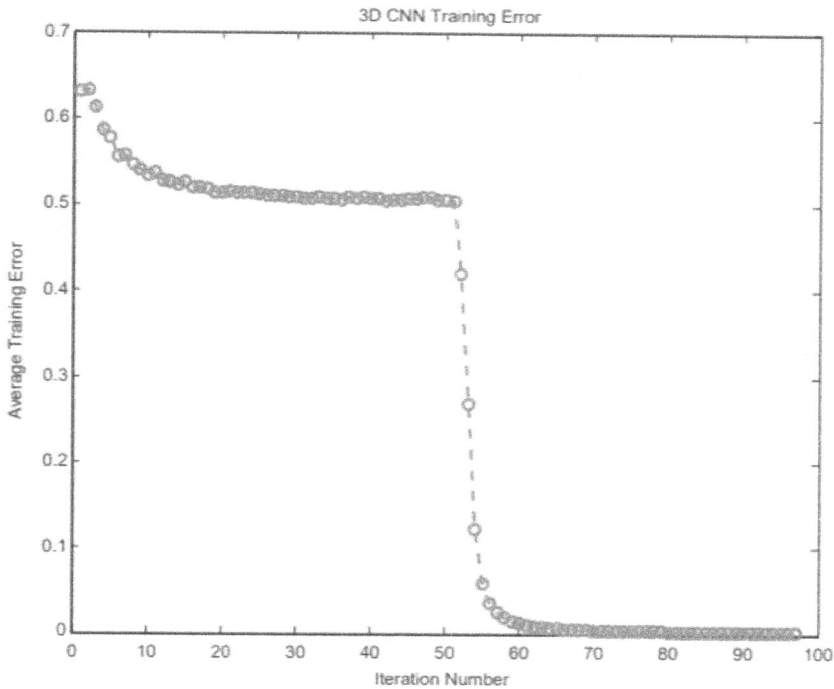

FIGURE 8.10   Training error.

Confusion Matrix of 3D CNN Using 30% Testing

	Predicted	
	Abnormal	Normal
Actual		
Abnormal	**0.853**	0.147
Normal	0.119	**0.881**

FIGURE 8.11   Confusion matrix.

The results are shown in figure 8.11 by confusion matrix achieved on the DSB dataset with 3D CNN. The accuracy of model is 86.6%, and the misclassification rate is 13.4%, the FP rate is 11.9%, and the false-negative rate is 14.7%. Almost all patients are classified correctly. Secondarily, there is an enhancement on accuracy due to efficient U-Net architecture and segmentation.

8.10 CONCLUSION

In this chapter we developed a deep CNN architecture to detect nodules in patients with lung cancer and detect the interest points using U-Net architecture. This is a prheprocessing step for 3D CNN. The 3D CNN model performed the best on the test set and produced results accordingly. While we achieve state-of-the-art performance in area under the curve (AUC) of 0.83, we performed well, considering that we use less labeled data than most state-of-the-art CAD systems. As an interesting observation, the first layers were a preprocessing layer for segmentation using different techniques. Identification of nodules is done through these variables: threshold, watershed, and U-Net. The network can be trained end-to-end from raw image patches. Its main requirements are the availability of training database, but otherwise no assumptions are made about the objects of interest or underlying image modality. Advancement could be possibly to extend our current model to not only determine whether or not the patient has cancer but also determine the exact location of the cancerous nodules. The immediate further improvisation on this architecture is to use watershed segmentation as the basic lung segmentation. Other opportunities for improvement include making the network deeper and more extensive hyper parameter tuning.

REFERENCES

[1] Y. Xu, T. Mo, Q. Feng, P. Zhong, M. Lai, and E. I. Chang, "Deep learning of feature representation with multiple instances learning for medical image analysis," in *IEEE International Conference on Acoustics, Speech and Signal Processing, ICASSP*, pp. 1626–1630, 2014.

[2] D. Kumar, A. Wong, and D. A. Clausi, "Lung nodule classification using deep features in ct images," in *2015 12th Conference on Computer and Robot Vision*, pp. 133–138, June 2015.

[3] Y. Bar, I. Diamant, L. Wolf, S. Lieberman, E. Konen, and H. Greenspan, "Chest pathology detection using deep learning with non-medical training," in *Proceedings—International Symposium on Biomedical Imaging*, vol. 2015, pp. 294–297, July 2015.

[4] W. Sun, B. Zheng, and W. Qian, "Computer aided lung cancer diagnosis with deep learning algorithms," in *SPIE Medical Imaging*, vol. 9785, pp. 97850Z–97850Z, International Society for Optics and Photonics, 2016.

[5] A. Chon, N. Balachandar, and P. Lu, *Deep Convolutional Neural Networks for Lung Cancer Detection*, Technical report, Stanford University, 2017.

[6] Y. LeCun, K. Kavukcuoglu, and C. Farabet, "Convolutional networks and applications in vision," in *Proceedings of the IEEE International Symposium on Circuits and Systems (ISCAS)*, pp. 253–256, IEEE, 2010.

[7] C. Chola, A. Y. Muaad, M. B. Bin Heyat, J. B. Benifa, W. R. Naji, K. Hemachandran, . . . T. S. Kim, "BCNet: A deep learning computer-aided diagnosis framework for human peripheral blood cell identification," *Diagnostics*, 12(11), 2815, 2022.

[8] K. Hemachandran, A. Alasiry, M. Marzougui, S. M. Ganie, A. A. Pise, M. T. H. Alouane, and C. Chola, "Performance analysis of deep learning algorithms in diagnosis of malaria disease," *Diagnostics*, 13(3), 534, 2023.

9 Breast Cancer Prediction Using Machine Learning Algorithms

Raja Krishnamoorthy, Arshia Jabeen, and Harshitha Methukula

9.1 INTRODUCTION

Cancer is a disease that occurs when there are changes or mutations in genes relating to cell growth. These mutations allow the cells to divide and multiply in an uncontrolled and chaotic manner. These cells keep increasing and start making replicas which end up becoming more and more abnormal [1]. These abnormal cells later on form a tumor. Tumors, unlike other cells, don't die even though the body doesn't need them.

The major cancer classifications are categorized two types, that is, malignant and benign. Malignant cancers are cancerous. These cells keep dividing uncontrollably and start affecting other cells and tissues in the body [2, 3]. They reach to all other organs of the body, and it is hard to cure this type of cancer. Chemotherapy, radiation therapy and immunotherapy are types of treatments that can be given for these types of tumors. Benign cancer is non-cancerous. Unlike malignant, this tumor shouldn't propagate to rest of the organs and hence is much less risky than malignant. In many cases, such tumors don't really require any treatment.

9.1.1 MACHINE LEARNING

The machine learning (ML) advancement contains better classification cum computation capability in various domains. ML has been seen many times by individuals while shopping on the internet. They are then shown ads based on what they were searching earlier on [4]. This happens because many of these websites use ML to customize the ads based on user searches, and this is done in real time as shown in Figure 9.1. ML has also been used in other various places like detecting fraud, filtering of spam, etc.

9.1.2 MACHINE LEARNING METHODS

Supervised learning—Here both the input and output are known. The training dataset also contains the answer the algorithm should come up with on its own [5].

DOI: 10.1201/9781003328414-9

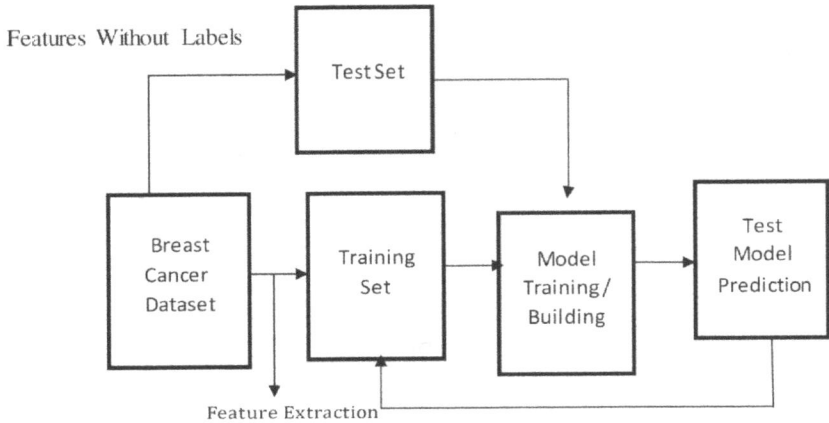

Train/Test loop: Test accuracy predictions matching Test Labels

FIGURE 9.1   System architecture.

Unsupervised learning—Here input dataset is known but output is not known. A deep learning model is given a dataset without any instructions on what to do with it. The training data contains information without any correct result. The network tries to automatically understand the structure of the model.

Semi-supervised learning—This type consists of labelled and unlabelled datasets.

Reinforcement learning—In this type, artificial intelligence (AI) agents are trying to find the best way to accomplish a particular goal. It tries to predict the next step which could possibly give the model the best result at the end.

9.2   EXISTING SYSTEM

Typically, all the researchers are following any one of the methods denoted here:

- **Breast exam**
- **Mammogram**
- **Breast ultrasound**
- **MRI of the breasts**
- **Removing a sample of breast cells for testing (biopsy)**

9.3   PROPOSED SYSTEM

In the proposed system we plan on using existing data of breast cancer patients which has been collected for a number of years and run different ML algorithms on them. It is done by taking the patients' data and mapping it with the dataset and checking whether there are any patterns found with the data. If a patient has breast cancer, then instead of taking more tests to check whether the cancer is malignant or benign, ML can be used to predict the case based on the huge amount of data on breast cancer. This proposed system helps the patients as it reduces the amount of money they need to spend just for the diagnosis. Also, if the tumor is benign, then it is not cancerous, and the patient doesn't need to go through any of the other tests. This saves a lot of time as well.

9.3.1 Module Description

9.3.1.1 Dataset

Wisconsin Diagnostic Breast Cancer methods have improved and it has been used in the United States. The features which are in the dataset are processed. These features describe different characteristics of the cell nuclei found in the image. There are 569 data points in the dataset: 212 for malignant and 357 for benign [1].

Each of these features has information on mean error, standard error, and "worst" or largest datasets computed. Hence, the dataset has a total of 30 features.

9.3.2 Proposed System-1 (Decision Trees)

This algorithm is used to predict the exact outcome or target variable based on many input variables. They are a collection of divide and conquer problem-solving strategies denoted in Table 9.1. It takes the shape of a tree-like structure. It starts with root nodes and these split into sub-nodes or child nodes [6]. These branches keep spitting until the outcome isn't reached, shown in Figure 9.2.

In this table have been compared with existing data and proposed system 1 like decision tree method, it represents the system analysis.

9.3.3 Proposed System-2 (Random Forest and Tree Methods)

Random forest [7] procedures for both classification and regression problems are shown in Figure 9.3. All together these methods use plenty of learning models to add better analytical results shown in Table 9.2. Random forest intends to reduce association concern by select only a sub-sample of the feature denoted in Figure 9.4.

Here, the proposed work in Table 9.2 denoted the random forest compilation outcome with an existing system.

TABLE 9.1

Proposed System – Decision Tree Method

S. No.	Existing System Data	Proposed System Data
1	4.2	7.0
2	2.6	4.3
3	3.4	6.0
4	4.8	7.9

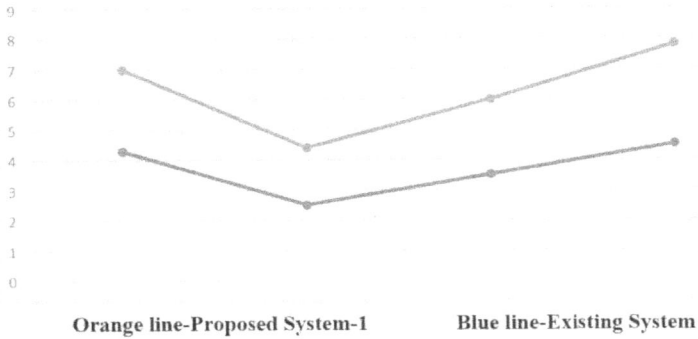

Orange line-Proposed System-1   Blue line-Existing System

FIGURE 9.2   Comparison between existing system and proposed system-1.

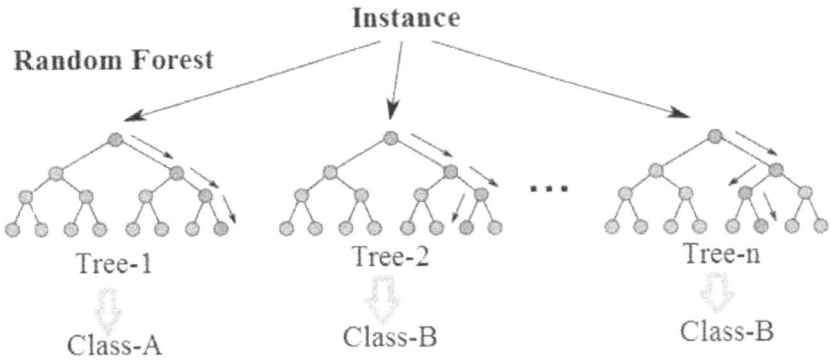

FIGURE 9.3   Random forest.

TABLE 9.2

Proposed System – Random Forest Method

S. No.	Existing System Data	Proposed System Data
1	3.1	4.0
2	4.0	7.0
3	6.0	8.0
4	8.1	9.0

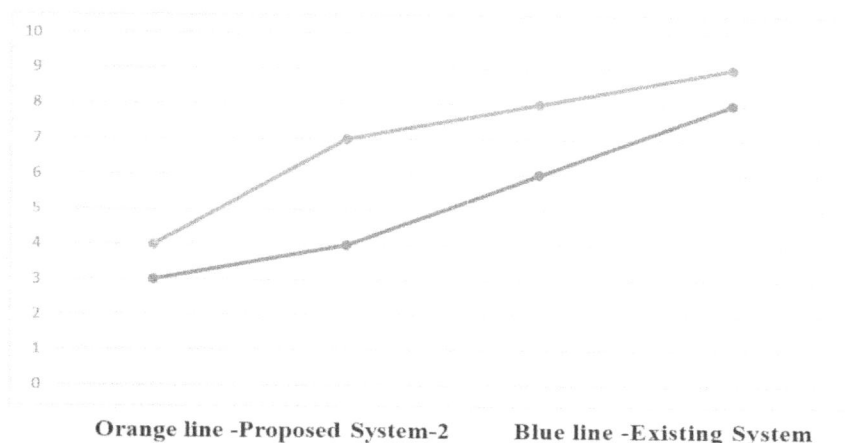

Orange line -Proposed System-2   Blue line -Existing System

FIGURE 9.4   Comparison between existing system and proposed system-2.

9.3.4   PROPOSED SYSTEM-3 (LOGISTIC REGRESSION)

Logistic regression is a supervised learning machine learning model that envisages the probability of a binary outcome [8–10], as shown in Figure 9.5. So, the resultant outcome should be an unconditional or distinct value discussed in Table 9.3.

The proposed logic regression system-3 has been implemented, and the result was analyzed through Table 9.4 then represented in Figure 9.6. In this work progress is evaluated with the necessary parameter.

Consequently, all the three proposed systems are investigated through the parameterwise comparison as mentioned in Table 9.4 and the analysis report mentioned in Figure 9.7.

9.3.5   DESIGN

A. DESIGN GOALS
Under this model, the goal of this project is to create a design to achieve the following:

B. ACCURACY
Only accurate outcomes can help make this model a good one. It can be reliable only when all the outcomes are correct and can be trusted. As this data is required for healthcare purposes, it is important that no errors occur.

C. EFFICIENCY
The model should be efficient as there is no requirement of manual data entry work or any work by doctors. It takes less time to predict outcomes after all the ML algorithms have been used on the data.

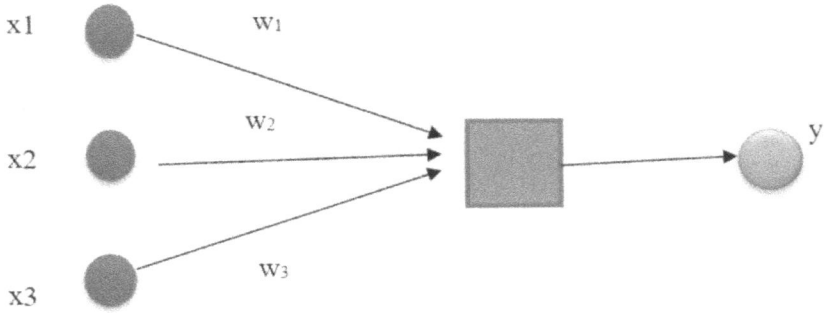

FIGURE 9.5   Logistic regression.

TABLE 9.3
Proposed System – Logistic Regression Method

S. No.	Proposed System Data	Existing System Data
1	2.0	2.0
2	2.5	3.0
3	3.5	3.0
4	4.5	2.9

TABLE 9.4
All Proposed System Comparison (Decision Tree, Random Forest and Logistic Regression)

S. No.	Proposed Sys-1	Proposed Sys-2	Proposed Sys-3
1	7.0	4.0	2.0
2	4.3	7.0	2.5
3	6.0	8.0	3.5
4	7.9	9.0	4.5

9.3.6   System Architecture

As this project does not have any user interface, the architecture is basically the dataset and the features of the dataset. It is trying to understand the dataset and make the system as simple and easy as possible. The dataset is first split into

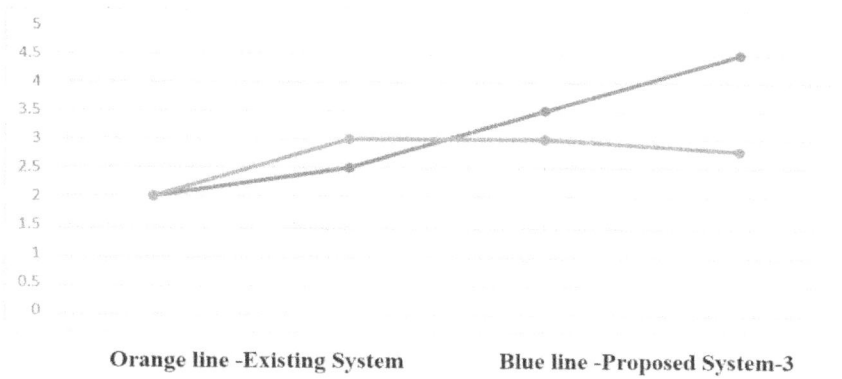

Orange line -Existing System   **Blue line -Proposed System-3**

FIGURE 9.6   Comparison between existing system and proposed system-3.

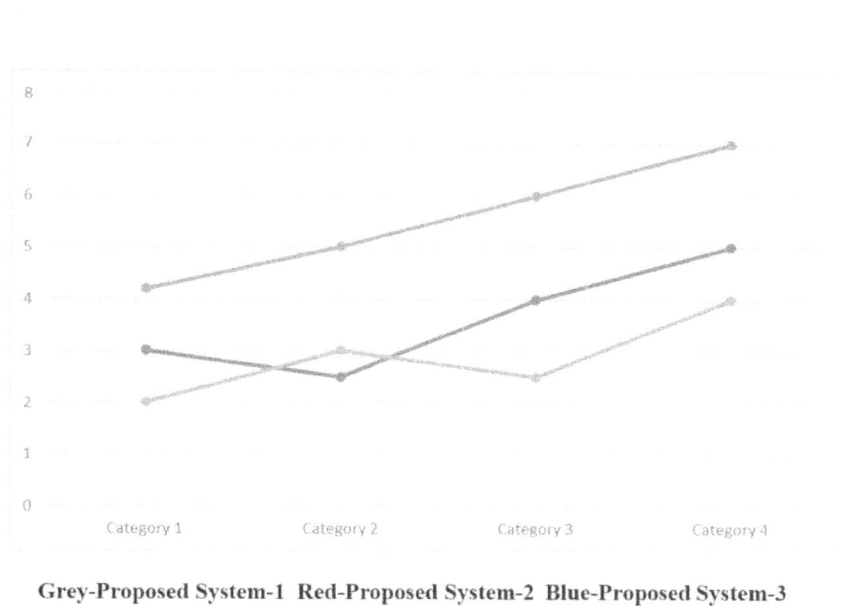

Grey-Proposed System-1   Red-Proposed System-2   Blue-Proposed System-3

FIGURE 9.7   All proposed systems comparison chart.

training and testing sets. The training set is first exposed to the ML algorithms so that the system understands what data gives what type of outcome. After the system is trained, the testing data is used to test whether the system can correctly predict the class of the data illustrated in Figure 9.8. It checks the percentage accuracy of the model.

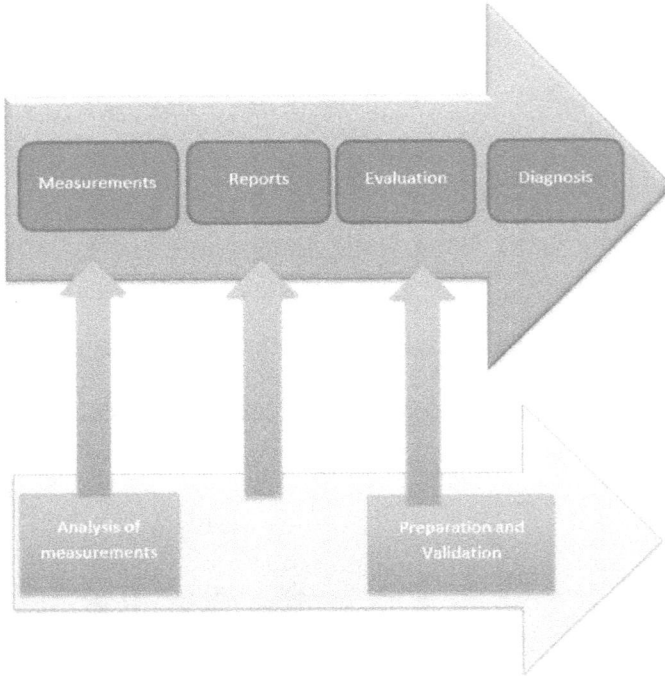

FIGURE 9.8   Dataflow diagram.

9.4   IMPLEMENTATION

PREPARING THE DATA

Step 1: The first step in the ML process is to prepare the data.

Step 2: After importing all the necessary packages, we need to load the dataset. We use the help of Pandas to load the dataset.

Step 3: We need to drop the first column of the dataset which consists of IDs, as this field will not help us in the classification process. This is done as follows:

Stage Four: Predict the number of DP points damaged and initiate.

diagnosis_all = list(data.shape)[0]

diagnosis_categories = list(data['diagnosis'].value_counts())

print("\n\t The data has{} diagnosis, {} malignant and {}benign.".format(diagnosis_all,diagnosis_categories[0],diagnosis_categor is[1]))

The data has 569 diagnosis: 357 malignant and 212 benign.

9.4.1   Visualizing the Data

The dataset will be reorganized by applying different methods and calculation sns.countplot(df['diagnosis'].label="count"

Relating the factors of the tissue for more understanding
sns.pairplot(df.iloc[:,1.:5],hue="diagnosis")
Get the correlation
df,iloc[:,1:32].corr()
The relation between the factors is calculated and demonstrated using a correlation matrix but it has all the numerical values in which they are related to get more understanding of the correlation matrix visualized using a heat map
features_mean= list(data.columns[1:11])

9.4.2 APPLYING THE ALGORITHMS

DECISION TREE

Steps to use in Jupyter:

1. Transfer the set of data to. csv file.
2. Bring the data into Rstudio.
3. Take away irrelevant columns and data.
4. Stabilize the data with different scales.
5. Crack the data into running out and appraise data (80–20).
6. With help of testing appraise data models and decision tree, get the envisaged outcome values.
7. Evaluate the observed values to actual values and calculate accuracy of model.

Accuracy Output: 0.964912280

9.4.3 RANDOM FOREST AND TREE METHODS

To utilize all the mean features contains in dataset, compile the dataset for predicting the precision-level outcome of breast cancer data. By training the model using the dataset and using prediction techniques, accuracy is found and the resultant outcome was disclosed in Figure 9.9.
Random Forest Accuracy: 0.99780

9.5 RESULTS

According to the compiled algorithm we have secured the following outcome

9.6 CONCLUSION

In this chapter, the suitable dataset is collected to help in this predictive analysis. This dataset is then processed to remove all the junk data. The predictive analysis method is being used in many different fields and is slowly picking up pace. It is helping us by using smarter ways to solve or predict a problem's outcome. This scheme was developed to reduce the time and cost factors of the patients as well as to minimize the work of a doctor, and we have tried to use a very simple and understandable

```
Model 0
                  precision    recall  f1-score   support

              0      0.96      0.99      0.97        67
              1      0.98      0.94      0.96        47

       accuracy                         0.96       114
      macro avg      0.97      0.96      0.96       114
   weighted avg      0.97      0.96      0.96       114

Accuracy :  0.9649122807017544
Model 1
                  precision    recall  f1-score   support

              0      0.94      0.96      0.95        67
              1      0.93      0.91      0.92        47

       accuracy                         0.94       114
      macro avg      0.94      0.94      0.94       114
   weighted avg      0.94      0.94      0.94       114

Accuracy :  0.9385964912280702
Model 2
                  precision    recall  f1-score   support

              0      0.96      1.00      0.98        67
              1      1.00      0.94      0.97        47

       accuracy                         0.97       114
      macro avg      0.98      0.97      0.97       114
   weighted avg      0.97      0.97      0.97       114
```

FIGURE 9.9   Accuracy of all the modes.

model to do this job. The ML algorithm of this work computed the training data and testing data that must be used to check accuracy of the outcome.

REFERENCES

[1] Wolberg, Street and Mangasarian, "Wisconsin Diagnostic Breast Cancer Dataset", http://archive.ics.uci.edu/ml.
[2] S. Palaniappan and T. Pushparaj, "A Novel Prediction on Breast Cancer from the Basis of Association Rules and Neural Network", *International Journal of Computer Science and Mobile Computing*, 2(4), (2013), 269–277.
[3] N. Khuriwal and N. Mishra, "A Review on Breast Cancer Diagnosis in Mammography Images Using Deep Learning Techniques", *Journal of Image Processing & Pattern Recognition Progress*, 5(1), (2018), 51–57.
[4] Mengjie Yu, *Breast Cancer Prediction Using Machine Learning Algorithm*, The University of Texas at Austin, 2017.
[5] Wenbin Yue, Zidong Wang, Hongwei Chen, Annette Payne, and Xiaohui Liu, "Machine Learning with Applications in Breast Cancer Diagnosis and Prognosis", *Designs*, 2(2), (2018), 13. https://doi.org/10.3390/designs2020013

[6] J. A. Cruz and D. S. Wishart, "Applications of Machine Learning in Cancer Prediction and Prognosis", *Cancer Informatics*, 2, (2006), 117693510600200030.

[7] Vikas Chaurasia and S. Pal, "Using Machine Learning Algorithms for Breast Cancer Risk Prediction and Diagnosis", *FAMS*, 83 (2016), 1064–1069.

[8] Y. Khourdifi and M. Bahaj, "Feature Selection with Fast Correlation-Based Filter for Breast Cancer. Prediction and Classification Using Machine Learning Algorithms," 2018 International Symposium on Advanced Electrical and Communication Technologies (ISAECT), Rabat, Morocco, 2018, pp. 1–6.

[9] R. M. Mohana, R. Delshi Howsalya Devi, and Anita Bai, "Lung Cancer Detection using Nearest Neighbour Classifier", *International Journal of Recent Technology and Engineering (IJRTE)*, 8(2S11) (September 2019).

[10] Ch. Shravya, K. Pravalika, and Shaik Subhani, "Prediction of Breast Cancer H. Using Supervised Machine Learning Techniques", *International Journal of Innovative Technology and Exploring Engineering (IJITEE)*, 8(6) (April 2019).

10 Breast Cancer Histopathological Images Classification Using Deep Learning

Laxminarayan Pimpdae, Tanvi Gorantla, Raul V. Rodriguez, and Jk. Dhivya

10.1 BACKGROUND

Cancer involves abnormal and uncontrollable cell growth with the potential to proliferate and advance to other body parts. This contrasts with benign cancers, which do not spread to other body parts. In the year 2018, with a death rate of 9.6 million, cancer ranks as the second most common cause of death worldwide. Low- and middle-income countries witness close to 70% of these deaths [1]. After skin cancer, breast cancer is the second most occurrent cancer to affect women. Mammograms are x-rays which aid in the early detection of breast cancer. Breast cancer is formed in the breast cells. Breast cancer occurs in both men and women, but it rarely occurs in men. Nearly 500,000 women around the world are killed due to breast cancer. In 2018, breast cancer took the lives of 627,000 women, which makes up 15% of all cancer-related deaths in women, according to estimates from the World Health Organization (WHO).

As a part of the screening process women are tested to find cancer cells before any symptoms arise. The use of mammography, professional breast examinations, and breast self-exams are only a few of the various techniques used for screening [2].

10.2 MAMMOGRAPHY

It employs low-energy x-rays to find breast abnormalities. In high-resource environments, it has been proven to reduce breast cancer mortality by about 20%. The WHO position paper on mammography screening deduced that in well-equipped environments women between the age group of 50 and 69 years should undergo organized, population-based mammography screening if prerequisite conditions on program implementation are satisfied. Mammography is not cost-effective in places with weak health systems and limited resources; thus early detection should be focused at diagnosis via increased awareness. The WHO advises systematic mammography

DOI: 10.1201/9781003328414-10

only in the context of thorough research and in locations with adequate resources for women aged 40–49 and 70–75.

10.3 CLINICAL BREAST EXAM

It is an examination of both breasts carried out by a trained healthcare professional. Based on the findings from ongoing research, the clinical breast exam (CBE) will be adopted since it appears to be a promising strategy for low-resource settings. Screening incurs a considerable demand for expenses and, consequently, makes it unaffordable to many; hence, the choice to proceed with screening should be pursued only after a basic breast scan has been conducted with further tests proven necessary [3].

The two types of breast cancer are benign, which is not hazardous to health, and malignant, which is potentially lethal. Benign cancers are not considered destructive since the cells in benign cancers look like normal cells in appearance, they grow slowly, and they don't affect nearby tissues or spread to other body parts. Malignant cells could be metastatic. See Figure 10.1.

10.4 BENIGN TUMOR

Benign tumor cells tend not to spread, and the growth rate of these cells is also slow. These cells are seen under a pathologist's microscope. In most cases, the outlook with benign cancers is optimal. However, benign cancers are considered serious and may require treatment if they press on important structures such as nerves or blood vessels [4].

10.5 CAUSES

Benign breast problems are generally brought on by a variety of factors, including breast composition, age, and hormone problems. External treatments such

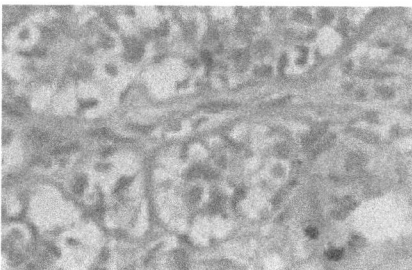
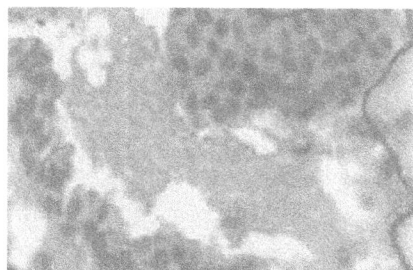

Benign   Malignant

FIGURE 10.1 Benign cancer and malignant cancer.

as birth control pills and hormone therapy can also be an active cause for it. Gynecomastia, or male benign breast disease, is brought on by an imbalance of hormones. In addition, certain diseases, hormone therapy, and being overweight contribute to it.

1. Adenosis:

Adenosis occurs once several lobules (milk-producing sacs) grow larger and contain more glands than usual. If the enlarged lobules contain scar-like fibrous tissue, this is often referred to as sclerosing gland disease. Adenosis might cause a lump that you or your doctor will feel (Figure 10.2). A diagnostic assay is required to assess the distinction between gland disease and cancer. With a real gland disease, any increase in carcinoma risk seems to be slight.

2. Fibroadenoma:

A fibroadenoma is typically felt as a lump within the breast that is smooth and moves easily below the skin. Fibroadenomas are sometimes painless; however, they'll feel tender or perhaps painful. See Figure 10.3.

Types of fibroadenoma:

1. Simple fibroadenoma:

Most fibroadenomas are 1–3 cm in size and are referred to as 'easy fibroadenomas'. Once checked out beneath a magnifier, simple fibroadenomas can look similar everywhere. Simple fibroadenomas don't increase the chance of developing carcinoma in the future.

2. Complex fibroadenoma:

Having a noticeable adenoma will increase the danger of developing carcinoma within the future.

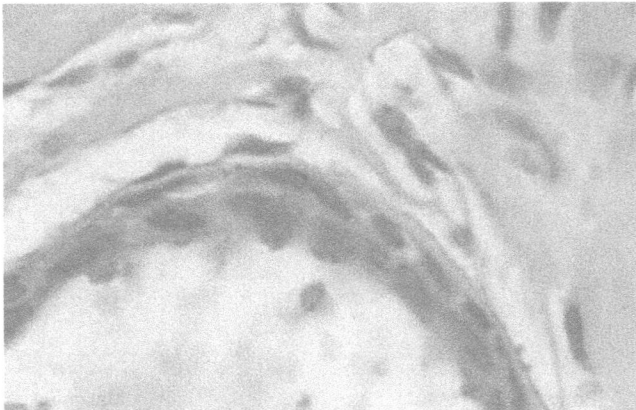

FIGURE 10.2 Adenosis.

FIGURE 10.3   Fibroadenoma.

3. **Giant or juvenile fibroadenoma:**
   Fibroadenoma will grow over 5 cm and will be referred to as a 'giant-fibroadenoma'. Those found in teenaged women are also referred to as 'juvenile fibroadenomas'.

3. **Phyllodes tumor:**
   Less than 1% of breast tumours are phyllodes tumours, which are extremely uncommon. The word "phyllodes," which translates to "leaf-like" in Greek, alludes to the established fact that tumour cells develop in a pattern like a leaf. These tumour cells also go by the labels phyllodes tumour and cystosarcoma phyllodes [5]. Phyllodes tumours rarely spread outside the breast but have a tendency to grow quickly. The majority of Phyllodes tumours are benign, although a small number are malignant or borderline. Phyllodes tumours can form at any age; however, they typically start to manifest once a woman reaches her forties. See Figure 10.4.

4. **Tubular adenoma:**
   A cannular non-malignant tumor may be a reasonably non-malignant tumor characterised by 'tubular' and 'acinar' -bud-like cell structures that are closely packed. An associated benign tumor may be a disordered neoplasm consisting of varying proportions of organ, fat, and fibrous tissue which will lead to a mass or pseudo-lesion [6]. Tubular adenomas are also referred to as 'pure' adenomas as a result of their inclination to not show excessive epithelial hyperplasia and tissue growth. However, they seem mixed with the generic 'fibroadenoma'. See Figure 10.5.

Malignant tumor: malignant tumor cells rapidly spread. These cells may have an abnormal shape.

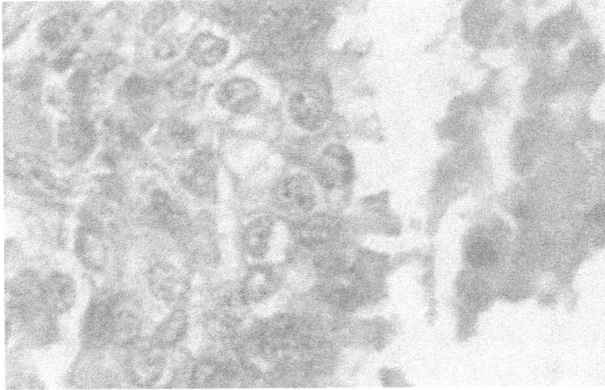

FIGURE 10.4   Phyllodes tumor.

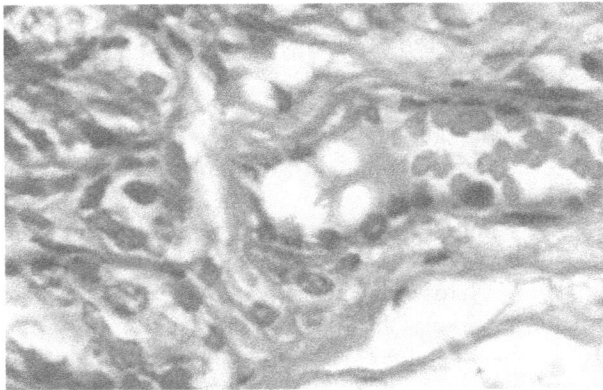

FIGURE 10.5   Tubular adenoma.

Malignant tumors are classified into:

1. Ductal carcinoma:
Ductal cancer is a non-invasive carcinoma that's restricted to the ducts in the breast. Accrued use of diagnostic techniques like screening has led to a rapid increase in ductal cancer detection (DC). Sixty-four thousand cases of DC are annually identified in the United States, and 90% of DC cases are classified as suspicious calcifications using diagnostic techniques having a linear, clustered, segmental, focal, or mixed distribution. See Figure 10.6.

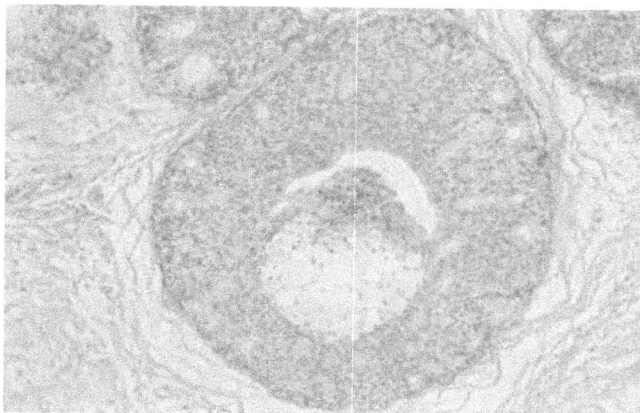

FIGURE 10.6 Ductal carcinoma.

DC is classified into three types, that is low grade (grade I), intermediate grade (grade II), or high grade (grade III) on the basis of growth rate of cells. Classifying DC as low and intermediate-grade indicates that the cancer cells' square measure is growing at a comparatively slow rate. Low-grade DC cells resemble atypical ductal dysplasia cells or conventional breast cells in many ways. Grade II DC cells develop more quickly than usual and seem less likely to be unaffected cells. Due to the rapid growth of grade III DC cells, grade III DC has a higher risk of developing invasive cancer in the first five years after diagnosis.

2. **Lobular carcinoma:**
   Lobular cancer in place (LC) develops from the lobe at the end of the duct and shows a widespread distribution throughout the breast, which explains its appearance like a non-palpable mass in most cases (as shown in Figure 10.7) [7]. The incidence of LC has doubled over the previous 25 years and now occurs at a rate of 2.8 per 100,000 females.

3. **Mucinous carcinoma:**
   Mucinous carcinoma is another uncommon microscopic anatomy type, seen in less than 5% of invasive carcinoma occurrences. It usually appears in septuagenarians as a palpable mass or appears on mammography as a poorly defined growth with uncommon calcifications and tends to restrict mucin production. Type A and type B are the two main kinds, while AB lesions can have either one [8, 9]. Type A glycoprotein malignant neoplastic disease is represented by selections with larger amounts of animate glycoprotein (as shown in the image), while the blood group could be a different variation with endocrine differentiation and microscopic anatomy showing additional granular living substance than type A carcinoma.

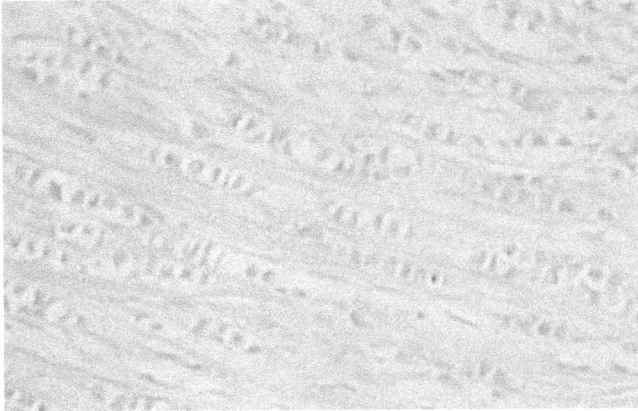

FIGURE 10.7   Lobular carcinoma.

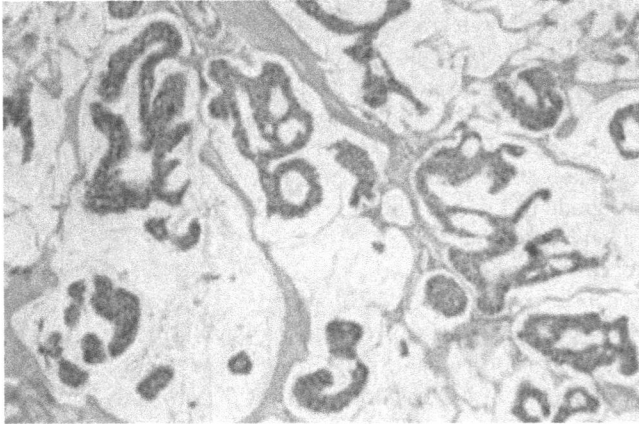

FIGURE 10.8   Mucinous carcinoma.

4. Papillary carcinoma:

Papillary cancer comprises a range of microscopic anatomy subtypes. The two common types are cystic (non-invasive form) and micropapillary ductal cancer (invasive form). Papillary carcinoma typically affects women over the age of sixty and makes up about 1%–2% of all breast cancer cases. Papillary carcinomas are located at the centre of the breast and might appear as discharge of blood from the breast. They are powerful sex hormones receptor (ER) and lipo-lutin receptor (PR). Low mitotic activity in cystic papillary carcinoma results in a prognosis and course that

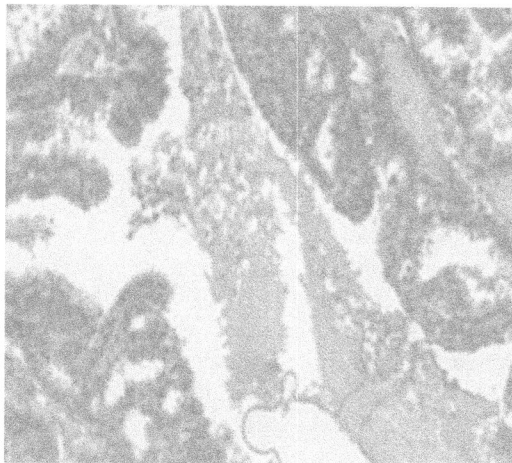

FIGURE 10.9   Papillary carcinoma.

are both slow-moving. Invasive micropapillary ductal cancer features an aggressive constitution, although the seventieth of cases are ER-positive. See Figure 10.9.

10.6 METHODS

Deep learning approaches are able to mechanically extract choices, retrieve data, and develop sophisticated abstract representations. Deep learning is also a set of machine learning algorithms with networks capable of differentiating unattended from unstructured information [7].

This chapter analyses the breast cancer images using transfer learning through pre-trained networks. We are classifying these histopathological images using Shufflenet by dividing the dataset according to the analysis. We are using convoluted neural network (CNN) for the classification problem, and we are comparing the CNN with the transfer learning for better results with increasing performance. A multi-class classification for diagnosis is also being conducted. The experiment aims to prove the Shufflenet provides the best accuracy results compared to CNN. In this experiment, the input size should be 224×224 for transfer learning. Our final goal is to provide a deep analysis using transfer learning on the BreaKHis Dataset. We also provide a confusion matrix for summarizing the performance of the classification method. We are utilizing a 400× sub-dataset.

10.7 DATASET DESCRIPTION

A total of 9,109 images of breast tumour tissue taken under various magnifications from eighty-two patients make up the BreaKHis dataset (40×, 100×, 200×, and

400×). So far it has 2,480 benign and 5,429 cancerous samples (700 × 460 pixels, three-channel RGB, 8-bit depth in every channel, PNG format) (https://web.inf.ufpr.br/vri/databases/breast-cancer-histopathological-database-breakhis/).

The two primary categories of the BreaKHis dataset are benign tumours and malignant cancers. A lesion that doesn't satisfy any criteria for malignancy, such as mitosis, significant cellular atypia, disruption of basement membranes, distribution, etc., is said to be histologically benign. Typically, benign tumours are relatively "harmless," show a modest rate of growth, and remain contained. A lesion that can spread to distant areas (metastasize) and invade and destroy nearby structures (locally invasive) resulting in loss of life is referred to as a malignant tumour. In the present iteration, samples in the dataset were gathered using the surgically obtained biopsy (SOB) methodology and were designated as excisional biopsy or partial mastectomy. Compared to other needle biopsy techniques, this type of operation extracts a larger sample of tissue, and it is carried out in a hospital under anaesthesia. Adenosis (A), phyllodes tumour (PT), fibroadenoma (F), and tubular adenoma (TA) are the four distinctive types of benign breast tumours currently present within the dataset, while ductal carcinoma (DC), mucinous carcinoma (MC), lobular carcinoma (LC), and papillary malignant neoplastic disease (PC) are the four malignant tumours shown in Table 10.1.

10.8   CONVOLUTIONAL NEURAL NETWORK

The ideal neural network model for image-related issues is a CNN which is capable of segmenting, classifying, and processing images. In 2012, CNN launched with Alexnet, and since its inception it has expanded to include other networks. It is also utilised in many other applications, including recommendation systems, object detection, natural language processing, and facial recognition. CNN has the benefit of automatically extracting features from images without any human intervention.

10.9   ARCHITECTURE

All of the architectural models in this CNN are comparable to one another. Convolution, pooling, and fully connected layers are included in these architectural models.

10.10   CONVOLUTION

Convolution involves applying filters to an input image before multiplying the result with additional filters to create a new image. Consider the 5 × 5 image matrix with pixel values of 0 and 1, for instance, and the 3 × 3 filter image matrix. Together, this 5 × 5 image matrix and 3 × 3 filter image matrix multiply the images. The output is referred to as 'a feature map'. Using a variety of filters and numerous convolutions, we generate various feature maps from the input images. To create a new image, we further merge all feature maps. Edge detection, blurring, and sharpening are just a few of the many processes that may be carried out using various filters.

TABLE 10.1

Images in Subclasses in 400× Magnification

MAGNIFICATION	BENIGN				MALIGNANT			
	Adenosis	Fibroadenoma	Phyllodes Tumour	Tubular Adenoma	Ductal Carcinoma	Lobular Carcinoma	Mucinous Carcinoma	Papillary Carcinoma
400x	106	237	115	130	788	137	169	138

10.11   STRIDES

Strides are used to indicate how many steps a convolution filter must move through in a single step. The default setting is 1. To minimise issues with field overlap, we can provide higher values. The feature map's size may be decreased as a result.

10.12   PADDING

When the filter does not exactly fit the input image, padding is a technique used to minimise the image dimensions. Padding can be performed in two ways: 1. To pad the bits with zeros, which is referred known as zero-padding. 2. Since only valid padding bits are used, we can remove the area of the image where the filter is not perfectly fit.

10.13   NON-LINEAR FUNCTION

We can employ a variety of functions, including the ReLu, tanh, and sigmoid functions. All negative numbers are converted to zeros or positive values using these functions. ReLu function, which converts all negative numbers to zeros, is the most accurate of these functions and also the best performing. ReLu produces the result $f(x)=\max(0, x)$.

10.14   POOLING LAYER

After performing all these convolution operations, the next layer is the pooling layer. In this layer, we will try to reduce the dimensions by reducing parameters. This spatial pooling could be done in three ways:

1. Minimum pooling
2. Average pooling
3. Sum pooling
4. Maximum pooling

It takes bigger value elements from the modified feature map output image. For instance, our output image size is 4×4 and after applying max-pooling, the matrix was converted to a 2×2 matrix where larger values were found and placed in the 2×2 matrix's first block. The process of max pooling is shown in Figure 10.10:

1	1	2	4
5	6	7	8
3	2	1	0
1	2	3	4

FIGURE 10.10   Max pooling.

10.14.1 AVERAGE POOLING

It will take the average of all values of each 2 × 2 matrix.

10.14.2 SUM POOLING

It will perform sum operation on each 2 × 2 matrix. Check the process in Figure 10.11:

1	1	2	4
5	6	7	8
3	2	1	0
1	2	3	4

FIGURE 10.11   Sum pooling.

10.14.3 FULLY CONNECTED LAYER

Similar to how a neural network would send information, this layer output matrix is converted into a vector and sent to all fully linked layers. Each neuron in this layer is given a vector, and the layer neurons are connected with weights and biases. In order to create the output image with its class name, it then generates the output matrix and the final layer. Our CNN output is produced using two layers: softmax and logistic. For performing binary classification, the logistic layer is utilised, and for multiclass classifications, the softmax layer. This is CNN's entire architecture, where we are running retrained networks for CNN utilising the transfer learning approach. See Figure 10.12.

10.15   TRANSFER LEARNING

Transfer learning is the process of reusing the gained knowledge for the same type of problems. It can reduce the time and also increase the performance of our model in the classification process. With transfer learning, the main advantage is that we can use large datasets for our experiments.

10.16   TRANSFER LEARNING STEPS

1. Select the pre-trained network for your model. For example, alexnet, googlenet, resnet50, vgg16, etc.
2. Classify the dimensions of the image according to your network.

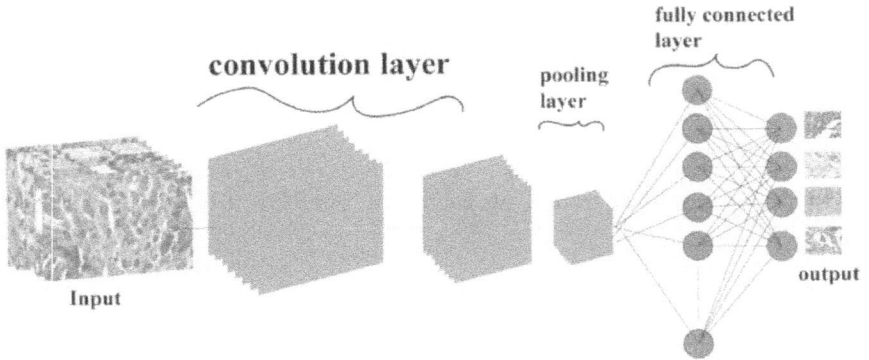

FIGURE 10.12   Convolutional neural network.

3. Fine-tune your model. In this fine-tuning we have several techniques: 1. We will replace the output layers and attach the newly defined output layers to the model and then train the network. For example, removing the SoftMax layer and replace it with a new SoftMax layer that is related to our model; 2. We will modify the learning rates that gives the best accuracy to our model; 3. We will freeze the weights in some layers. This makes our model more accurate.

10.17   DATA AUGMENTATION

In deep learning, data augmentation is the technique used to virtually expand the size of training images accordingly to the pre-trained network input size.

10.18   EVALUATION PROCESS

We are using Shufflenet as a pre-trained network with a learning rate of 0.002 and a maximum of epochs 20 where we freeze 25 layers' weights.

Firstly, we divided our dataset into three parts: train, validation, and test with rates of 70%, 10%, and 20%, respectively. Then we shift the data into augmented data. Now we started training the data and acquired the accurate values for validation. We also loaded the separated test data to the trained model and got testing accuracy. The testing accuracy results was approximately 89±%. By training the model repetitively, the accuracy may increase. For the purpose of measuring the performance of our model, we plotted the confusion matrix. We determined the sensitivity, specificity, recall, and f_score.

10.19 CONFUSION MATRIX

A confusion matrix is calculated for the performance evaluation of our model. It describes how our model is predicting the correct and incorrect samples. Sensitivity, which also refers to the percentage of tests that are correctly identified, is also known as a true positive rate [10]. Specificity, also called the true negative rate, indicates the discovered false tests.

Recall values give the rate of correctly classified positive samples.

Precision shows how often the results are true.

F-measure helps us comparing the precision and recall by using the harmonic mean.

The formulas for these metrics are:

$$Accuracy = \frac{TP+TN}{TP+TN+FP+FN}$$

$$Recall = \frac{TP}{TP+TN}$$

$$Precision = \frac{TP}{TP+FP}$$

$$F\_measure = \frac{2*Recall*Precision}{Re\,call + Precision}$$

10.20 SHUFFLENET

Shufflenet is one of the pre-trained networks provided by MATLAB, and it is a CNN that is trained on a large number of images from the ImageNet datastore. MATLAB Shufflenet is available in the Deep Learning Toolbox Model package. If the support packages are not installed, then the Shufflenet function automatically provides a link to download the package. Shufflenet input image size is 224×224. In this network, there are 173 layers connected to lower layers [11]. This pre-trained network is mainly used for mobile devices. It contains residual blocks called Shuffleunits.

10.21 RESULTS

In this section of the chapter, we are presenting the classification results on the breast cancer histopathological images from the BreaKHis dataset. In this experiment, we are using MATLAB as a platform for execution and programming language. For this training progress, we have used graphics processing unit (GPU).

TABLE 10.2

Performance Metrics for Each Class

	Accuracy	Sensitivity	Specificity	Recall	F_score
Adenosis	0.967213	0.636363636	0.988372093	0.636363636	0.7
Ductal carcinoma	0.907104	0.898734177	0.913461538	0.898734177	0.893081761
Fibroadenoma	0.967213	0.875	0.981132075	0.875	0.875
Lobular carcinoma	0.945355	0.5	0.982248521	0.5	0.583333333
Mucinous carcinoma	0.956284	0.764705882	0.975903614	0.764705882	0.764705882
Papillary carcinoma	0.967213	0.714285714	0.98816568	0.714285714	0.769230769
Phyllodes tumour	0.956284	1	0.953488372	1	0.733333333
Tubular adenoma	0.994536	0.923076923	1	0.923076923	0.96

xThe results are on the augmented dataset which performed the training process. The results are displayed in Table 10.2 on the 400X dataset. The images are taken from eighty-two patients.

The metrics for each class and overall performance metrics for each class are described in the table.

We divided the dataset into train, validation, and test. The training options used are at 0.002 as initial learning rate with 20 epochs. After the training is done, the validation accuracy is up to 89±% and the test accuracy is 90±%. The confusion matrix is provided in Table 10.3.

Table 10.4 includes the confusion matrix values for each class.

The overall performance metrics graph is shown in Figure 10.13. The average accuracy of all classes is 95±%.

The resultant images are the random images displayed after testing is done. The images display how much the probability percentage of the data is matched with the training data. See Figure 10.14.

10.22   DISCUSSION

All these results are provided by using the transfer learning method with the help of pre-trained networks on the histopathological images of breast cancer. It is very useful for doctors to find out which type of cancer can be diagnosed from images. It reduces the time of clinical testing and allows doctors to provide an immediate diagnosis.

This analysis is the base for researchers to further explore histopathological images of breast cancer in the future.

10.23   CONCLUSIONS

We have proposed our methods of using pre-trained networks in the transfer learning process using MATLAB on histopathological images of breast cancer in this chapter. All

TABLE 10.3

Correlation Matrix for Each Class

	Adenosis	Ductal carcinoma	Fibroadenoma	Lobular carcinoma	Mucinous carcinoma	Papillary carcinoma	Phyllodes tumor	Tubular adenoma
Adenosis	7	0	0	0	1	0	3	0
Ductal carcinoma	0	71	1	3	1	2	1	0
Fibroadenoma	0	0	21	0	1	0	2	0
Lobular carcinoma	0	5	0	7	1	0	1	0
Mucinous carcinoma	2	1	1	0	13	0	0	0
Papillary carcinoma	0	3	0	0	0	10	1	0
Phyllodes tumor	0	0	0	0	0	0	11	0
Tubular adenoma	0	0	1	0	0	0	0	12

TABLE 10.4

Confusion Matrix for Each Class

	TP	TN	FP	FN
Adenosis	7	170	2	4
Ductal carcinoma	71	95	9	8
Fibroadenoma	21	156	3	3
Lobular carcinoma	7	166	3	7
Mucinous carcinoma	13	162	4	4
Papillary carcinoma	10	167	2	4
Phyllodes tumor	11	164	8	0
Tubular adenoma	12	170	0	1

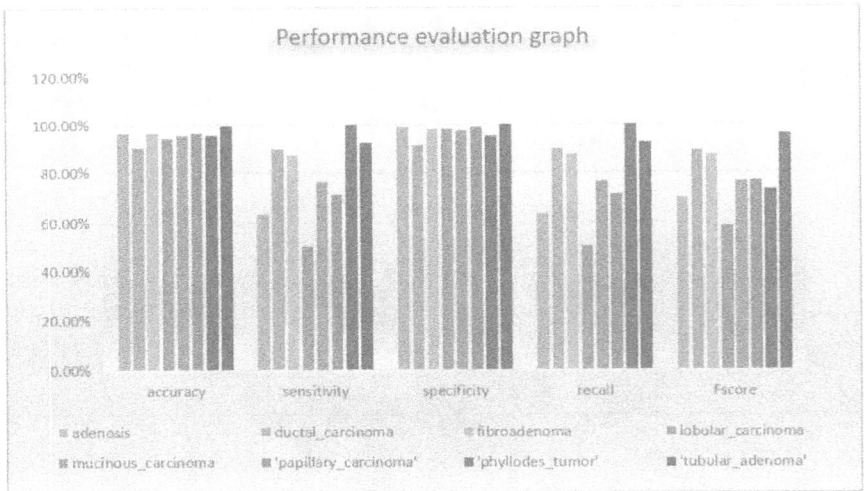

FIGURE 10.13 Performance metrics graph for the classification model.

results shown are on an augmented dataset in multi-class classification. Our confusion matrix also provided the valid results about how our model performance is. Researchers could use different networks with other training options and different parameters. They could use regular CNN without transfer learning and pre-trained networks.

10.24 ABBREVIATIONS

A—Adenosis
CNN—Convolutional Neural Networks
DC—Ductal carcinoma

ductal carcinoma, 100%

ductal carcinoma, 98.9%

ductal carcinoma, 100%

fibroadenoma, 100%

ductal carcinoma, 98.4%

fibroadenoma, 100%

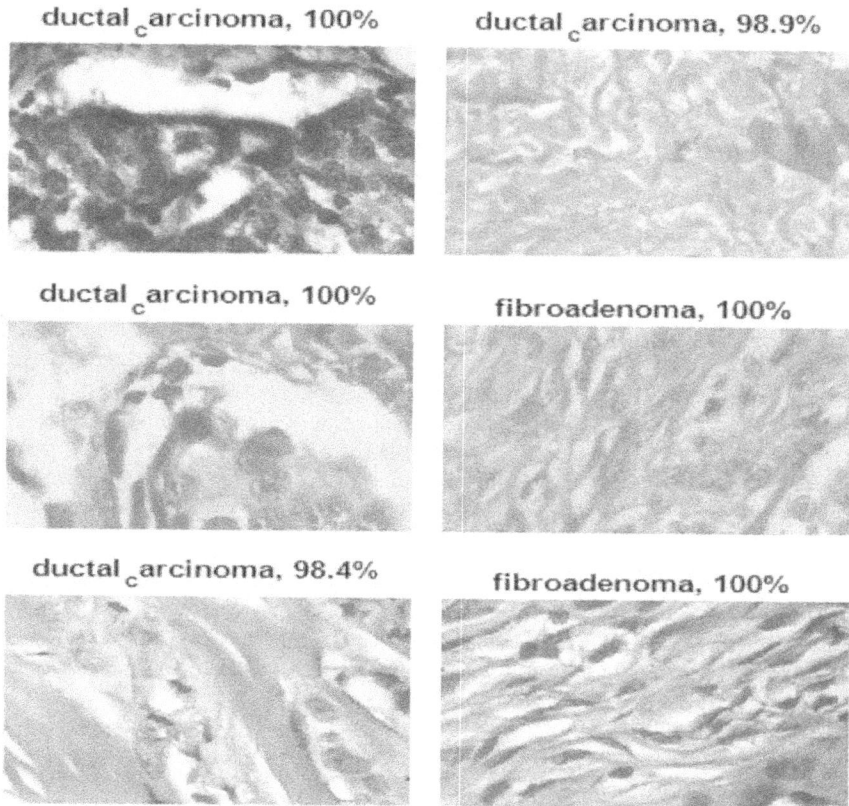

FIGURE 10.14   Classification model.

ER—Estrogen receptors
F—Fibroadenoma
FN—False negative
FP—False positive
GPU—graphics processing unit
LC—Lobular carcinoma
MC—Mucinous carcinoma
PC—Papillary malignant neoplastic disease
PNG—Portable Network Graphics
PR—Progesterone receptors
PT—Phyllodes tumor
RGB—Red, Green, Blue, concerning computer display
SOB—Surgically obtained biopsy
TA—Tubular adenoma
TN—True negative
TP—True positive

REFERENCES

[1] American Cancer Society: Phyllodes Tumors of the Breast. www.cancer.org/cancer/breast-cancer/non-cancerous-breast-conditions/phyllodes-tumors-of-the-breast.html. Accessed 14 Jan 2020.

[2] American Cancer Society: Adenosis of the Breast. www.cancer.org/cancer/breast-cancer/non-cancerous-breast-conditions/adenosis-of-the-breast.html. Accessed 14 Jan 2020.

[3] F. Bray, J. Ferlay, I. Soerjomataram, R. L. Siegel, L. A. Torre, and A. Jemal. Global Cancer Statistics 2018: GLOBOCAN Estimates of Incidence and Mortality Worldwide for 36 Cancers in 185 Countries. www.wcrf.org/dietandcancer/cancer-trends/breast-cancer-statistics.

[4] Breastcancer.org: Invasive Ductal Carcinoma (IDC). www.breastcancer.org/symptoms/types/idc. Accessed 14 Jan 2020.

[5] Breastcancernow: Breast Cancer Causes. https://breastcancernow.org/information-support/have-i-got-breast-cancer/breast-cancer-causes. Accessed 14 Jan 2020.

[6] Breastcancer.org: IDC Type: Mucinous Carcinoma of the Breast. www.breastcancer.org/symptoms/types/mucinous. Accessed on 14 Jan 2020.

[7] D. Sarkar. A Comprehensive Hands-on Guide to Transfer Learning with Real-World Applications in Deep Learning. 2018. https://towardsdatascience.com/a-comprehensive-hands-on-guide-to-transfer-learning-with-real-world-applications-in-deep-learning-212bf3b2f27a. Accessed 14 Jan 2020.

[8] Boubacar Efared et al. Tubular Adenoma of the Breast: A Clinicopathologic Study of a Series of 9 Cases. Clinical Medicine Insights. 2018; doi:10.1177/117955571875749.

[9] E. Roth and D. Weatherspoon. Lobular Breast Cancer: What Are the Prognosis and Survival Rates? 2018 www.healthline.com/health/breast-cancer/lobular-breast-cancer-prognosis-survival. Accessed 12 Jan 2020.

[10] Abhishek Sharma. Confusion Matrix in Machine Learning. www.geeksforgeeks.org/confusion-matrix-machine-learning/. Accessed 14 Jan 2020.

[11] X. Zhang, X. Zhou, M. Lin, and J. Sun. Shuffle Net: An Extremely Efficient Convolutional Neural Network for Mobile Devices. *2018 IEEE/CVF Conference on Computer Vision and Pattern Recognition.* 2018; doi: 10.1109/CVPR.2018.00716.

11 Machine Learning and Signal Processing Methodologies to Diagnose Human Knee Joint Disorders
A Computational Analysis

*Murugan R, Balajee A, Senbagamalar L, Mir Aadil,
and Shahid Mohammad Ganie*

11.1 INTRODUCTION

One of the critical disorder analyses in the medical field trending nowadays is the diagnosis of knee joint disorders which are very complex due to the structure of the knee joints. At present there are two different methods of diagnosing in practice. The invasive and the non-invasive methods are adopted for the knee joint disorder diagnosis. An invasive method is similar to arthroscopy, as it is not only expensive but also not suitable for regular diagnostics. The other disadvantage of this method is it is entirely prone to infection [1]. Non-invasive methods are those which use systemized methods to diagnose without clinical surgery methods. This includes computed tomography (CT), x-rays, magnetic resonance imaging (MRI), ultrasonography (US), etc.

Vibroarthrography (VAG) is a non-invasive method that involves analyzing vibrations in the knee joint. These vibrations are obtained from the relative movement of the articular surfaces of the synovial joints. In a healthy state, the outer surface of the joint is covered by a smooth and slimy hyaline ossein, which detects optimal arthro-kinematic movement quality. Osteoarthritis is again and again observed by using the patello-femoral joint (PFJ), a portion of the knee joint. VAG signals exhibit the ability of non-linearity, are multi-component, and are non-stationary in nature. Thus, the analysis of VAG signals would not be

DOI: 10.1201/9781003328414-11

preferred for digital signal processing using conventional methods. The greatest awareness of the VAG test results is those recorded from the skin-deep position of the knee.

VAG signals play a vital role in discriminating the different levels of joint disorders and they could also act as a database for future reference [2]. There are a number of models that are proposed for handling the binary classification, whereas the abnormal level-based multiple classifications remain unaddressed [3]. The binary classifications are directly performed with the feature vector that consists of data lacking in performance in terms of sensitivity [4–6]. Combining the optimization methods with abnormality-level identification could induce the overall performance of the VAG systems [7].

Thus, the initial goal of this chapter is to analyze the materials and methods used for performing the binary classification of VAG data samples using signal processing and machine learning methods [8]. In the signal processing criteria, we considered the recently proposed CEEMDAN method for analysis. This method performs not only the decomposition of the raw VAG signals but also acts as a pre-processing method for creating the original data matrix that could act as input data to feature extraction and the classification methods that could induce the overall performance of the VAG-based disorder diagnosis. Two major machine learning methods are adopted for analyzing the classification performance of data samples. LS-SVM and SVM-RFE are used, where each method supports the performance improvement of the VAG system.

The VAG method is a dynamic measurement system that achieves higher performance metrics when compared to the other non-invasive modalities. The results achieved by the VAG methods using the considered classification systems are projected in the results section. Visualizing the knee joint as x-rays at the time of recording the signal samples also used US [9, 10]. Invasive methods are commonly performed via image review, and it could not provide information about the early bone disorder [11]. On the other hand, for the early diagnosis of bone joint disorders, a non-invasive method was employed. The surface of the bone joint disorder was analyzed using LabVIEW software with the aid of an acquired VAG signal.

11.2 METHODOLOGIES

This section covers the major methodologies focused on the process of analyzing knee joint disorders using VAG signals. Earlier decomposition techniques like variational mode decomposition lack the successful mode splitting that is calculated as intrinsic mode functions (IMFs), whereas the recently proposed CEEMDAN method [7] achieves the better split of mode functions so that the noise can be removed. The noise removal also utilizes the minor addition of added noise via transformation metrics, which also achieved better pre-processing of available VAG signal samples. The alternate mechanism of including the existing decomposition strategies also could improve the overall signal pre-processing results. By including the CEEMDAN, VMD, and also the timing decomposition techniques, an empirical signal processing method can be adopted as shown in Figure 11.1.

Figure 11.1 explains the overall block diagram of the feature extraction methods. This is designed to work well for data that is non-stationary and non-linear frequency images that will give input to the time-frequency representation. The features extracted from time-frequency images are classified using SVM, LS-SVM, and SVM-RFE machine learning algorithms to analyze the efficiency of each of the algorithms in clinical classification performance. Finally, the classification system identifies the healthy and unhealthy samples as shown in Figure 11.1.

11.2.1 Complete Ensemble Empirical Mode Decomposition with Adaptive Noise (CEEMDAN)

CEEMDAN adopts a different strategy when compared to other empirical methods in segregating the signal samples into a number of mode functions. CEEMDAN decomposes the VAG signal into IMFs. The mode overlapping problem was successfully reduced, and it also induced performance improvement using empirical methods. This adaptive measure reduces the noise experienced during the signal extraction process. The parameter tuning and the algorithmic descriptions of CEEMDAN are provided in the following section.

Let us assume that S_n is a function to produce the nth mode for creating the continuous mode functions that is denoted as $S(I) = X - S(I)$ where 'I' denotes the input signal and S denotes a cumulative occurrence of input signals. The single-mode function can be denoted as follows:

$$S_1(I) = X - S(I) \tag{11.1}$$

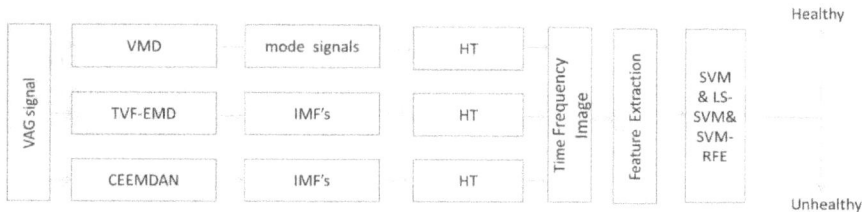

FIGURE 11.1 Block diagram for VAG signal-based classification of healthy and unhealthy knee joints.

ALGORITHM: CEEMDAN

Input Time-frequency signal images $i, i_2, i_3 \ldots i_n$

Output Pre-processed signal data

Begin

Step 1: Remove the white noise using the following equation:

$$I^P = I + \beta_0 X_1 \left(W^P \right) \tag{2}$$

Where W^P denotes the white noise included in each of the signal samples.

Step 2: Calculation of mean value along with the R real values is performed using the following notation:

$$R_1 = S \left(I^P \right) \tag{3}$$

Step 3: The outliers that are left unaddressed till the mean value calculation are removed using the following notation:

$$O_1 = I - R_1 \tag{4}$$

Step 4: The overall residues that are associated with the original samples are calculated by finding the iterative average of the local means as follows:

$$O_2 = R_1 - S(R_1 + \beta_0 X_2 \left(W^P \right) \tag{5}$$

End

11.2.2 Feature Extraction Using Different SVM Algorithms

The signal processing mechanism produces the outcome of a time-frequency image which is used as input data to the classification system. As mentioned earlier, we had utilized two supervised classification algorithms, namely LS-SVM, and SVM-RFE, which are modifications of traditional SVM. This section discusses the feature extraction and classification mechanism that are considered for analysis.

11.2.2.1 Support Vector Machine

Support Vector Machine (SVM) is effective and mostly used for classification problems such as binary class classifications. Maximal margin using a separating hyperplane between the positive and negative classes are constructed by SVM. It can handle a quadratic programming problem that contains inequality constraints with a linear cost on the slack variables, though the dual space is solved and it is more feasible. In fine-tuning the results of the classification system, kernel methods play a vital role. Different statistical kernel metrics, including radial-basis, Gaussian, etc., could support the traditional SVM classifier to improve the overall performance for the dataset considered for analysis. Thus, the SVM performance-based analysis requires the following:

i. Select the appropriate kernel metric for the SVM
ii. Choose the appropriate regularized parameter C

iii. Compute the quadratic programming using the following notation:

$$\sum_{i=1}^{n} \alpha_j - \frac{1}{2} \sum_{i=1}^{n} \sum_{j=1}^{n} \alpha_i \alpha_j y_i y_j K(x_i, x_j)$$ (11.6)

$$0 \leq \alpha_i \leq C \qquad \sum_{i=1}^{n} \alpha_i y_i = 0$$

The difference in the regularized parameters can be calculated through the bias and variance for the considered identical data sample.

iv. The discriminant function can be constructed as follows:

$$g(x) = \sum_{i \in SV}^{n} \alpha_i K(X_i, X) + b$$ (11.7)

Finally, detect the knee joint disorder using the SVM classifier.

11.2.3   SVM–RECURSIVE FEATURE ELIMINATION (SVM-RFE)

The RFE-SVM is a feature-elimination algorithm that iteratively reduces the number of irrelevant features available in the dataset by removing the duplicated attribute in a recursive manner. This algorithm is used in most clinical data analysis, especially in genetic data analysis [12, 13]. This strategy can also be adopted for removing the irrelevant features from the feature vector for performing the efficient identification of knee joint disorders. In the knee joint dataset, some of the sample details might hold redundant information and outliers, which affects the process of identifying the target class. These iterative attributes can be removed via a similar strategy that is adopted for gene expression data. Thus we considered the same flow of eliminating the feature that are followed in genetic data analysis, as shown in Figure 11.2.

ALGORITHM: SVM-RFE

Input	Original dataset that has number of samples $s, s_2, s_3 \ldots s_n$
Output	Feature vector with absolute set of features
Begin	
Step 1:	Provide input data subset
	$S_i = \{1, 2, \ldots m\}$
	Identify the minimal weight as $W = \{\}$
Step 2:	Iterate the steps 3 to 8 till S_i is filled
Step 3:	Perform the training for traditional SVM using S_i

(Continued)

(Continued)

Step 4:   Weight vector can be calculated as follows:

$$V = \sum_{i=1}^{m} \beta_i p_i q_i$$

Here i denotes the entities and p and q denote two identical target classes relevant

to the subset. β_i is the Lagrangian multiplier estimated from the training set.

Step 5   Calculate the ranking criteria by performing V^2

Step 6   Rank the attributes based on the calculated criteria

Step 7   Update the rank list after every iteration

Step 8   Features with minimal ranking criteria are eliminated from the vector

Step 9   Once the criteria reach the smallest rank (i.e. <0.01) stop performing the ranking

End

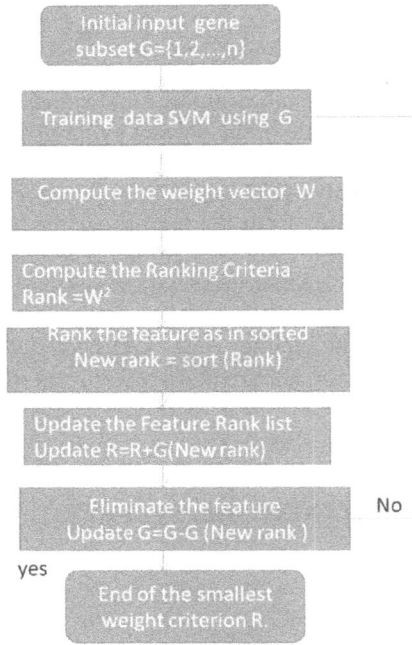

FIGURE 11.2   SVM-RFE feature elimination algorithm.

11.3 RESULTS AND DISCUSSIONS

11.3.1 RESULTS FOR VARIATIONAL MODE DECOMPOSITION (VMD)

The VAG signals are given as input to the VMD algorithm. The VAG signals have composite input values, and the input values are passed into the different frequency levels that are represented in Figure 11.3.

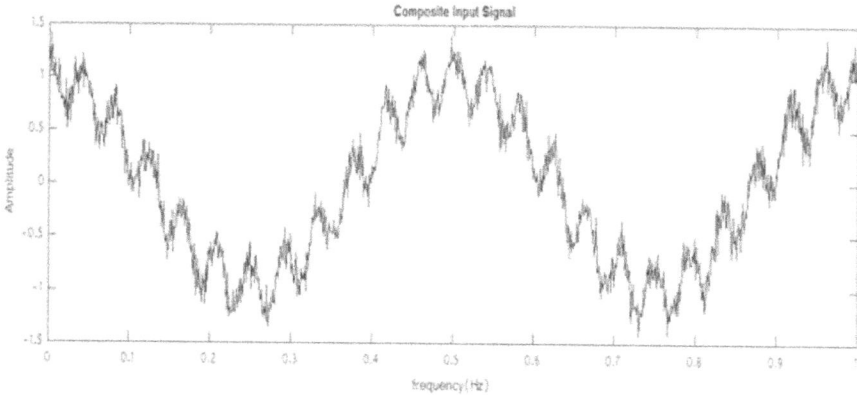

FIGURE 11.3 Composite input signal for the VMD method.

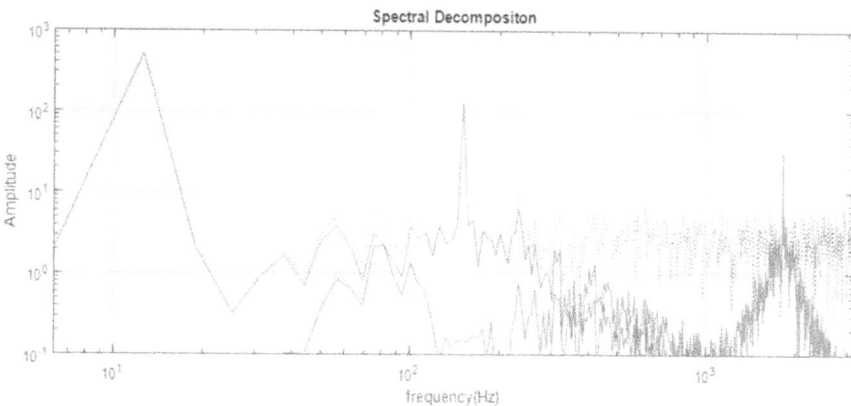

FIGURE 11.4 Spectral decomposition for VMD method.

The VAG signals are decomposed into the IMF's signal. The VMD signals have to perform mode signal operation. The VMD method examines the total number of modes and their center frequencies, and the mode has reproduced the input signals as smooth and demodulated to the baseband of spectral decomposition shown in Figure 11.4.

Spectrum-based decomposition of the VAG signal can be applied to any band with a number of modes. The modes are extracted concurrently using a non-recursive variation mode decomposition model. The model is shown for an ensemble of modes, then their corresponding center frequencies, such as the modes collected, and reproduces the input signal of each as smooth after being demodulated into the baseband as shown in Figure 11.5.

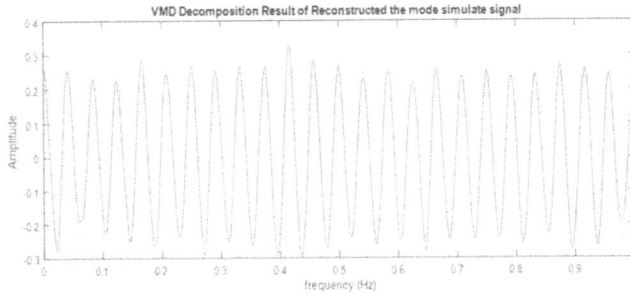

FIGURE 11.5   Reconstructed mode of input signal for the VMD method.

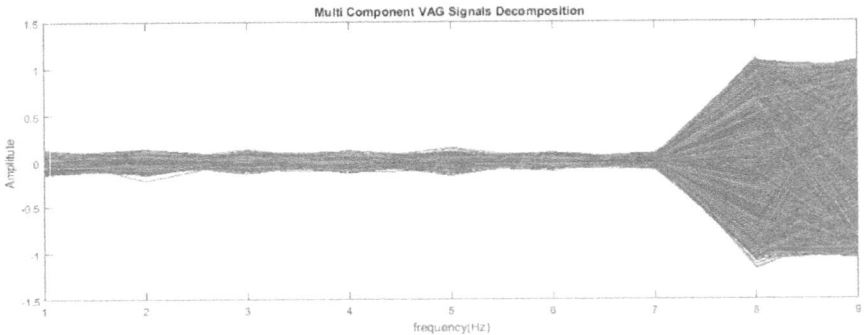

FIGURE 11.6   Multi-component VAG signal decomposition using TVF-EMD.

11.3.2   RESULT FOR TIME-VARYING FREQUENCY EMPIRICAL MODE DECOMPOSITION (TVF-EMD)

Sampling rate and frequency are used to analyze the signal and add the noise signal to the linear and non-stationary VAG signal. Time-varying filtering-based EMD results in the intrinsic mode function using Hilbert transformation to extract the instance of amplitude and instance of frequency. Computing the basic cut-off frequency to deal with the mode mixing problem to locate the maxima of the input signal after pre-processing found the all intermittences. The scenario, while iterative units produce only a single maximum value, will be considered and it is extracted using Gabriel Rilling. Multi-component VAG signal with decomposition using TVF-EMD is shown in Figure 11.6.

11.3.3   RESULT FOR CEEMDAN

The combination of modes results in overlapping which is a major issue that must be handled through the decomposition. The mode functions are denoted as IMFs where empirical mode decomposition (EMD) is calculated for each of

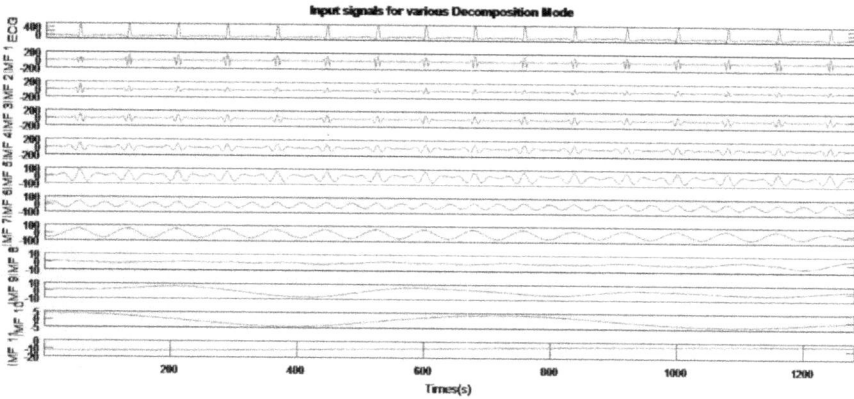

FIGURE 11.7 Input signal for various decomposition modes using the CEEMDAN method.

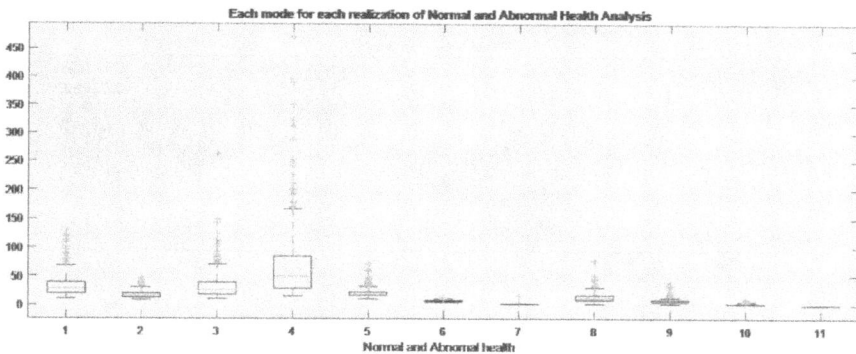

FIGURE 11.8 Shifting iteration of each mode for normal and abnormal samples.

the original vectors of the considered sample. This mode avoids the overlapping issue, and it combines the functions to create a new vector. Whenever the individual vector has a higher value that could not be accommodated within the normal mode function, then the bivariate EMD can be adopted for creating the mode functions that are equivalent to the dual step of shifting the sequence taking place in the traditional decomposition schemes. Signal masking techniques are adopted by CEEMDAN to avoid the noise that occurs with each of the data samples shown in Figure 11.7.

The overlap that occurs in modes is eliminated through the scaling and partitioning schemes of decomposition methods. The scale partition capabilities of EEMD enable the elimination of the mode mixing problem. The CEEMDAN method improves the shifting process for complete shifting in multiple-mode operation shown in Figure 11.8.

11.3.4 TIME-FREQUENCY IMAGE OF INPUT SIGNAL

In VAG signals of waveforms after decomposition, different frequencies of the input signal of VMD, TVF-EMD, and CEEMDAN methods are given as an input to the time frequency image as shown in Figure 11.9.

11.3.5 FEATURE EXTRACTION OF SPECTROGRAM OUTPUT

The time-frequency image is used for the feature detection method of scale invariant feature transformation for computer vision. In machine learning, using pattern recognition, various statistical parameters can be calculated from the time-frequency image which contributes to segregating the samples into accurate target classes. It will detect the local features in the image as a histogram-oriented gradient. After feature detection, we extracted statistical features classified as the pattern of signals healthy and unhealthy using SVM and LS-SVM algorithms as shown in Figure 11.10 and Figure 11.11.

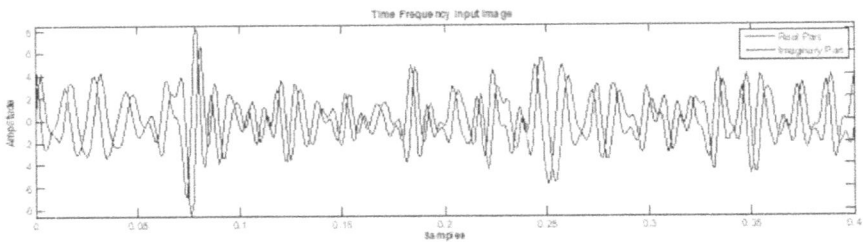

FIGURE 11.9   Input signal of time frequency image.

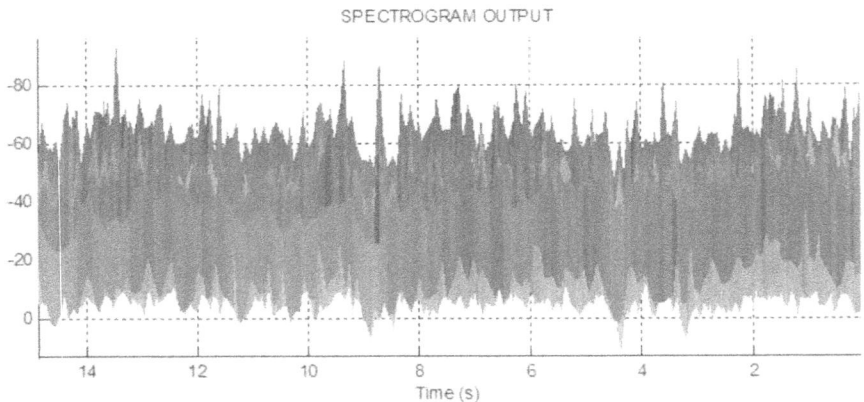

FIGURE 11.10   Feature extraction of spectrogram output using LS-SVM.

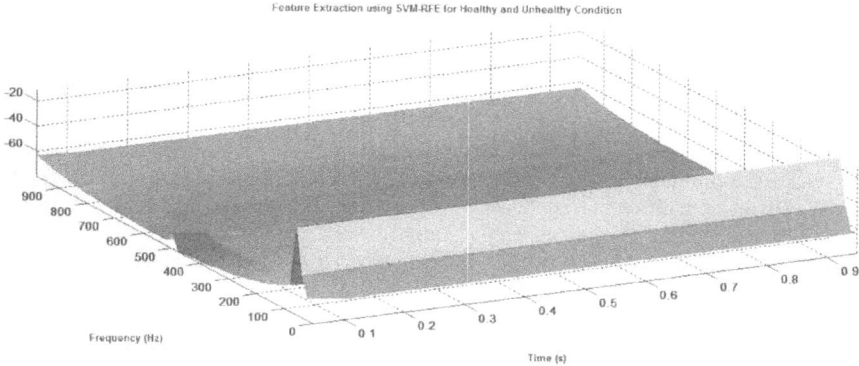

FIGURE 11.11   LS-SVM using feature extraction for healthy and unhealthy.

11.4 CONCLUSION

In this chapter, we analyzed various features that are key factors in discriminating between the normal and abnormal data samples. The analysis also contains the conversion techniques that are used to interact between the mode functions and the original class of the data. The data analysis had been implemented for the dataset of non-stationary and non-linear signals using different processing techniques. The Hilbert transformation method has been performed for the different mode functions of TVF-EMD, VMD, and CEEMDAN methods. To add the noise performs the shifting process reconstructs the input signal of time-frequency images. The time-frequency data is fed to perform the pattern classification of LS-SVM and SVM-RFE algorithms for feature extraction where SVM-RFE extraction shows the best optimum results. Finally, VAG signals are analyzed and classified for both healthy and unhealthy knee joint samples.

REFERENCES

[1] G. Rajalakshmi, C. Vinothkumar, A. Anne Frank Joe, and T. Thaj Mary Delsy. "Vibroarthographic signal analysis of bone disorders using Arduino and piezoelectric sensors". International Conference on Communication and Signal Processing, April 4–6, 2019, India.

[2] A. C. D. Faria, G. R. C. Pinheiro, J. Neri1, and P. L. Melo. "Instrumentation for the analysis of changes in the knee joint of patients with rheumatoid arthritis: Focus on low-frequency vibrations". *Journal of Physics: Conference Series*, 1044, 2018, conference 1.

[3] Manish Sharma and U. Rajendra Acharya. "Analysis of knee-joint vibroarthographic signals using bandwidth-duration localized three-channel filter bank". DOI: 10.1016/j.compeleceng.2018.08.019.

[4] Saif Nalband, Amalin Prince, and Anita Agrawal. "Entropy-based feature extraction and classification of vibroarthographic signal using complete ensemble empirical model decomposition with adaptive noise". *IET Science, Measurement & Technology*, 12 (3), 2018, pp. 350–359 © The Institution of Engineering and Technology 2017.

[5] Jawad F. Abulhasan and Michael J. Grey. "Anatomy and physiology of knee stability". *Journal of Function Morphology and Kinesiology*, 2 (4), 2017, pp. 34.

[6] F. Picard, A. Deakin, N. Balasubramanian, and A. Gregori. "Minimally invasive total knee replacement: Techniques and results". *European Journal of Orthopaedic Surgery & Traumatology*, 28, 2018, pp. 781–791.

[7] Aditya Sundar Nalband, A. Amalin Prince, and Anita Agarwal. "Feature selection and classification methodology for the detection of knee-joint disorders Saif". 2016. DOI: 10.1016/j.cmpb.2016.01.020.

[8] K. Hemachandran, S. Khanra, R. V. Rodriguez, and J. Jaramillo (Eds.). *Machine Learning for Business Analytics: Real-Time Data Analysis for Decision-Making*. CRC Press, 2022.

[9] M. T. Hirschmann and W. Müller. "Complex function of the knee joint: The current understanding of the knee". *Knee Surgery, Sports Traumatology, Arthroscopy*, 23, 2015, pp. 2780–2788.

[10] Dawid Bączkowicz, Edyta Majorczy, and Krzysztof Kręcisz. "Age-related impairment of quality of joint motion in vibroarthrographic signal analysis". 2015. DOI: 10.1155/2015/591707.

[11] Yunfeng Wu, Pinnan Chena, Xin Luoa, Hui Huangc, Lifang Liaoa, Yuchen Yao, Meihong Wu, and Rangaraj M. Rangayyan. "Quantification of knee vibroarthrographic signal irregularity associated with patello femoral joint cartilage pathology based on entropy and envelope amplitude measures". 2016. DOI: 10.1016/j.cmpb.2016.03.021.

[12] Mei-Ling Huang, Yung-Hsiang Hung, et al. "SVM-RFE based feature selection and Taguchi parameters optimization for multiclass SVM classifier". *Hindawi Publishing Corporation ⊡e Scientific World Journal*, 2014, Article ID 795624.

[13] Jianchen Wang, Ganlin Shan et al. "Improved SVM-RFE feature selection method for multi-SVM classifier". 2011 International Conference on Electrical and Control Engineering. DOI: 10.1109/ICECENG.2011.6058060.

12 Diagnostics and Treatment Help to Patients at Remote Locations Using Edge and Fog Computing Techniques

T. Sunil, J. Gladson Maria Britto, and K. Bharath

12.1 INTRODUCTION

In the current scenario it is very important to see that diagnosis and treatment happen at the location where the patient is available. The framework is designed to see that a provision is made to reach the patient in the remote location and to see that the patient is monitored 24/7 by making use of technology like the fog and edge computing [1].

The basic idea is to see that the patient is monitored all the time with the assistance of different layers like the fog and edge layers, which are connected in the network, where all the required devices like the sensors will be connected to the patient in order to monitor them and based on the data captured by the various sensors.

This data which is captured by the various sensors will be the input for the edge computers, and then after performing the process of filtration, the data is then transferred or uploaded to the fog systems, where again the process of filtration happens, and ultimately the data is transferred to the cloud system or the server [2].

Here the basic use of edge computing is to get the data from the sensors and other devices which are used to monitor the patient and then to send the same to fog computers; care is taken to see that the required treatment is also provided to the patient in case of any emergency [3].

The basic advantage is that the required data will be sent to the cloud server whereby the traffic is controlled and reduced. The nodes which are connected to the edge of the network, which are referred as edge systems, will help in generating

DOI: 10.1201/9781003328414-12

faster results as the nodes will process and transfers the data to the next layer. Fog systems which are connected between the edge and the cloud systems help in the filtration process and thereby reduce the amount of the data which is getting transferred to the cloud server [4].

Here we also make use of the concept of Internet of Things, which will be used in order to connect the devices to the patient and to capture the data from those devices. All the sensors used will be connected to the edge system, and then after collecting and filtration of data, it will be sent to the fog computer. The data collected at the edge will be localized, and based on the need the data will be escalated to the fog system for further processing and finally will be sent to the cloud server. The overall concept is to see that the traffic on the cloud is reduced [5].

The system is designed with the help of multiple systems which are connected in order to ensure that the data which is transferred to the cloud is minimal, which helps in reducing the hassles of the bandwidth required. Basically the devices like the sensors and the actuators which are connected to the body of the patient in order to monitor the health generate lot of data. So instead of directly sending the data which is generated, the data is first transferred to the first layer in the system, which is the edge computer. The edge computer is responsible for handling the data that is generated by the various devices connected.

The next layer in the system is the fog system. This system will be responsible for accepting all the data that is given or transferred by the edge systems. The task of the fog system is to see that the data is correctly analyzed, and then it has to check for the relevance of the data, and it has to pull out all the data which is not relevant to transfer to the cloud system.

By placing the fog system, the advantage is that only the data which is relevant for monitoring the health of the patient will be transferred to the cloud and not all the data that is generated. It means that the data will be voluminous when it is generated by the devices connected to the network which will be reaching the cloud after multiple layers of processing and filtration processes.

By introducing multiple layers in the system, the system will be transferring only the required data, which helps to keep the cloud server to help in making decisions by the experts in a faster manner, as the data to be searched will be less and will only be useful data.

12.2 PROPOSED METHODOLOGY

The system proposed will ensure that the data sent to the cloud server will be minimal, and this is achieved by placing or working with multiple layers in the system. The different layers are fog computing layer and edge computing layer. As we know the devices which are connected to the body of the patient will generate data every second, and the amount of data which is generated will be huge and voluminous. If transferred directly to the cloud, this will require more bandwidth, and the process will be slow in order to make decisions because of the amount of data which is transferred. The data is generated by the various devices connected every second on a 24/7 basis. So in order to reduce the data which is to be transferred to the cloud system, these layers are introduced. These layers will help to reduce the quantity of data which needs to be on the cloud.

The data generated will be sent to the various nodes connected at edge of the network, which are referred as edge system or computers. The data which is given as input to the edge systems or the nodes are then processed at local level so that some amount of data is reduced at this level.

The data is processed at every edge of the network while taking into consideration only the important data which is to be sent to the cloud. So by processing the data and collecting the data at every edge, this will help the system to transfer only the required data to the cloud server.

The next layer in the system is the layer where fog systems are available. These systems will receive the stream of data from the first layer, which is the edge layer or edge nodes. With the introduction of a fog system in between the edge and cloud, data latency is reduced and the data efficiency is increased. As these systems, which are referred as fog systems, receive the stream of data from the edge systems, the data is then processed on the basis of some parameters which are important to monitor the health of the patient, and then the data is also filtered in order to see that only relevant and important data is transferred to the cloud server.

The system consists of edge and fog systems, which basically create platforms to see that the data is collected at a place where they are generated. So in order to maintain the efficiency of the system along with the security and the system capacity, these layers are introduced.

The system can replace or can avoid using the fog systems but cannot afford to skip the edge nodes which are connected to the edge of the network. By processing the data which is collected right near to the point where the data is generated, this will help in the reduction of data transferred to the cloud system. Collection of the data and storing it a point where it is generated will help to reduce the cost of the system as the same need not to be transferred to the cloud server.

The purpose of reducing data sent to the cloud server is achieved when the edge and fog layer is introduced. The concept is to see that the data is gathered at a point where it is generated by various Internet of Things (IoT) devices like the sensors and so on. Bringing the storage and processing near to the devices connected or the application used will help to minimize the time required for processing of the data and also helps in using less Internet bandwidth [3].

So in order to see that the information pertaining to the patient is available on the cloud server in order to take the decision on the health of patient and also to provide the right treatment, the system makes use of edge and fog computing layers. The advantage is improvement in efficiency of the data and reduction in bandwidth along with latency.

Figure 12.1 shows how different layers are used in order to control traffic on a network. So by making use of different layers in the system, we can see that the bandwidth required is reduced along with an increase in system efficiency and speed.

The concept is to make sure that the application present on the cloud which is used by the end user has to be simple and clear in terms of providing information about

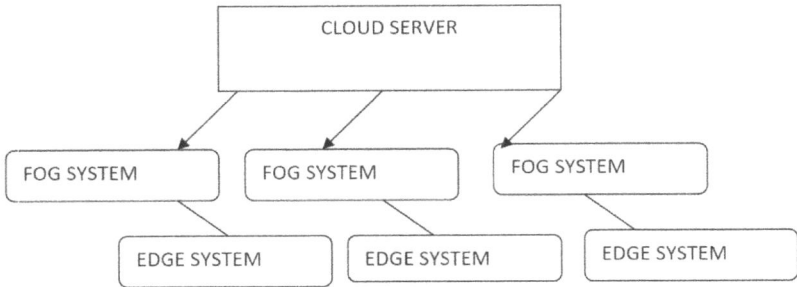

FIGURE 12.1　Different layers are used in order to control traffic on a network.

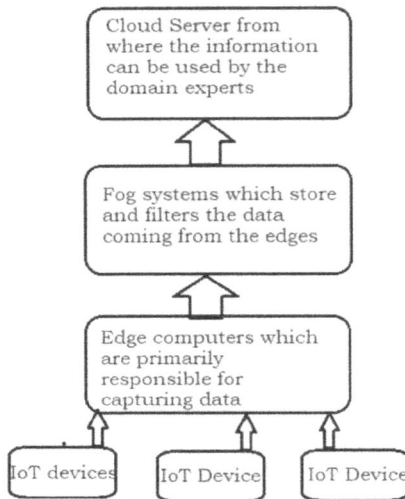

FIGURE 12.2　Flow chart to show how the data is captured and sent to the cloud.

the patient. The system is designed so that it collects the data from the various IoT devices, which are wearable devices and which are connected to the patient's body and capable of generating data that should be processed as shown in Figure 12.2.

If we make all the devices which are connected to generate and send the data to the cloud, the system will be clunky and costly and the decision-making process will take a back seat. Introduction of edge nodes at the edges of the network will help in collecting the data at the point of generation thereby providing more security to the data which is collected. The edge computers will store the data which is generated and will perform the task of processing the data in order to reduce the load on the next layer. This means the edge nodes will store and buffer the data and will also process the same before the data is sent to the fog systems in the network.

The fog systems, which are connected between the edge and the cloud server, are responsible for processing the data which is captured, and then on the basis of certain parameters the data is analyzed, and the data is also filtered so that only the relevant and important data reaches the cloud server.

The system will provide multiple advantages in terms of improved efficiency along with reduction in bandwidth required. The system will also help in reducing the congestion of the data while it is transferred to the cloud system.

12.3 RESULTS/OUTCOME OF THE DESIGNED SYSTEM

The designed system has reduced the amount of time required for the diagnosis of the patient and also the time required for providing the treatment; this has helped to save the lives of multiple patients when an emergency exists. The amount of time required by an expert is drastically reduced, as the monitoring of the patient was possible from the remote location. This system was very useful during the pandemic, as travel was restricted in multiple places, and because of the checks on the roads the time required for permitted travel was high. So in this situation the designed system has played a major role in providing proper treatment to the patient on time every time [5]. Results of the system which is designed with the fog computing and the modeling of graph is shown in Figure 12.3.

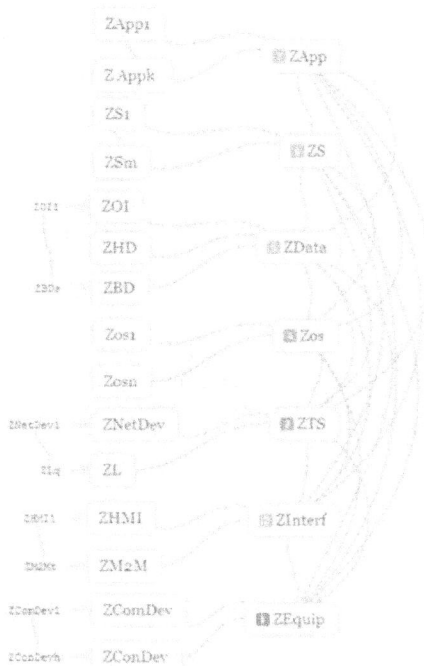

FIGURE 12.3 Fog computing graph model.

We have shown the equipment in the first level (Zequip,ZconDev).

This vertex-Zequip will decide the problems which are solved at this level.

Vertex ZconDev are devices to perform the task of calculations.

Next level: M2M and HMI are inter-machine interaction systems and human-machine interface.

Zinterf specifies the task at this level.

The third level pertains to the transport level.

The fourth level relates to the OS level.

The fifth level pertains to real-time analysis.

The sixth can be for online services.

12.4   CONCLUSION

The designed system was implemented by making use of various devices like the sensors connected to the patient present at a remote location and then capturing the data related using the edge computers and then transferring the same to the fog computers, so that the data required can be extracted and the irrelevant data can be filtered at this level in order to see that the traffic and congestion on the cloud server are reduced. The data from the cloud server was properly utilized by the various domain experts as per the need in order to provide right and perfect treatment for the patient in need. This designed framework has helped the patients to get the experts' advice on time every time. As the system makes use of the edge and fog computing concept, it is very easy for the computers connected at the edges to capture the data at various locations and then to store the same at the local level and also to filter and analyze the data so that only the relevant data from all the edges will pass on to the fog system, and the edges will receive the data from the various devices connected at every time interval fixed. This has to be filtered and then transferred to the fog, which is connected between the edges and the cloud server. The system provides a lot of advantages in terms of required bandwidth and congestion on the cloud server, as the process of filtration and removing irrelevant data will be done at this level also.

REFERENCES

[1] Mora-Sánchez, O.B., López-Neri, E., Cedillo-Elias, E.J., Aceves-Martínez, E. and Larios, V.M., 2020. Validation of IoT infrastructure for the construction of smart cities solutions on living lab platform. IEEE Transactions on Engineering Management, 68(3), pp. 899–908.

[2] Fortino, G., Fotia, L., Messina, F., Rosaci, D. and Sarné, G.M., 2020. Trust and reputation in the internet of things: State-of-the-art and research challenges. IEEE Access, 8, pp. 60117–60125.

[3] Goudarzi, M., Wu, H., Palaniswami, M. and Buyya, R., 2020. An application placement technique for concurrent IoT applications in edge and fog computing environments. IEEE Transactions on Mobile Computing, 20(4), pp. 1298–1311.

[4] Ali, B., Pasha, M.A., ul Islam, S., Song, H. and Buyya, R., 2020. A volunteer-supported fog computing environment for delay-sensitive IoT applications. IEEE Internet of Things Journal, 8(5), pp. 3822–3830.

[5] Sarkar, S., Wankar, R., Srirama, S.N. and Suryadevara, N.K., 2019. Serverless management of sensing systems for fog computing framework. IEEE Sensors Journal, 20(3), pp. 1564–1572.

13 Image Denoising Using Autoencoders

Mursal Furqan Kumbhar

13.1 INTRODUCTION: BACKGROUND AND DRIVING FORCES

Denoising of images is the process of eliminating unwanted and extra noise from images and photos. Here, unwanted noise is defined as everything that can be eliminated by a well-designed preprocessing filter in visual processing and computer vision. With the use of machine learning and artificial intelligence approaches, a novel image denoising strategy based on soft-threshold and edge enhancement is presented to reduce noise while keeping picture edge and texture quality.

In order to enhance edge detection, the author applies Canny edge detection operators to identify image features, flattens the edges, and employs stationary wavelet transformation to process both the edges and the associated noise images. Subsequently, the processed edges are added to the final output. These wave coefficients represent the corresponding level, allowing us to use the soft threshold denoise technology to remove the noise. Because the characteristic is reinforced before soft threshold denoising, it can be preserved. Figure 13.1 shows how an encoder works.

The term "noise" here could be:

- JPEG compression artefacts
- Poor paper quality
- Quantization noise
- Image perturbations

Figure 13.2 illustrates how a denoiser autoencoder (DAE) works.

13.2 LITERATURE REVIEW

X. Li et al., in their paper [1], explain how to reduce noise while maintaining image edge and texture quality using a visual denoise strategy that lies upon a soft threshold on images and edge enhancement of given images. To improve the edge, first they detect the image feature using the Canny edge detector. Then flatten the edge image and perform the extraction of the edge image and the noise image using a stationary wavelet transformation. And finally, by adding these wave coefficients to the corresponding levels, they eliminate noise using soft threshold noise elimination technology.

DOI: 10.1201/9781003328414-13

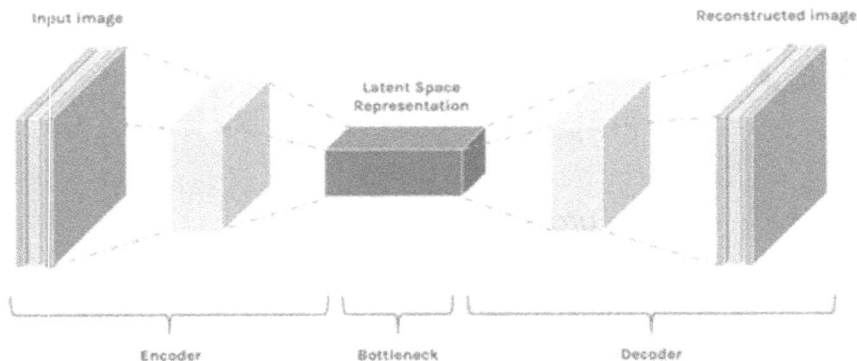

FIGURE 13.1   Encoder-decoder architecture.

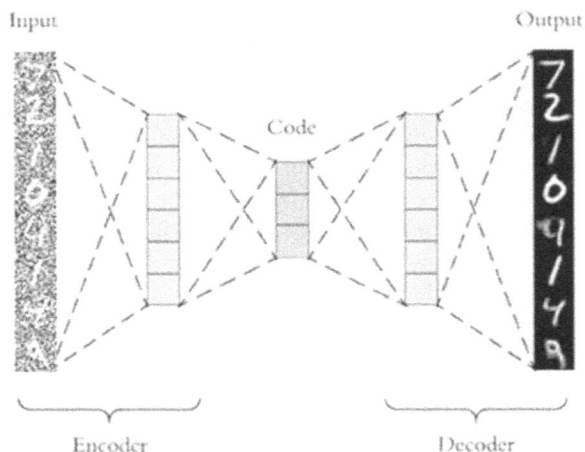

FIGURE 13.2   An overview of how the entire procedure works to denoise the noisy images.

In their paper [2], Wu, C. and Gao, T. describe the concept of image/photo denoising, share a complete list of common image noises, and discuss several classic algorithm discussions used in traditional denoising procedures. Additionally, the shortcomings of old methodologies are assessed. Then we summarize the deep learning–based image denoising techniques, including the image denoisation methods based on several structures, including GAN, DnCNN, REDNet, Noise2Noise, and CBDNet, along with the concepts and structures of different methods. Finally, the limitations of picture denoising are investigated, as well as potential future study fields.

The authors, Chen, J., Chao, H. et al. in [3] of the study used image blind deformation to study methods for removing unknown distortions from loud images. We all know that discriminatory learning algorithms like DnCNN can produce cutting-edge results in denoisation. They are not suitable for our applications, however, as there are no pairing training data. To solve this problem, Chen J. et al. proposed a new and efficient two-step approach. The generation contrast network (GAN) is a noise prediction and noise sampling network on noise input images. Second, the noise blocks of the previous phase were used to create training pairs and then trained in deep convolutional neural networks (CNNs).

As described by Guo, S. in [4], in recent years, gaussian noise analysis algorithms have been successful, especially if gaussian noise is produced or encountered at regular intervals. While, on the other hand, original noise or real-world noise is chaotic and complex. In order to properly decrease noise and capture a high-quality image, improved and high-quality hardware devices are required. Furthermore, the generated image may be distorted, hazy, or low resolution. As a result, figuring out how to recover the latent clean picture from the superposed noisy image is critical. Furthermore, although deep learning algorithms require the use of ground truth to collect features, the created real-world noisy images do not. These are significant issues that academics and researchers must solve [4].

In [5] the researcher Li, X.X. created a revolutionary wavelet denoising approach based on an unsupervised learning model. The technique constructs an unsupervised dictionary learning algorithm for noise reduction dictionary synthesis using wavelet transform properties such as sparsity, multiresolution structure, and proximity to the human visual system. We develop an adaptive dictionary by learning the wavelet decomposition of the noisy picture using the K-singular value decomposition (K-SVD) approach. Our suggested technique surpasses state-of-the-art denoising algorithms in terms of peak signal to noise ratio (PSNR), structural similarity index (SSIM), and visual effects with varying noise levels, according to experimental results on benchmark test images.

13.3 METHODOLOGY

As mentioned earlier, image noise is caused by several intrinsic or external reasons, and it is difficult to deal with them. Denoising the image is a major problem in the processing and visualization of images. This makes it useful in various industries where the acquisition of original images is crucial to strong performance. Figure 13.3 shows the appearance of noisy images.

13.3.1 Encoder-Decoder Network (Autoencoders)

Autoencoders are unsupervised artificial neural networks that copy inputs and outputs. When working with image data, the autoencoder first encodes the image into a lower-dimensional representation and then decodes it to the original image. The encoder-decoder consists of two structures:

- Encoder: This network consists of a low-dimensional sampling of data.
- Decoder: This network reconstructed the original data of the lower-dimensional representation.

FIGURE 13.3   An example of an image with noise.

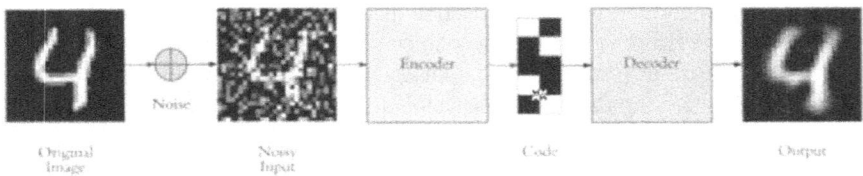

FIGURE 13.4   A thorough overview of the autoencoding process.

The reduced-dimensional representation generated by the network, often referred to as a hidden space representation, is typically obtained through an autoencoder. It is crucial for the autoencoder to consider that it can only compress data similar to what it has been trained on. Additionally, it incurs a loss, resulting in a decrease in output compared to the original input. Backpropagation trains them in the same way as an artificial neural network. The overview of autoencoding process is shown in Figure 13.4.

13.3.2   Implementation

Because actual research implementations are done in Python, libraries and other relevant information are simply called Python programming languages.

13.3.2.1   The Libraries Used

The first step is to import the required Python libraries. Importing libraries such as NumPy improves matrix multiplication, and Matplotlib displays data as graphical diagrams first [6]. The Keras sequence model provides an empty box in which we design different density layers. Dense builds complete network connections, and MNIST quickly imports Keras datasets. An example of a MNIST dataset is shown in Figure 13.5.

FIGURE 13.5 An example of a MNIST dataset.

13.3.2.2 Dataset

In this study, the updated dataset of the National Institute of Standards Technology (MNIST) is used. The MNIST datasets are the foundations of computer vision. It is composed of labelled greyscale 28 × 28 handwritten number images. The MNIST dataset is divided into two portions for the optimal model optimization:

- Training dataset: There are 60,000 training datasets.
- Test datasets: There are 10,000 points in the test dataset.

13.3.2.3 Plotting Images as Grayscale Images

Following the separation of the train and test datasets, as shown in Figure 13.4, some of the images in the dataset were plotted using the MATPLOTLIB package, specifically, the subplot function of the MATPLOTLIB package, which allows you to plot many pictures at the same time.

13.3.2.4 Adjusting Data with Respect to Keras

Following that, the two-dimensional picture array was reduced to 2828 × 784 integer vectors. There is no difference in how it is flat as long as the array format is constant across all photos. MNIST pictures, from this perspective, are essentially points in 784-dimensional vector spaces. Data must, however, always be represented as follows: date numbers, data point dimensions. The training data in this example are in the 60,000 × 784 format.

13.3.2.5 Adding Noise to Images

When trying to solve the issue statement, keep in mind that the goal is to construct a model capable of removing noise from photographs. To create this, existing pictures were used and then random noise was injected into the images shown in Figure 13.6. In this phase, models (autoencoders) learn how to clean up noisy images. The input is the original photo, and the output is the loud image.

The first hyperparameter is the noise factor provided. A random matrix 0.0 is multiplied by a noise factor and standard deviation 1.0. This matrix sample is generated using a normal distribution (gaussian). When adding noise, the shape of the regular random array is considered equivalent to the data of the noise addition. Finally, you can use the np.clip function to check if the last image array has 0–1. The Numpy Clip function replaces the value outside the Min-Max range with a user-assigned value.

13.3.2.6 Defining an Encoder-Decoder Network

The encoder-decoder network contains 784 neurons in the input layer because of the 828-pixel size.

FIGURE 13.6   Images with noise added to them.

13.3.2.7   Compiling the Model

After a specific parameter is specified, the model must be compiled. The optimizer and all measurements are provided when building the loss function that will be used. To achieve the best results in our model, Adam optimizers and mean squared errors are used for problem statements.

13.3.2.8   Model Training/Fitting

The model is completed and ready for training. The network receives training data, defines validation data, and only uses it to validate models.

13.3.2.9   Evaluating the Model

Finally, the training model is evaluated with the test datasets from the previous round. The first shot is the test image, the second is the loud image, and the third is the cleaned image (reconstruction) as shown in Figure 13.7.

13.4   RESULTS

DAEs bypass the identity function and, unlike traditional noise reduction filtering methods, do not produce too smooth images and may be computed rapidly. DAEs, in general, use an updated autoencoder approach for operation, which is essentially based on the injection of noise into the input and reconstruction of the output from the damaged picture. This change to the general autoencoder approach prohibits DAEs from simply copying the input to output, forcing DAEs to eliminate noise from the input before extracting valuable data.

CNN was used in our DAE approach because of its effectiveness in denoising and retaining spatial links within images. Furthermore, when arbitrary-sized images are utilized as input, the employment of CNN fulfils the goal of lowering dimensional and computational complexity. An example of different images from the MNIST dataset is shown in Figure 13.8.

13.5   CONCLUSION

In this study, an autoencoder model is developed that effectively cleans very noisy pictures that it has never seen before, i.e., those on the dataset. These images do, however, include certain unnoticed abnormalities. However, it can greatly clean distorted and noisy images, demonstrating that our method is effective in recovering damaged images. In the future, this autoencoder model might be expanded and integrated into a picture enhancement software to increase the clarity and crispiness of the images.

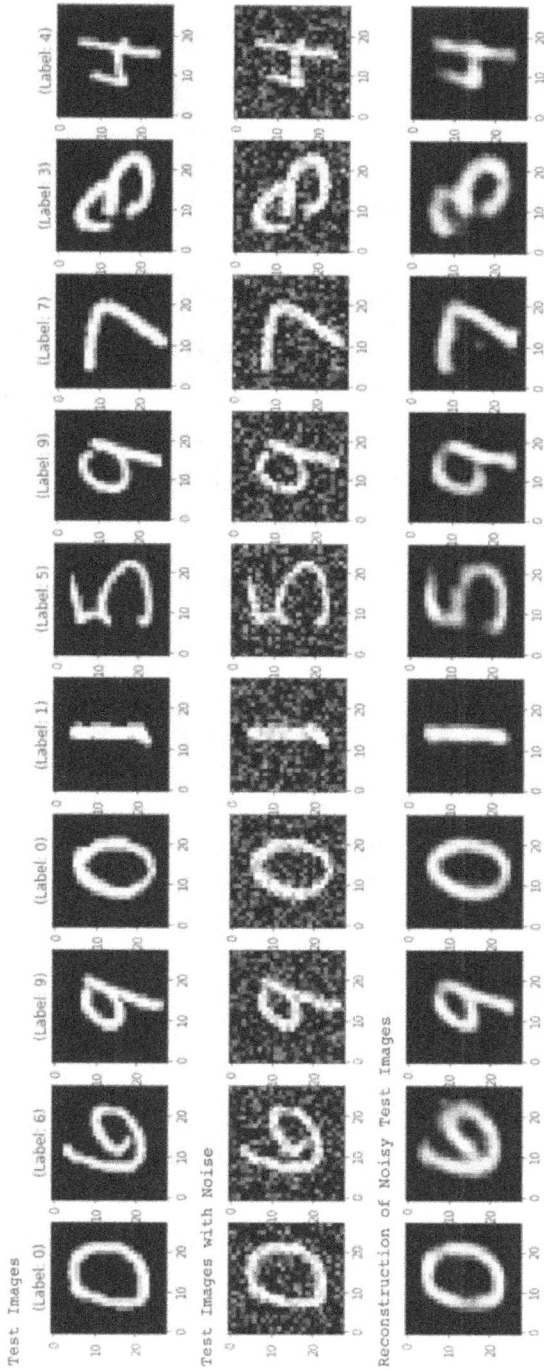

FIGURE 13.7 Denoised images.

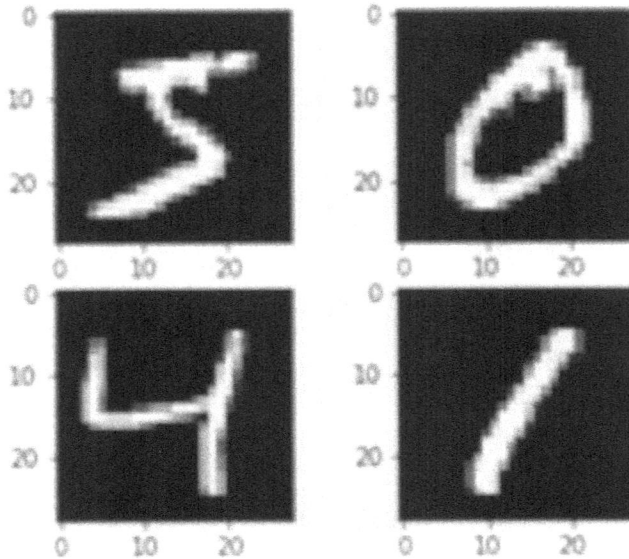

FIGURE 13.8   An example of different images from the MNIST dataset.

REFERENCES

[1] Li, X., Yan, G., Mei Li, X.-and Chen, L. 2007. Image denoise based on soft-threshold and edge enhancement. Second Workshop on Digital Media and its Application in Museum & Heritages (DMAMH 2007).

[2] Wu, C. and Gao, T. 2021. Image denoise methods based on deep learning. *Journal of Physics: Conference Series*, 1883(1), p. 012112.

[3] Chen, J., Chen, J., Chao, H., et al. 2018. Image blind denoising with generative adversarial network-based noise modelling [C]. Proceedings of the IEEE Conference on Computer Vision and Pattern Recognition. Salt Lake City, Utah, pp. 3155–3164.

[4] Guo, S. 2019. *A Study on Real Camera Image Denoise Based on Convolutional Neural Network [D]*. Harbin Institute of Technology.

[5] Li, G.X. 2006. A study on image wavelet denoise based on wavelet shrinkage method [J]. *Journal of Remote Sensing*, 5, pp. 697–702.

[6] Hemachandran, K., Khanra, S., Rodriguez, R. V. and Jaramillo, J. (Eds.). 2022. *Machine Learning for Business Analytics: Real-Time Data Analysis for Decision-Making*. CRC Press, New York.

14 Genetic Disorder Prediction Using Machine Learning Techniques

Sirisha Alamanda, T. Prathima, Rajesh Kannan K, and P. Supreeth Reddy

14.1 INTRODUCTION

An illness known as a genetic disorder is typically brought on by abnormalities in the DNA or alterations to the chromosomes' size or structure in general. Hereditary gene mutations are linked to a number of prevalent disorders. Studies show that the frequency of hereditary diseases has increased rapidly in parallel with population growth. Hereditary illnesses are becoming more common because people are less aware of the value of genetic testing. Genetic testing must be done when children are young because many children die from these conditions.

Existing systems for diagnosing genetic disorders are very expensive and time-consuming. State-of-the-art equipment and cutting-edge technology are needed, and it can only be provided by a few top-level institutes and hospitals, which are out of reach for most people. Even after the process of genome sequencing, there are a lot of difficulties in finding the abnormalities in the gene. There are no direct tests to find out the genetic disorder a person has. Genetic illnesses are still a mystery to scientists, and the notion is still in the study stage.

There are two types of disorders that are genetically determined: "single gene inheritance diseases" and "mitochondrial gene inheritance disorders". A single-gene illness results from genetic changes to one gene. Single-gene diseases are highly diverse and can have an impact on all aspects of functioning because they can arise in any gene. The majority of inborn metabolic illnesses are mitochondrial gene inheritance abnormalities [1], and 1.6 out of every 5000 persons are affected according to [2].

In this study, the authors used machine learning and deep learning models to predict genetic disorders. Attributes used in this study don't require complex equipment and don't need genome sequencing. The authors have used features like white blood cell count, history of radiation exposure, substance abuse, birth asphyxia, history of previous pregnancies, number of previous abortions, birth

DOI: 10.1201/9781003328414-14

defects, (institute masked) tests, assisted conception IVF/ART, folic acid detail (peri-conceptional), etc.

The patients in the dataset considered for the study are in the age group of 0–15 years, which gives time to detect, acknowledge, and prevent further deterioration of the patient at an early stage. All the data for the various attributes mentioned earlier is retrieved from the patient and their biological parents. In this study, an autoencoder model was used for feature extraction and machine learning models. For classification, XGBoost and artificial neural networks were used.

14.2   RELATED WORK

Numerous studies have been conducted to predict genetic diseases (Y & M, 2015) [3] and (Menche et al., 2015) [4]. Genetic data collecting has increased as a result of improved genotyping and testing techniques. Despite this growth, it remains unclear how genetic variations contribute to illness development. The bulk of genomic alleles and malignant variations still lack genetic information, despite ongoing mapping efforts [5].

The study by (N.G., 2012) [6] forecasts the potential for cardiovascular disease by merging the benefits of genetic algorithms and neural networks. The principles of natural genetics are imitated by an optimization algorithm known as a "genetic algorithm". The deep learning model used is a multi-layer perceptron (MLP). The genetic algorithm is used to choose the optimal initial weights for the MLP model. To determine the likelihood of developing cardiovascular disease, the final weights given by the neural network model are used. (K. R. Gray et al., 2013) [7] had studied classification of Alzheimer's disease based on a random forest. Similar methods [8, 9] have been used to anticipate disorder modules, which present a similar difficulty; clusters of these modules can contain disturbed genes. Genetic risk prediction has been shown to have some short-term effects on people and communities; however, only significant improvements in high-density genotyping technology have made it possible to determine genetic risks. Alternate outcomes including dyslipidaemia, hypertension, or even smoking may also be caused by genes associated with cardiovascular disease.

The authors of (Hoskins et al., 2017) [10] used deep learning techniques and Illumina exome genotyping array data to generate a precise human Down syndrome (DS) prediction model. The authors presented a convoluted neural network (CNN) architecture. Three algorithms support vector machine (SVM), decision tree, and random forest, are used to develop DS prediction and screening models. The genotyping array is the input for other models, whereas the single-nucleotide polymorphism (SNP) map is the input for CNN. However, traditional machine learning methods treat each SNP individually as a feature. The considerable computing costs associated with obtaining certain features for subset characterization, and optimization further limited the discriminative abilities of these methods. As a result, geographical patterns and feature correlations were difficult for conventional machine learning methods to identify.

Improvements in sequencing technology have made it possible to investigate so-called missing heritability, and many affected were examined to find substantial connections between uncommon variations and human diseases (Kim W, 2017) [11]. However, given the high cost of genome sequencing, a robust statistical method for purposeful sampling would be helpful. To identify patients based on family history, the authors have suggested a new statistical strategy. Their proposed approach was effectively used in genome-wide association studies (GWAS) for type 2 diabetes, and the analytic findings demonstrated the approach's usefulness.

Despite the enormous successes of GWAS, (Gim et al., 2017) [12] developed a technique for risk prediction analysis with penalized regression analysis that takes into consideration both a substantial number of genetic variants and clinical risk factors. In a research study on lung cancer prediction using analysis of hundreds of gene expression profiles, advanced machine learning classifiers were used to assess the probability of genes causing cancer (Patil, 2018) [13]. Out of 7129 genes, the authors identified 72 genes as having the highest likelihood of being linked to lung cancer. (Frangly et al., 2018) [14] had worked on an ensemble approach for predicting genetic disease, and (Psomagen 2020) [15] has done a study on how can genetics help predict diseases.

Machine learning (ML) techniques have been used by (Romagnoni, A., et al, 2019) [16] to categorize healthy and diseased individuals based on their genomic information. The study demonstrated that gradient boosted trees or artificial neural networks (ANNs) may offer strong complementary ways to detect and categorize genetic markers in comparison to logistic regression. (Duc-Hau Le, 2020) [17] conducted a study in which ML-based methodologies for disease gene prediction and evaluation were examined and assessed. (Bracher-Smith et al., 2021) [18] thoroughly examined ML techniques for psychiatric disease prediction based solely on genetics and assessed their discrimination, bias, and application.

Medical histories of patients have been employed by (Nasir, M. et al., 2022) [19] to get over the restriction on predicting genetic disorders using genomic sequential data. This study has devised a genetic disease prediction algorithm in order to determine if a patient is prone to a genetic disorder. Different machine and deep learning approaches, including ANN, SVM, and K-nearest neighbours (KNN), are trained to predict and identify patients with genetic diseases. Their results indicate that the proposed model has the highest performance when combined with the ANN.

Many of the previous studies have mainly used microarray gene data, which is hard and expensive to get and operate on, so this has inspired the authors of the current study to use simpler attributes such as age, father's age, mother's age, blood cell count, and others to try to predict the genetic disorders. The earlier feature extraction methods are mainly statistical methods, so the authors have decided to use an autoencoder for feature extraction, which is a deep learning technique.

14.3   PROPOSED FRAMEWORK

In this study, the authors take input data and preprocess the data using various pre-processing techniques. The authors only select a subset of the attributes using feature extraction with autoencoders and trained XGBoost and ML models for performance comparison. The framework proposed for genetic disorder prediction is shown in Figure 14.1.

ALGORITHM 1: AEML-GDP (AUTOENCODER AND MACHINE LEARNING–BASED GENETIC DISORDER PREDICTION)

Input: Genetic disorder dataset.
Output: Optimal model for genetic disorder prediction
Process:
Step 1: Load the dataset
Step 2: Perform data preprocessing.

    a. For each feature, fill the missing values if any.
      i) If the feature is continuous, use random sample imputation
      ii) If the feature is categorical, use mode.
    b. For each continuous feature, perform discretization.
    c. Perform MinMax normalization of data
    d. For each categorical variable, apply one-hot encoding.

Step 3: Perform feature extraction using the autoencoder to retrieve the most relevant and significant features.
Step 4: Train the OvR-SVM, XGBoost, and ANN models on the resulting dataset.
Step 5: Compare the performance of the models for precision, recall, accuracy, and output of the optimal model.

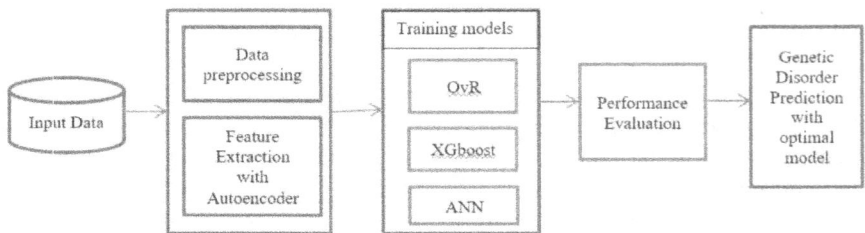

FIGURE 14.1   Proposed framework for genetic disorder prediction.

The proposed study is presented in the earlier algorithm AEML-GDP. The following four phases make up the current study:

1. Data preparation
2. Feature extraction
3. Classification model development
4. Performance evaluation of models

14.3.1 DATA PREPARATION

Data preparation is a method of putting raw data into a format that is comprehensible. Real-world data is unstructured and is often produced, processed, and saved by a variety of people, business operations, and applications. Because of this, a dataset can be incomplete, have manual input errors, have duplicate data, or use several names to refer to the same object. People are frequently able to see and fix these issues in the data they use in their line of work, but data that is used to train deep learning or ML algorithms must be automatically preprocessed.

The current study is based on the dataset taken from Kaggle (Amit Kumar, 2021) [20] to predict the genetic disorder. This dataset contains medical data of children with genetic disorders. The dataset consists of 22,083 records and 44 columns. There are three different classes of genetic disorders, and the distribution of these classes in the raw training data is shown in Table 14.1. As the study is only based on genetic disorder prediction, the disorder subclass column is ignored, and 14% of data has a missing value for the Genetic Disorder column.

The features "Patient#Id", "Family_Name", "Patient_First_Name", "Institute_Name", "Father's_name", "Place_of_birth", "Location_of_Institute", "Parental_consent", "Test_5", and "Status" were removed from the dataset as such features don't quite help understand the reason for inheriting a disorder. Test 5 and Parental Consent are removed as they contain only a single value. Furthermore, the following steps are performed on the data to make it ready for training.

1. The missing values are filled using random sample imputation for continuous variables and mode for categorical.
2. Some continuous variables, such as mother's and father's ages, are discretized.

TABLE 14.1

Distribution of Genetic Disorder Class Label in the Training Data

Type of Disorder	% of Tuples
Mitochondrial genetic inheritance disorder	44
Single-gene inheritance diseases	33
Other (multifactorial genetic inheritance disorders + missing label)	23

3. The data is normalized using MinMax scaling.
4. The target variable and categorical variables have been converted into one-hot encoded values.

14.3.2 FEATURE EXTRACTION

There are usually too many factors (features) used to make the final prediction in real-world ML issues. It becomes more difficult to visualize the training set and subsequently work on it as the number of features increases. These features can occasionally be redundant or connected to one another. Algorithms for dimensionality reduction are useful in this situation. By identifying a collection of primary or significant features, one can reduce the number of random features that must be taken into account.

Feature extraction is a process for transforming unprocessed data into usable numerical features while preserving the original dataset's information. Compared to using ML on the raw data directly, it produces better outcomes. Features can either be extracted manually or automatically. Identification and description of the characteristics that are pertinent to a particular problem are necessary for manual feature extraction, as is the implementation of a method to extract those features. Automated feature extraction eliminates the need for human intervention by automatically extracting features from signals or images using specialized algorithms or deep networks.

As not all input features are suitable for prediction, in this study autoencoders are used to choose only a subset of useful features. A neural network called an autoencoder is trained to reproduce its input to its output. The encoder and the decoder are both components of an autoencoder. It has a hidden layer on its interior that specifies the encoder that is used to represent the input. The encoder picks up on the input's semantics and learns to condense it into a representation provided by the bottleneck layer. The decoder makes an effort to reconstruct the input using the encoder's output (the bottleneck layer). In this situation, after the model is fitted, the reconstruction portion can be dropped, and the model can be used all the way up to the bottleneck. A fixed-length vector that offers a reduced depiction of the input data is models output at bottleneck layer is used as the characteristic vector in a supervised model. The dataset for the current study has been reduced to 37 features following the feature extraction using autoencoders.

14.3.3 MODEL DEVELOPMENT

The authors applied the ML models to the compressed data in this section. The authors have trained the XGBoost, ANN, and One vs. Rest classification model with SVM on the training data and estimated the models parameters.

14.3.3.1 One-vs-Rest Classifier (OvR)

OvR is an experimental technique for employing two-class classification algorithms for multi-class classification. In this technique a multi-class label prediction problem is carried out as multiple binary classification problems. Then, a binary classifier is trained and the confident model makes predictions for every two-class labelling problem.

Given a multi-class classification problem, for instance, with examples for the classes *"red"*, *"blue"*, and *"green"*. The following three binary classification problems might be created from this:

1. Binary Classification Problem 1: Red vs. [Blue, Green]
2. Binary Classification Problem 2: Blue vs. [Red, Green]
3. Binary Classification Problem 3: Green vs. [Red, Blue]

This method's potential drawback is that it necessitates the creation of one model for each class. The authors have employed the OvR Classifier with SVM.

An SVM can be handled using both regression and classification. SVM can address regression concerns, and categorization is appropriate term. The SVM method looks for a hyper-plane in an N-dimensional space that categorize the data points with precision. The size of the hyper-plane is proportional to the size of the features. When there are two input features, the hyper-plane effectively looks like a 2D line. The hyper-plane is a 2D plane even for three input features. Visualizing is challenging when there are more than three features. In this study the authors have achieved an accuracy of 85.27% using the OvR Classifier with SVM.

14.3.3.2 XGBoost

By combining several weak classifiers, the ensemble modelling technique known as "boosting" aims to create a powerful classifier. It is accomplished by using weak models in series to develop a model. First, a model is created using a training set of data. A second model is then created in an effort to fix the previous model's flaws. Models are added in this manner until either the full training dataset is properly predicted or the maximum numbers of models have been added.

In the XGBoost approach, the decision trees are generated in a sequential manner. Weights are significant in this approach. Independent variable weight is computed before considering it to be inserted into the decision tree to perform prediction. Variables that the tree incorrectly predicted are given more weight before being placed into the second decision tree. These different predictors are then joined to produce an accurate and robust model.

A gradient boosted decision tree implementation designed for speed and performance is called XGBoost. Gradient boosting is a technique that uses new models to predict the errors or residuals of older models, which are then merged to provide the final prediction. The method is called "gradient boosting" because it uses a gradient descent methodology to lessen loss while introducing new models. The authors got an accuracy of 88.9% from XGBoost.

14.3.3.3 Artificial Neural Network

An ANN is a computational network that makes an effort to imitate the network of neurons that make up the human brain in order to perceive things and make decisions in a way that is human-like. The various layers that can be used in an ANN include:

Input layer: This accepts inputs from the user in various formats.
Hidden layer: This stays in between the input and output layers. It makes all the computations necessary to uncover patterns and hidden features.
Output Layer: This layer is used to communicate the output after the input has undergone a number of alterations in the hidden layer.

The most popular ANN model in the deep learning field is an MLP. The MLP has multiple layers of neurons, and each layer multiplies the features with weights and applies an activation function such as tanh, ReLU, or sigmoid. In the final layer of the MLP, a softmax function is used, which produces probability for each class of disorder.

In this study, the authors used an ANN model to include a basic sequential model followed by a sequence of dense layers and dropout layers. Preprocessed data with 37 columns is given as input to the input layer, which is followed by five hidden layers of neurons with a rectified liner unit as the activation function used and an output layer with three neurons corresponding to the three classes of genetical inheritance disorders: a. multi-factorial, b. mitochondrial, and c. single. The ANN model utilizes categorical cross entropy loss and the Adam optimizer. The training set achieves an accuracy of 92.62%, while the test set achieves an accuracy of 94.13%. Analyzing Figure 14.2 reveals that the ANN model stabilizes after approximately 70–75 epochs, with no signs of over-fitting as both the train and test loss exhibit similar patterns.

The performance of all these models is evaluated, and the comparison of the results is discussed in detail in this next section.

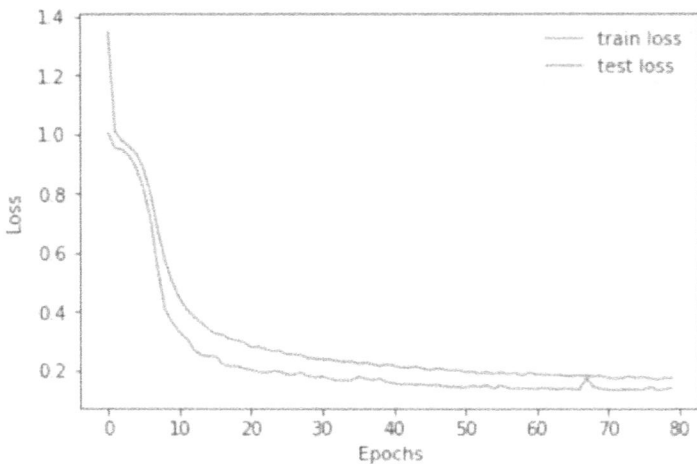

FIGURE 14.2   The graph of epochs vs loss of ANN.

14.4 RESULTS AND DISCUSSION

This section compares the trained ML model's performance on the test data using the metrics precision, recall, F1 score, and accuracy. The definition of the parameters used for the computation is provided in Table 14.2.

The evaluation metrics: recall, accuracy, precision, and F1-score are computed by using the following formulae:

$$\text{Precision} = \frac{T.P.}{T.P. + F.P.} \tag{14.1}$$

$$\text{Recall} = \frac{T.P.}{T.P. + F.N.} \tag{14.2}$$

$$\text{F1-Score} = \frac{2\left(Precision * Recall\right)}{Precision + Recall} \tag{14.3}$$

$$\text{Accuracy} = \frac{T.P. + T.N.}{T.P. + F.P. + T.N. + F.N.} \tag{14.4}$$

From Table 14.3, it can be observed that the ANN has better accuracy and can be chosen as the best model for genetic disorder prediction. But as the study is on genetic disorder prediction, the cost of false negatives is much higher than the cost of false positives. So false negatives should be given more importance. The lower the recall, the more false negatives the model predicts and can be considered a bad model.

From Table 14.3, it can be observed that the OvR Classifier is performing well for mitochondrial genetic disorders but for multifactorial and single-gene genetic disorders its performance is not good in terms of precision. From Table 14.3 it can be seen that the recall is very low for multifactorial genetic inheritance disorders with the OvR Classifier so this is not a good model. On the other hand it can be observed that the XGBoost model is performing well for mitochondrial genetic disorders and single-gene genetic disorders but not so well for multifactorial genetic disorders, whereas it can be observed that the ANN model is performing well for all classes of disorders. So it can be considered an optimal model for the genetic disorder.

TABLE 14.2

Parameter Description

Parameter	Definition
True Positives (TN)	Number of concerned (target) class tuples, labelled as such.
False Positives (FN)	Number of other (non-target) class tuples, labelled as concerned (target) class tuples
True Negatives (TN)	Number of other (non-target) class tuples, labelled as such.
False Negatives (FN)	Number of concerned (target) class tuples, labelled as other (non-target) class tuples.

TABLE 14.3

Comparison of Prediction Models (SVM, XGBoost and ANN) Used in the Current Study

Prediction Model Used	Type of Disorder	Precision	Recall	F1-score	Accuracy
OnevsRest classifier with SVM	Mitochondrial genetic inheritance disorder	0.97	0.88	0.92	85%
	Multifactorial genetic inheritance disorder	0.71	0.43	0.54	
	Single-gene inheritance diseases	0.76	0.93	0.84	
XGBoost	Mitochondrial genetic inheritance disorder	0.90	0.96	0.92	89%
	Multifactorial genetic inheritance disorder	0.83	0.66	0.74	
	Single-gene inheritance diseases	0.89	0.87	0.88	
ANN	Mitochondrial genetic inheritance disorder	0.90	0.99	0.95	94%
	Multifactorial genetic inheritance disorder	1.00	0.82	0.90	
	Single-gene inheritance diseases	0.99	0.91	0.95	

After the detailed analysis from the obtained results, the framework is compared with the parallel research results on the same dataset in Table 14.4. From the table it is evident that feature extraction with the autoencoder helps to extract more significant features and thereby predicts the genetic disorder more efficiently.

14.5 CONCLUSION

Many genetic disorders are hard to detect. They can be very dangerous if not detected. Many people are suffering from them. Many people die due to them. Detecting them early can be very helpful. The existing systems for detecting genetic disorders can be very expensive, and some disorders are hard to identify. So, this has inspired the authors to create a system that will be able to identify the disorder quickly in the people who are suffering from it. The authors have trained the genetic disorder data from Kaggle on the OvR Classifier with SVM, XGBoost, and ANN models and compared their performance. The results have shown the best model is ANN. The authors were able to address the limitation identified in genetic disorder prediction sequential data by using patient medical history information. Many lives can be saved with the use

TABLE 14.4

Comparison of Prediction Models Used in [19] and the Current Study

Author, Year	Pre-proceeding	Feature Extraction Method	Prediction Models Used	Accuracy Observed
Nasir et al., 2022 [19]	yes	Linear Regression	*SVM*	60.1 %
			Artificial Neural Network (ANN)	84.9 %
Proposed framework, 2022	yes	Autoencoder	*One-vs-Rest Classifier with SVM*	85.29 %
			Artificial Neural Network (ANN)	94.13 %

of this method to predict the genetic adversity early based on medical data. In the future, this system can be made accessible to everyone in the world, anywhere and anytime, by creating web applications and mobile applications. The system's accuracy can be increased even more by working on large amounts of data.

REFERENCES

[1] C. R. Ferreira, C. D. M. van Karnebeek, J. Vockley and N. Blaue, "A proposed nosology of inborn errors of metabolism", *Genetic Medicine*, vol. 21, no. 1, pp. 102–106, 2019.

[2] J. Tan, M. Wagner, S. L. Stenton, T. M. Storm, S. B. Wortmaan et al., "Lifetime risk of autosomal recessive mitochondrial disorders calculated from genetic databases", *Lancet*, vol. 54, pp. 111–119, 2019.

[3] Y. Park and M. Kellis, "Deep learning for regulatory genomics", *Nature Biotechnology*, pp. 825–826, 2015.

[4] J. Menche, A. Sharma, M. Kitsak, S. Ghiassian, M. Vidal et al., "Uncovering disease-disease relationships through the incomplete human interactome", *Science*, vol. 347, no. 6224, pp. 1257601, 2015.

[5] S. Won, H. Choi, S. Park, J. Lee, C. Park et al., "Evaluation of penalized and nonpenalized methods for disease prediction with large-scale genetic data", *BioMed Research International*, p. 605891, 2015.

[6] N.G, B. A et al., *Cardiovascular Disease Prediction using Genetic Algorithm and Neural Network*. IEEE, 2012.

[7] K. R. Gray et al., "Alzheimer's disease neuroimaging initiative. Random forest-based similarity measures for multimodal classification of Alzheimer's disease", *Neuro Image*, vol. 65, pp. 167–175, 2013.

[8] Y. liu, D. A. Tennant, Z. Zhu, J. K. Health, X. Yao et al., "Dime: A scalable disease module identification algorithm with application to glioma progression", *PloS One*, vol. 9, no. 2, pp. 866–876, 2014.

[9] S. D. Ghiassian, J. Menche and A. L. Barabasi, "A disease module detection (diamond) algorithm derived from a systematic analysis of connectivity patterns of disease proteins in the human interactome", *PloS Computational Biology*, vol. 11, no. 4, pp. 1004120, 2015.

[10] W. Hoskins, Y. Zhang, Y. Guo, and J. Tang, *Down syndrome prediction/screening model based on deep learning and Illumina genotyping array*. IEEE, 2017.

[11] W. Kim, D. Qiao, M. H. Cho, S. H. Kwak, K. S. Park et al., "Selecting cases and controls for DNA sequencing studies using family histories of disease", *Statistics in Medicine*, vol. 36, pp. 2081–2099, 2017.

[12] Jungsoo Gim, Wonji Kim, Soo Heon Kwak, Hosik Choi, Changyi Park, Kyong Soo Park, Sunghoon Kwon, Taesung Park, and Sungho Won, "Improving disease prediction by incorporating family disease history in risk prediction models with large-scale genetic data", *Genetics*, vol. 207, no. 3, pp. 1147–1155, 1 November 2017.

[13] Jaydeep Patil, *Gene Expression Analysis for Early Lung Cancer Prediction Using Machine Learning Techniques*. IEEE, 2018.

[14] Frangly Francis and T. N. Namitha, *Ensemble Approach for Predicting Genetic Disease*. IEEE, 2018.

[15] Psomagen, "How can genetics help predict diseases?" *Psomagen*, 31 August 2020. [Online]. Available: https://psomagen.com/how-can-genetics-help-predict-diseases-psomagen/.

[16] A. Romagnoni, S. Jégou, K. Van Steen et al., "Comparative performances of machine learning methods for classifying Crohn Disease patients using genome-wide genotyping data", *Scientific Reports*, vol. 9, pp. 10351, 2019.

[17] Duc-Hau Le, "Machine learning-based approaches for disease gene prediction", *Briefings in Functional Genomics*, vol. 19, no. 5–6, pp. 350–363, 2020.

[18] M. Bracher-Smith, K. Crawford, and V. Escott-Price, "Machine learning for genetic prediction of psychiatric disorders: A systematic review", *Molecular Psychiatry*, vol. 26, pp. 70–79, 2021.

[19] M. U. Nasir, M. A. Khan, M. Zubair, T. M. Ghazal, R. A. Said, and H. Al Hamadi, "Single and mitochondrial gene inheritance disorder prediction using machine learning", *CMC – Computers Materials & Continua*, vol. 73, no. 1, pp. 953–963, 2022.

[20] Amit Kumar, "Predict the genetic disorders dataset-of genomes: Dataset of genomes and genetics", *ML Challenge Hackerearth*, 2021. www.kaggle.com/datasets/aibuzz/predict-the-genetic-disorders-datasetof-genomes.

15 Bayesian Models in Cognitive Neuroscience

Gabriel Kabanda

15.1 INTRODUCTION

Big Data analytics, artificial intelligence (AI) and robotics, machine learning (ML), cybersecurity, blockchain technology, and cloud computing are some of the most revolutionary technologies available today. In essence, the Fourth Industrial Revolution (4IR) is about cyber-physical technologies that allow the physical and virtual worlds to intersect. In order to optimize production chains as part of the 4IR, which drives scientific and technological developments, cyber-physical systems (CPS) are built on the Internet of Things (IoT) and its supporting technologies. To secure information systems, computers, devices, programs, data, and networks against internal or external threats, injury, damage, attacks, or illegal access, a mix of technologies, processes, and operations are referred to as cybersecurity [1]. In order to stop an attack from happening, cybersecurity combines the confidentiality, integrity, and availability of computing resources, networks, software, and data into a coherent set of policies, technologies, processes, and strategies.

A combination of laws, methods, technologies, and procedures is needed to protect the availability, confidentiality, and integrity of computing resources, networks, software, and data against attack. Ref. [2] asserts that when computers, associated telecommunications equipment, and other components that allow for the fast transport of vast amounts of data are connected, a human-created information environment called "cyberspace" is created. The use of IP addresses reveals the virtual nature of cyberspace. Unlike addresses in the physical domain, IP addresses give users navigational information without necessarily relating to a physical place. The networked devices and data that make up cyberspace were all created by humans. Cyberspace is divided into three layers: the physical layer, the intellectual layer, and the social layer.

The escalating usage of the Internet and the dangers it poses has given rise to network intrusion detection systems (NIDSs). A sort of computer program called NIDS monitors system activity to find behavior that violates security rules and can discriminate between malicious and legitimate network users [3]. Misuse network detectors and anomaly detectors are the two types of NIDSs. Misuse detection

DOI: 10.1201/9781003328414-15

systems maintain an extensive attack base and monitor all incoming traffic for any sequences that might be present there. Anomaly detection systems, on the other hand, concentrate on identifying fresh, unidentified threats. Ref. [3] asserts that research on network anomaly detection has utilized a number of well-known AI paradigms, including support vector machines, fuzzy logic, genetic algorithms, finite automata, neural networks, and genetic algorithms. The best tool for achieving this integration of misuse network detectors and anomaly detectors is a set of Bayesian networks. Following a period of training, the Bayesian network learns how the model behaves and can then predict its conclusion.

Based on observations and prior assumptions, Bayesian statistics provide a framework for drawing conclusions about the fundamental nature of the universe. A set of potential causes for the observed data are taken into account, and each is given a probability using the Bayesian approach to data analysis. The variance of a probability density function (PDF), or the width of the PDF, used to convey a belief about the state of the world, is a measure of how uncertain that belief is. The fact that Bayesian systems take this uncertainty into account and use it to weight various sources based on their varying degrees of accuracy is a critical component of the Bayesian methodology. In Figure 15.1, a sample of a Bayesian network is displayed. The directed graph's nodes correspond to the variables that make up the problem, while its edges show their conditional dependencies.

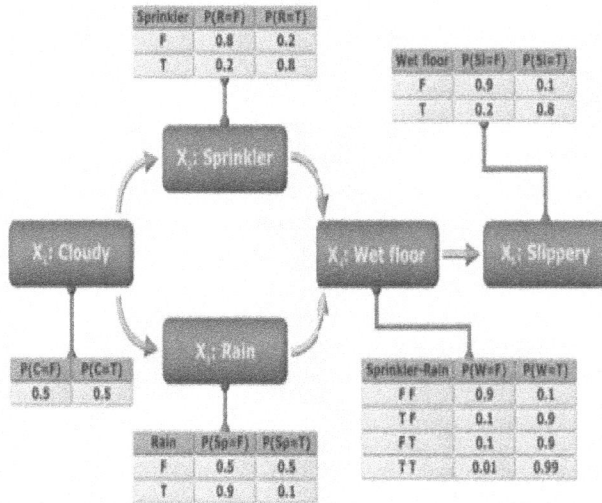

FIGURE 15.1   An example of a Bayesian network. (Source: [3])

The study looks into how the healthy population is affected by the dysfunctions in the hierarchical Bayesian inference process from a perceptual and belief fixation standpoint [4–8]. These hierarchical Bayesian models were sparked by an influential theory in cognitive and computational neuroscience that models the brain as a "probabilistic prediction machine" striving to minimize the mismatch between internally generated predictions of its sensory inputs and the sensory inputs themselves [9–13]. This theory models the brain as a "probabilistic prediction machine" that strives to minimize the mismatch between internally generated predictions of its sensory inputs and the sensory. Computational psychiatry has emerged as a paradigm for understanding how changes in brain functions can lead to the onset of severe psychiatric symptoms, and this understanding is largely due to growing knowledge of the brain as an organ of predictive inference [14].

Researchers in the newly developed discipline of computational psychiatry have recently made an effort to address these problems by using dysfunctions in the hierarchical Bayesian inference process that are hypothesized to underlie perception and belief fixation in the healthy population [4–8]. These hierarchical Bayesian models have been greatly influenced by predictive processing, also known as hierarchical predictive coding, a significant theory in cognitive and computational neuroscience that conceptualizes the brain as a "probabilistic prediction machine" aiming to reduce the discrepancy between internally generated predictions of its sensory inputs and the sensory inputs themselves [9–13].

15.1.1 Background

Bayesian networks (BNs) are directed acyclic graphs with a corresponding probability distribution function that are described as graphical probabilistic models for multivariate analysis [3]. BNs are a widely diverse family of models, according to [15], that may be used to depict nested, acyclic statistical models of almost any type of non-pathological joint probability distribution. A Bayesian network is easily described by [16] as a directed acyclic graph (DAG) with nodes and arcs, where the nodes stand in for random variables (RVs) and the directed arcs between nodes show dependencies between the RVs. The edges of the directed graph show conditional relationships between the variables that make up the problem. The probability function also shows how strong these connections are in the graph.

Let's formalize the definition of a Bayesian network B as a pair, $B = (D, P)$, where D is a DAG, $P = \{p(x_1|\Psi_2), \ldots, p(x_n|\Psi_n)\}$ is the set made of n conditional probability functions (one for each variable), and I is the set of parent nodes of the node X_i in D. The joint probability density function is used to define the set P [3].

$$P(x) = {}^n\prod_{i=1} p(x_i \mid \Psi_i)$$

A Bayesian network G is a probabilistic graphical model that uses conditional dependencies to describe a joint probability distribution over a set of variables $X = \{X_1, X_2, \ldots, X_n\}$. A direct probabilistic connection between the two connected nodes is indicated by an edge in this DAG, which also represents each node as a random variable [17].

The joint probability distribution is faithfully represented by the BN as

$$P(X_1, X_2, \ldots, X_n) = {}^{n}\prod_{i} = 1\, p\left(X_i \mid \mathrm{PaG}(X_i)\right)$$

where $\mathrm{PaG}(X_i)$ denotes the set of parent nodes of X_i in G, and $p(X_i \mid \mathrm{PaG}(X_i))$ describes the conditional probability distribution (CPD) for X_i given $\mathrm{PaG}(X_i)$ [17].

The ability of BNs to calculate the likelihood that a particular hypothesis is true, given a history dataset (for example, the likelihood that an email is spam or authentic) is their most crucial feature. The Bayesian technique is the one that combines knowledge under uncertainty with the most accuracy, as may be demonstrated mathematically. However, to determine whether people mix information in similar ways, we must first determine whether these models are effective in cognitive neuroscience. Both humans and animals have been known to exhibit behavior that is somewhat optimal as predicted by Bayesian theory. The human brain can be viewed, at its most basic level, as a machine whose function it is, at least in part, to infer the state of the environment. The application of Bayesian theory suggests that individuals possess a prior belief that influences their information processing. By employing the iterative Bayes' rule, it becomes possible to model the process of learning from a sequence of observations. Ref. [15] cites the following benefits of using BNs for this kind of research:

1. They are graphical models, able to show relationships simply and intuitively.
2. They can reflect cause-and-effect interactions since they are directed.
3. BNs are able to deal with ambiguity.

Theoretical neuroscience makes use of computer simulations and mathematical modeling to comprehend the brain and the behavior it produces. The computational, the algorithmic, and the level of implementation are all possible in Bayesian modeling of perception.

15.1.2 THE BAYESIAN BRAIN

The two-part hypothesis being considered is as follows:

1) To help us make decisions and direct our actions in the world, the brain uses Bayesian inference.
2) Probability distributions are how the brain represents sensory data.

Bayes Theorem states that:

$$p(h/e) = p(e/h)p(h)/p(e)$$

Under a set of plausible assumptions, Bayes Theorem defines the optimum calculus for updating beliefs in the face of uncertainty. Bayes Theorem specifically asserts that the likelihood of the hypothesis given the evidence, or $p(e/h)$, is proportional to its prior probability, or $p(h)$, which is the probability of the hypothesis evaluated independently of the evidence. If e is a piece of evidence and h is a potential explanation for this evidence, then in light of this equation, one should modify their beliefs in accordance with Bayes Rule. The study of cognitive phenomena has seen "a boom in research employing Bayesian models" in recent years [18–20]. At least two important factors have contributed to this "revolution" [21]: first, a growing understanding of the ways in which Bayesian statistics and decision theory can be used to capture the solutions to such problems in mathematically precise and empirically illuminating ways; and second, a growing understanding that across many psychological domains, the main challenge that the brain faces is inference and decision-making under uncertainty [19].

Although Bayesian models have been used to explore a wide range of cognitive phenomena, including categorization, causal learning and inference, language processing, abstract thinking, and more, they have been most frequently utilized in perceptual and sensorimotor psychology [18, 19]. Sometimes, without any pretension of being sufficiently descriptive, these models are designed just as normative "ideal observer" models. They seek to record cognitive task performance at its best. The Bayesian models used in this study, however, are focused on providing descriptive accounts of genuine cognitive systems. The success of such descriptions served as inspiration for the "Bayesian brain hypothesis" [22], which proposes that some knowledge is stored in the brain.

The Bayesian brain hypothesis is hindered by at least two significant challenges. First of all, accurate Bayesian inference requires a lot of time and is frequently computationally difficult. As a result, algorithms for approximate Bayesian inference have received a lot of attention in statistics and AI, with sampling and variational methods being the most common [23]. Second, scientists must explain how the approximation algorithms they have selected are implemented in the brain's neural networks. In other words, to be descriptively realistic about Bayesian cognitive science, Ref. [24]'s three-tiered schema for computational explanation requires researchers to find credible hypotheses at both the "algorithmic" and "implementational" levels.

Numerous studies have been conducted on these problems. Predictive processing, also known as hierarchical predictive coding, is the most well-known of these systems [9–13, 25, 26, 27]. There are excellent overviews of predictive processing available elsewhere in Refs. [10, 13], and [27]. Predictive processing has been extensively studied in both scientific and philosophical literature. Here, the study solely focuses on three issues: how to formulate approximate Bayesian inference in terms of precision-weighted prediction error minimization, how to make this process of minimization hierarchical, and how to employ predictive coding as a message-passing technique for doing so.

15.1.3 Scope of Research

The purpose of the study is to determine whether or not the brain can be viewed as a Bayesian machine and whether perception can be regarded as a form of Bayesian inference.

The objectives of this research are to:

a) analyze directional BNs that can be visualized as graphical models to describe cause-and-effect interactions;
b) Develop a BN model that can handle cybersecurity complexity.
c) Determine whether or not the brain can be viewed as a Bayesian machine and whether perception can be regarded as a form of Bayesian inference.

The research questions are:

i. How may cause-and-effect interactions be visually represented using BNs?
ii. What role do Bayesian models play in theoretical neuroscience?
iii. How can we discover that perception is Bayesian inference or that the brain is a Bayesian machine from the usage of Bayesian models in theoretical neuroscience?

15.2 LITERATURE REVIEW

15.2.1 Bayesian Network Concepts

BNs, which have numerous uses in a variety of walks of life, can be viewed as causal models created from observational data. Ref. [28] contends that whereas each causal mechanism in a given individual's joint set of mechanisms is frequently shared with a large number of other individuals, the joint set as a whole is essentially exclusive to that individual. Building the proper set of mechanisms for that person based on the characteristics we are aware of about that person and a training collection of data on numerous other people is the goal of causal learning for that particular person. This instance-specific causal learning approach is applicable to different causal systems even outside of human biology.

A graphic depiction of the probabilistic interactions between a number of variables is called a BN [28]. The graphical model structure G of a BN is composed of a DAG and a set of parameters for the DAG. The nodes of a DAG (G) represent variables, and the directed edges represent the conditional dependent relationships between those variables. A set of parameters is used to parametrize the relationships in the DAG. Thus, a state-of-the-art method for determining a BN structure from observational data is the greedy equivalence search (GES). According to Ref. [28], GES is a two-phase score-based system that includes a forward phase. The two-phase score-based method known as GES includes both forward equivalence search (FES) and backward equivalence search (BES), in accordance with Ref. [28]. Since each forward and backward step in GES requires scoring a single node given its parents, a node-wise decomposable score is required.

The creation of a Bayesian method for learning a BN structure entails searching a dataset for a structure with a high posterior probability. Let D be a dataset containing n discrete variables $X = \{X_1, X_2, \ldots, X_n\}$, where each variable X_i can take r_i values and its parents $Pa(X_i)$ can take q_i distinct instantiations [28].

The structure we want to score is G. The posterior probability of graph G given data D is as follows, according to Bayes Theorem:

$$P(G|D) = P(D|G) \cdot P(G) \, / \, P(D)$$

where P(G) is the structural prior, P(D) is the probability of the data, and P(D|G) is the marginal likelihood of the data. We define the model G score as given in Ref. [28] because P(D) is an independent normalization constant that does not depend on the model:

$$score(G) = P(D|G) \cdot P(G)$$

where we can compute P(D|G) by integrating over all unknown parameters θ as follows:

$$P(D \mid G) = \int_{\theta} P(D \mid G, \theta) \cdot P(\theta \mid G).$$

Bayesian techniques demand the specification of a generative process using a probability function that, given certain parameters, produces the observed data x. We may use the Bayes formula to invert the generative model by declaring our prior belief and then derive inferences about the likelihood of the parameters. The formula shown earlier shows that to compute the marginal likelihood P(D|G), integration (or summation in the discrete case) over the full parameter space is required. Remember that most of the time, this integral cannot be solved analytically. Markov chain Monte Carlo (MCMC), which provides a way to create samples from the posterior distribution, is one sampling strategy that avoids this problem. These methods, which will be discussed in more detail later, have been quite effective.

Despite certain limitations, the instance-specific greedy equivalence search (IGES) method proposed by Ref. [28] can be improved and expanded in the following ways:

a) For the IGES algorithm to learn instance-specific models iteratively for each instance in the training set, then use an average of those models to create the population-wide model;

b) Better understand the reason for the instance-specific BN models' significantly lower recall and try to raise it while maintaining precision;

c) To train BN structures with different types of variables (such as continuous or a combination of continuous and discrete variables) to create an instance-specific score;
d) Expand the experimental evaluations and create a more informative structure and parameter prior.
e) Show that IGES will always find the causal model that generates the data for a test instance within the big sample limit.

According to the work by Ref. [28], the suggested IGES method is a potential strategy to develop a BN structure that more accurately captures the interactions between variables of a particular instance T as opposed to a population-wide model.

Ref. [29] examined the effects of family characteristics, socioeconomic status, the biophysical environment, institutional support, and farm features on local inhabitants' decisions on reforestation. They did this by using the BN. The BN was effectively used to pinpoint the key variables impacting landowners' planted forest area, their interactions with one another, and the restrictions on tree planting. The belief network, commonly known as the BN, is a well-liked technique for managing and evaluating actual data. BN may study scenario-based subjects, incorporate qualitative variables with quantitative and spatially explicit data, and make robust predictions with high accuracy even with relatively small sample sizes without over-fitting [29]. A BN is useful in finding the pertinent aspects for qualitative reasoning in order to decide how to encourage or discourage particular options among a complicated group of interrelated elements. BN is a suitable method to address our study concerns because it is known to better capture such interdependencies without penalizing significant factors that do not have the strongest influence. Making decisions can be a difficult process impacted by many interconnected variables [29]. According to Ref. [29], the BN was successfully used to identify important variables influencing tree planting decisions and was useful in highlighting the complexity of decision-making. Causal, belief, and probabilistic networks are other names for BNs.

A BN G is a probabilistic graphical model that uses conditional dependencies to describe a joint probability distribution over a set of variables $X = \{X_1, X_2, \ldots, X_n\}$. A direct probabilistic connection between the two connected nodes is indicated by an edge in this DAG, which also represents each node as a random variable [17]. The joint probability distribution is faithfully represented by the BN as

$$p(X_1, X_2, \ldots, X_n) = \prod_{i=1}^{n} p(X_i \mid \mathrm{Pa}_G(X_i))$$

where $\mathrm{PaG}(X_i)$ defines the CPD of X_i given $\mathrm{PaG}(X_i)$, and $\mathrm{PaG}(Xi)$ denotes the set of parent nodes of Xi in G [17].

There are different types of graphical models, each with its own properties, structure, and benefits; these types can be classified into three categories: undirected graphical models, factor graphical models, and directed graphical models as shown in Figure 15.2. The probabilistic graphical models are also shown in Figure 15.2.

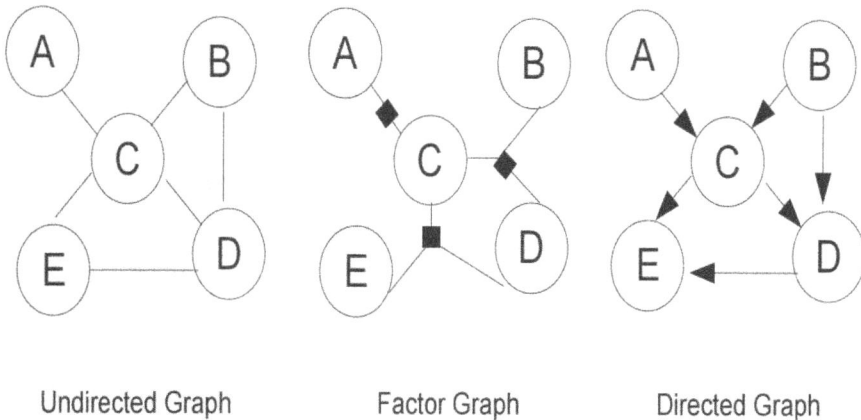

FIGURE 15.2   Probabilistic graphical models.

The *undirected graphical model* is a model of the (full) joint probability distribution of a collection of random variables. It is also known as the Markov network or Markov random field. It can express some dependencies that a BN is unable to describe (such as cyclic dependencies), but it is unable to represent some dependencies that a BN is capable of (such as induced dependencies).

A second class of graphical models that resemble undirected graphs is a factor graph, which is an undirected bipartite graph connecting variables and factor nodes. Each element represents a probability distribution over the variables it is linked to. A Bayesian network, often called a belief network, is a probabilistic DAG model that depicts a set of variables and their probabilistic connections. For example, a Bayesian network may show the likelihood of relationships between diseases and symptoms. The network can be used to determine how likely it is given a collection of symptoms that a particular group of diseases will be present.

Sequential data modeling is done using *dynamic Bayesian networks* (DBNs). There is sequential data everywhere. For instance, in speech recognition, eye tracking, or financial forecasting, temporal data describes a system that is dynamically changing or evolving through time. Sequential data can also be used in text processing or biosequence analysis to depict system changes, such as state changes. These types of data have problems with classification, segmentation, state estimation, problem diagnosis, and prediction. DBNs facilitate reasoning in fields where variables' values change over time. We collect data in regular time slices and reproduce the network structure for each slice (i.e., it is assumed that the relationship between variables in the same time slice is stationary).

Matrix representation is the most practical approach to represent graphs. An n 2n matrix is used to represent a DBN, with the number of variables being n. The observation model is represented by the first submatrix (1..n, 1..n) and the transition model is represented by the second submatrix (1..n, n+1..2n), as shown in Figure 15.3.

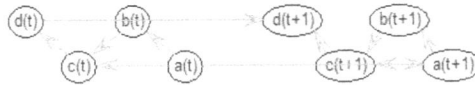

FIGURE 15.3   An example for a DBN (a) drawn using BNT, its matrix representation (b) and its equivalent initial PrDBN (c).

The chance that an edge exists between nodes I and j is represented by a real integer between 0 and 1 in the DBN(i, j). Initial probabilities for all candidate edges acquired from the constraint step are 0.5. Edges that the search space restriction phase removes from the candidate set correspond to entries that are set to 0. The *probabilistic DBN* is the name of this matrix (PrDBN).

(a)
     0 1 1 0 1 0 0 0
(b) 0 0 1 0 0 0 0 0
     0 0 0 1 0 0 0 0
     0 0 0 0 0 0 0 1
     0 0.5 0.5 0 0.5 0 0 0
(c) 0 0 0.5 0 0 0 0 0
     0 0 0 0.5 0 0 0 0
     0 0 0 0 0 0 0 0.5

BN analysis is extended in time by DBN. These models belong to a broad class that can depict intricate temporal stochastic processes. For the analysis of microarray data, BNs are incredibly powerful, but they only function when there are no circular dependencies. A DBN is a kind of BN that depicts a series of random variables. It is a directed stochastic temporal graphical model. These series are frequently time series or symbol sequences. In a graphical model, random variables stand in for nodes, while edges indicate direct probabilistic relationships between nodes. A set of conditional independence between variables is represented by a network structure. Finding the structures that are best supported by the data is the core objective of structure learning algorithms. The most popular and efficient technique for doing this activity is called "search and score." These approaches frame learning as an optimization problem and employ a variety of conventional heuristic search methods. By utilizing techniques like greedy climbing and simulated illumination, you can identify structures that have received high marks based on structural fitness criteria. These local search techniques occasionally succeed but frequently become stuck at local maxima rather than discovering new global maxima.

15.2.2 Predictive Processing

We can first formulate Bayesian inference by presuming a Gaussian distribution (i.e., a density function), as shown later. Calculate the prediction error by comparing the mean m of the historical distribution to the mean e of the evidence. The prediction error is the difference between these two numbers. As a result, the prediction error is inversely proportional to the likelihood of the hypothesis. Less prediction error is produced by m the better it anticipates the evidence. The posterior for Bayesian inference is therefore provided by the amount to update the former in response to this prediction mistake. This should be determined in relation to prior probability and likelihood uncertainty, according to Bayes Theorem. We can calculate this attribute for the Gaussian distribution using two distribution accuracies (reciprocal variances). In particular, when the prior probabilities closely align with the evidence, the adjustment of m in response to the predicted error becomes less probable, and vice versa. Consequently, the Bayesian learning rate is contingent upon the relative accuracy of the two distributions. The extent to which the previous probabilities are revised in response to an unsuccessful forecast plays a crucial role [30].

Applying this straightforward Bayesian inference procedure recursively will reduce prediction errors and arbitrarily correct prior probability in a universe where evidence is a linear function of its causes. This is due to the fact that human predictions properly represent the current situation [30]. In actuality, though, this would be unacceptable. The brain receives sensory information from settings that are dynamic and unstable and have complex hierarchical causal structures [31] and [13]. An inference system must react to this structure in order to successfully minimize prediction errors under such circumstances. In particular, hidden causes at various spatio-temporal scales must be uncovered by consistently reversing the organized causal processes that generate sensory inputs [32, 33, 34]. Therefore, such a system creates a hierarchy of hypotheses spanning several levels of spatio-temporal scales and abstractions, where hypotheses at level L+1 are L, rather than matching a single hypothesis space H to the evidence. Lower levels are given priority, and proof is given for prediction errors calculated at levels higher than L+2. For further information on this hierarchy's notion, see [32] and [35]. These hierarchical models enable Bayesian systems to adapt learning rates in a flexible and contextual way using high-level assumptions [30, 36, 37]. For instance, such a system might change its learning rate in accordance with its expectations of high-fidelity visual evidence in daylight and noisy evidence in fog [10].

The formalization of hierarchical Bayesian inference in terms of a recursive procedure of precision-weighted prediction error minimization has been briefly described thus far. How can a neural network be used to implement this process? Utilizing hierarchical predictive coding and predictive processing.

Only the signal's unpredictable components are sent on to the following stage of information processing when using predictive coding [10]. This coding technique for predictive processing captures the essential characteristics of hierarchically structured communications in the neocortex [10, 25, 27, 38]. In particular, this approach contends that only the unpredicted component of the signal—the prediction error—can be transmitted and that postsynaptic connections from higher cortical regions (such as the frontal or temporal) match activity and prediction at lower levels. The brain must minimize these internally accessible variables provided by these prediction mistakes. Predictive processing states that the brain constantly strives to reduce this quantity, which leads to the installation of a rich, hierarchically structured generative model of the environment and real-time parameter adjustments. An influential theory in the literature contends that changes in the "postsynaptic gain" of such superficial pyramidal cells at the "implementation level" result in changes in the prediction and prediction error carrying capacities of deep and superficial cortical pyramidal cells, respectively. It offers accuracy weighting. Second, neuromodulators like dopamine, serotonin, and acetylcholine have a role in at least some of the pathways that affect postsynaptic strengthening [4].

Some individuals may find that this method of hierarchical predictive coding only captures their preferences for the visual cortex when it comes to information processing [26]. A foundation for comprehending all cortical information processing has been added by others [10, 13, 25, 38]. A potential explanation for "perception and action, and everything in between, the mind," in particular, has been advanced as a component of this comprehensive theory of brain activity [13]. According to this perspective, which is conceptually connected to the "free energy principle" as it is stated by [11], "the brain is an organ for minimizing prediction mistakes" [39]. In other words, all brain activity is structured to minimize long-term prediction mistakes [10, 12, 40]. Even if the previous description of predictive processing was incomplete and left out many crucial details, it is nevertheless adequate for the purposes of this chapter to gauge its applicability in explaining the emergence and persistence of illusions.

15.3 HIERARCHICAL BAYESIAN MODELS OF DELUSION

One of the main goals of psychiatry is to relate psychopathologies and the symptoms they cause to malfunctions in the mechanisms underlying psychological processes in the general population [41].[1] The recent field of computational psychiatry uses these computational models to explain the psychiatric and neurological disorders at the root of mental illness because it is widely believed that our best explanations of cognitive mechanisms treat them as computational mechanisms involved in information processing [42, 43]. As stated by [44]:

> In the modern brain sciences, the most powerful models are computational in nature. . . . The relatively novel approach [of computational psychiatry] harnesses these powerful computational models and applies them to psychiatric and neurological disorders.

A flood of research using the Bayesian brain hypothesis and predictive processing has emerged in recent years to shed light on psychopathologies such as autism [45], anxiety disorders [46], and depression [47]. It is crucial to clarify how dysfunctions in the hierarchical Bayesian inference procedure described in Section 15.2 lead to the development and maintenance of delusional beliefs. It's crucial to be aware of the following:

- the argument that *precision-weighting* dysfunctions are the primary pathology underlying the development and maintenance of delusional beliefs; and
- the way that such hierarchical Bayesian models are intended to undermine accounts of delusions that separate perception from cognition.

However, it would be helpful to start by giving a brief overview of the conventional theoretical approaches to illusions.

The beginning point of this discourse is autism [45], anxiety disorders [46], and depression [47] in light of this explanatory technique. We first concentrate on recent attempts to explain the development and maintenance of delusional ideas in terms of limitations in the hierarchical Bayesian reasoning process that have been put forth. With this strategy, it is intended to pinpoint structural problems [4, 48]. The two points are specifically as follows:

a) A request for precision-weighted dysfunction to be recognized as the primary pathology driving the genesis and maintenance of delusional beliefs.
b) How such hierarchical Bayesian models are claimed to be insufficient for delusion-related explanations that make a distinction between perception and cognition.

15.3.1 DELUSIONAL BELIEFS AND THE TWO-FACTOR FRAMEWORK

Both a malfunction of the agent's perceptual system and a deficiency in her logical skills have typically been the focus of traditional theoretical approaches to delusions. It makes reference to traditional research [47] and [49], which is consistent with the traditional separation of perception and cognition in both cognitive psychology [50] and philosophy [51]. Delusional beliefs are accounted for in the first scenario as sensible reactions to anomalous events [52]. For instance, it could be plausible to assume that other people can hear the patient if they can be heard speaking aloud [6]. In the second instance, it has been argued that the agent's capacity for reasoning—or, more generally, the mechanisms underpinning hypothesis evaluation—is malfunctioning [53]. The systems that underlie the conclusions agents take from their perceptual experiences are generally healthy, yet they are damaged or malfunctioning. Recently, Ref. [47] asserted that both components must be flawed for an explanation of delusions to be tenable. This points out one error [54].

1. The first is a neuropsychological condition that gives patients new (erroneous) information, and delusional beliefs are what, when true, best account for that information.
2. The second is a cognitive defect in the belief assessment system, which stops patients from disproving newly formed ideas in spite of overwhelming evidence to the contrary.

Consider the Capgra craze as an illustration. People with this illness develop the erroneous belief that a visually indistinguishable (or almost indistinguishable) impostor has taken the place of someone dear to them (often a spouse or loved one). This deception's likely cause, according to Ref. [55], is injury to the area of the brain that generates the variety of autonomic reactions that agents often experience while visually recognizing faces. In these conditions, the agent sees the face but does not pick up the typical autonomic signals connected to that face. This suggests the following explanations:

This person is not my spouse. You are a scammer.

However, as Ref. [47] notes, this reliance on visual anomalies is insufficient to account for delusional ideas. The relevant evidence appears to have a far better explanation, to start. Second, some individuals with the proper neurological abnormalities do not develop delusions. As a result, the agent's "belief rating system" must be dysfunctional, which constitutes a second impairment or deficit [47]. In particular, a deficiency in this system, considered to result from injury to the frontal right hemisphere, is to blame for the agent's inability to let go of the delusion in the face of overwhelming evidence to the contrary.

The Capgras delusion is an example of a monothematic delusion that focuses on one unique theme. However, the two-factor framework has recently been tentatively extended to include the case of polythematic delusions, which can occur in conditions like schizophrenia and in which the person develops delusional beliefs about a variety of themes that are only tangentially related to one another [56]. There is general agreement that detecting harms or dysfunctions in the first category has been the framework's greatest achievement. Even its supporters admit that the second factor's specification is "too underspecified" [57] at the algorithmic level (a lack of a "belief evaluation system") and extremely coarse-grained at the physiological level ("frontal right hemisphere damage") to produce actual testable theories or mechanistic models. As Ref. [58] says:

> All two-factor accounts are based on a conceptual distinction (and an empirical disconnect) between perception and cognition. The first factor is cognitive abnormalities and the second factor is deficits in cognitive assessment.

The hierarchical Bayesian deception model under investigation aims to primarily achieve this frame-related property.

15.3.2 HIERARCHICAL BAYESIAN MODELS OF DELUSION

The most notable characteristic of hierarchical Bayesian models in the context of theoretical approaches to illusions is their support for a single deficit framework [6]. However, advocates of the hierarchical Bayesian model view perceptual and cognitive processes as remnants of folk psychology that are absent from modern cognitive neuroscience [7, 49] rather than attributing this dysfunction to the agent's perceptual system or to their capacity for reasoning. They specifically contend that claims about specific "belief assessment systems" are neurobiologically implausible [6] and that this can be explained by a single malfunction of hallucinations and delusions that can be understood in terms of a chaotic hierarchical Bayesian framework without taking experience and belief separately [6]. Central atypicality, a shift in the balance of Bayesian reasoning within hierarchically organized information processing systems, can be used to explain abnormal perceptual experiences and beliefs in psychosis [44].

What is the atypicality's proposed nature? The confusion of "error-dependent updating of inferences and views about the world" is cited by all advocates of hierarchical Bayesian models [6]. For instance, certain theories localize psychotic diseases with excessively exact sensory data [4], and others pinpoint them with excessively accurate predictions [59]. Refs. [4, 8] present the core of this proposal:

> Pervasive psychotic symptoms can be explained by the inability to express the accuracy of beliefs about the world. The delusional system can become elaborate when sensory evidence permeates it too precisely. There are primary pathologies here which show substantial metacognitive properties. In the sense that it is based on beliefs, is about beliefs, and more importantly, there are no obstacles necessary for the formation of predictions or prediction errors. [4]

Here, traditional statistical inference [4] and [13] serve as a useful parallel. Consider contrasting the data's mean with the null hypothesis that it is zero. The prediction error is determined by the difference between these two values. The null hypothesis is refuted by the prediction error. However, we must take into account the precision of the prediction error in order to calculate the amount of evidence we have. You shouldn't reject the null hypothesis if your data are highly variable (poor precision). Most likely, the prediction inaccuracy is just noise reflecting. On the other hand, the null hypothesis ought to be disproved if the precision-weighted prediction error is sizable enough. This illustration demonstrates how flaws in some second-order statistics can result in significant inference errors. That is, mistakes that happen during the weighting of the resulting prediction error rather than when comparing forecasts to the available data, or dependability. This latter calculation inaccuracy might very easily determine whether a novel medicine is regarded as safe or dangerous to the broader populace.

This deficiency is cited by proponents of predictive processing theories of delusions as a major pathology underlying the development of delusional beliefs in diseases like schizophrenia [4, 7, 48]. It has been proposed, for example, that errors in the accuracy-weighting procedure overstate the dependability of sensory evidence in comparison to earlier, "higher-level" beliefs. The agent is informed that its worldview is incorrect via persistent, high-weight prediction errors, which prompts quick rectification of the world model. However, because messaging in the prediction processing architecture is bidirectional, these updated predictions are transmitted back to affect how we interpret the sensory data we receive. Even worse, high-precision prediction errors necessitate updating both learning and inference (the construction of models based on those conclusions), which ultimately necessitates significant revisions to our understanding of the agent world. A possibility exists. To explain this phenomenon, Ref. [7] mentions a psychologist who talks about his own experience with paranoid schizophrenia.

> I had to make sense of all these macabre coincidences. I did so by radically changing my conception of reality. [60]

They write,

> In our terminology, these uncanny coincidences were false hypotheses engendered by prediction errors with inappropriately high precision. . . . To explain them away [60] had to conclude that other people, including radio and television presenters, could see into his mind. [7]

This hypothesis, which is currently being offered in the literature like most delusional notions, is therefore very rudimentary. It does, however, provide a number of well-known attractions.

1) The great persistence of delusional ideas is thought to be particularly difficult to explain by disruption of the brain's capacity to correctly attribute accuracy, according to good evidence [13].
2) Second, it has long been recognized that dopaminergic dysfunction contributes significantly to schizophrenia, and as we mentioned earlier, dopaminergic activity is regarded to be crucial for appropriate weighting within predictive processing [4].
3) Third, a remarkable collection of computer simulations of precision-weighting errors' consequences on reasoning appear to corroborate the notion that these errors lead to psychotic symptoms [4, 61].
4) Another strong corpus of neuroimaging research has linked false predictive error signals to delusions and psychosis [48, 62].

15.3.3 Hierarchical Bayesian Models and the Two-Factor Framework

The two-factor model presented in Section 3.1 is meant to be challenged by a hierarchical Bayesian model of delusions [5, 6, 44, 49]. We contend that two-factor theory and hierarchical Bayesian models "are at distinct levels of explanation." More specifically, we contend that while the two-factor theory "works well at the descriptive level," perception and cognition are not distinct "at a deeper level" and two components can be found at two levels in the same hierarchy [48]. There are three significant differences between the two frameworks, and it is crucial to identify what those differences are.

i. First and foremost, it should be noted that, in contrast to the two-factor paradigm, hierarchical Bayesian models of delusion only posit one deficit—namely, the aberrant encoding of accuracy in a predictive coding hierarchy. It is crucial to emphasize that this one component is nevertheless intended to disturb all levels of the hierarchy, creating a "multifactor psychopathology from a single pathophysiology," as a reviewer who wishes to remain anonymous points out.

ii. Second, proponents of hierarchical Bayesian delusion models emphasize the importance of bidirectional message flow. Standard uses of the two-factor concept indicate that information traveling between cognitive and perceptual systems goes in a single direction: perceptual information is transmitted onto systems underlying belief fixation, which are tasked with explaining this information. Perceptual systems are thus informationally contained [51]:

> what the agent believes is responsive to what she perceives, but not vice versa. In contrast, advocates of hierarchical Bayesian models stress the influence of top-down predictions (i.e. priors) on the agent's perceptual experiences.

iii. In particular, it is not merely that abrupt modifications to the agent's "higher-level" worldview are caused by exceptionally high-precision prediction errors. A recurrent process known as a "insulated self-confirming loop" [63] is when these revisions act as the prior factors the agent uses to interpret (predict) her own experience. This results in false perceptions that appear to confirm the extremely high-level beliefs that they are intended to test.

iv. The third and most significant difference, however, is that hierarchical Bayesian models of delusion disavow the very notion that perceptual experiences and beliefs—and thus hallucinations and delusions—should be treated "as distinct entities" from the perspective of psychiatric explanation in the first place. This difference does not stem from the use of a single deficit or the emphasis on bidirectional message passing as such. Refs. [6, 7, 49] [49], for example, write that positing two factors "is only necessary in so far as there is a clear distinction between perception and inference: a distinction which is not actually compatible with what is known about how the brain deals with the world."

This third distinction is significant in that it is more fundamental than the first two. For example, a dopaminergic abnormality at the center of predictive processing may be the primary flaw that causes the genesis and maintenance of delusional beliefs. Despite this, the dysfunction serves two purposes. Similarly, while acknowledging the distinction between perception and belief, we might also concede that an actor's beliefs might affect their perspective of the world. Although there is a two-way communication between them, this does not negate the fact that people are unique systems.

The third distinction goes far beyond pointing out the possibility of isolated errors and cognitive penetration. According to the two-factor model, the perceptual system's main responsibility is to choose the most plausible explanation for the agent's sensory information, which is then communicated to the belief-fixation system. This latter system determines what to believe and how to act by fusing this perceptual data with prior knowledge and reasoning skills [51]. In contrast, there is only one system in a hierarchical Bayesian model. In other words, there is just one unified hierarchy of inferences, and only the obscure parts of that hierarchy are tracked by our intuition about the distinction between perceptions and beliefs [9, 13]. This is a fundamental departure from conventional delusional theory and much of the relevant cognitive psychology studies [50]. This is a possibility, according to proponents of hierarchical Bayesian models; however, we contend in the following section that this notion is unwarranted.

15.3.4 INTEGRATIVE CONCLUSION

As a strong challenge to the popular two-factor paradigm, hierarchical Bayesian models of delusion are put forth. Such models specifically reject the separation of vision and cognition in favor of a unified inferential hierarchy where bidirectional message flow resembles Bayesian inference. Then, it is asserted, delusional ideas result from flaws in this information-processing architecture. This method is based on current developments in computational neuroscience, and it is supported by a strong body of theoretical and empirical arguments. The end result is an incredibly outstanding effort to shed light on the enigmatic process by which people lose their sense of reality, an effort that justifies its reputation as one of the most promising and fascinating by-products of the new discipline of computational psychiatry [44]. However, even those who support these models admit that hierarchical Bayesian models of illusion have limitations [48]. For instance, how can a single malfunction explain the disparity between those who have Capgras delusion and others who are not delusional yet share the same experience [64]? Advocates of hierarchical Bayesian models of delusion have yet to respond to this challenge [47], which applies to all monothematic delusions. This suggests that the application of such models may at best be limited to polythematic delusions of the type that are present in conditions like schizophrenia [65, 66].[2]

There are still more unanswered questions.

a) How, in the first place, can such theories explain the normal social components of delusional beliefs [48]? A temporal sequence by which one malfunction transforms into the other [67]?
b) Does the underlying dysfunction consist of unusually exact sensory evidence, abnormally precise priors, some combination of the two, or neither of these?

Hierarchical Bayesian models of delusion have the following two main summary features:

1. Hierarchy: In a single inferential hierarchy, delusions appear at the upper or highest parts.
2. Bayesian: This hierarchy's information processing employs an approximation of Bayesian inference.

15.4   THE INFERENTIAL HIERARCHY

In favor of a single inference hierarchy, hierarchical Bayesian delusion models reject the distinction between perception and cognition, contending that information processing in this hierarchy takes the form of approximative Bayesian reasoning. The argument in this paragraph and the one that follows is that both parts have significant issues.

The first component will be highlighted in this section. This claim makes the case that a unified hierarchy of thinking, rather than a difference between perception and cognition, is a better explanation for delusions. It can be claimed that the methods used to describe this hierarchy of thinking in the literature are incompatible with the variety of experiences that may signify delusions.

15.4.1   Understanding the Inference Hierarchy

One of the most defining and damaging aspects of hierarchical Bayesian theories of delusion is the appeal to a single hierarchy of reasoning rather than a substantial theoretical separation between perception and cognition [6, 44, 49]. The concept of a hierarchy of reasoning must be able to describe the kinds of phenomena that take place in delusional thinking for this extreme departure from conventional theory to be acceptable. We must therefore comprehend precisely how inference hierarchies are understood in hierarchical Bayesian models and where illusions are meant to be in it in order to determine whether this criterion is met.

First, it is frequently suggested in the literature that representations at "higher" levels of the inference hierarchy correspond to what we intuitively think of as beliefs.

> There is likely a hierarchy of such reasoning devices in the brain, with the lower levels of the hierarchy being more relevant to cognition and the higher levels being more relevant to beliefs.

It is obviously incorrect to think of the proposed hierarchy of inferences as resembling a stepladder. Instead, think of it as command hierarchies organized so that a level X is above a level Y only if an agent at level X can command an agent at level Y through the neocortex. The temporal organization and regularities of events determine how the brain is arranged hierarchically. Higher parts of the cortical hierarchy will correspond to progressively longer time scales. Basic sensory characteristics and causal regularities are typically predicted at early levels of the hierarchy at extremely short, millisecond time scales, whereas higher levels deal with more complicated regularities at progressively slower time scales. These hierarchies are arranged with the slowest time scale at the top in accordance with the temporal scales of the representations. The fundamental hypothesis is that temporal hierarchies in the environment are translated into structural hierarchies in the brain, with high-level cortical areas encoding slowly changing contextual states of the world and low-level areas encoding quick trajectories. According to this theory, a level X in the inferential hierarchy is superior to a level Y only if it captures events at a higher spatiotemporal scale. The fact that state transitions happen at various rates at various levels of the hierarchy is a key feature of this generative model. At higher levels, inference is more abstract. Due to this, we propose that the cerebral cortex's unified inference hierarchy is structured according to the spatiotemporal scales of the information it represents, or according to the content represented at distinct levels.

15.4.2 EVALUATION OF INFERENCE HIERARCHY

You are able to think quite abstractly about little, large, slow, and quick objects. Assumptions can also be made about events without a spatiotemporal scale, as number 42 and Bayes Theorem. In fact, this characteristic of thought is sometimes used by cognitive psychologists and philosophers to distinguish between perceptual systems and belief-fixing architectures [51].

Therefore, reasoning hierarchies can explain this characteristic of cognition given that our capacity to reason extremely abstractly about things at any spatiotemporal scale or abstraction is a distinguishing property of human cognition. It is unknown if such beliefs entail phenomena at all, as is the case with delusions in general. It is unknown whether such beliefs involve phenomena at any spatio-temporal level, much like with delusions in general. When the hierarchical links between levels are defined in terms of spatio-temporal scales or levels of invariance, it is crucial to comprehend Ref. [7]'s claim that schizophrenia delusions dwell at the "head" of the hierarchy. Do these delusions always represent the most significant, prevalent, or enduring phenomenon? What level of the chain of reasoning does this belief fall under?

It may be argued that this belief is consistent with those who have this delusion's anomalous experiences, which exist at a lower level of the inference hierarchy. As a result, it should be expected that many delusional beliefs will entail events that have the same amount of spatio-temporal granularity as the information in a person's perceptual experience. If this is the case, it is obvious that delusions do not necessarily depict events that are constant, bigger, or moving more slowly than those phenomena depicted "below" in the inference hierarchy. The issue emerges when such hierarchies are abandoned in favor of a unified hierarchy of reasoning where the levels and connections between them are defined in terms of the spatiotemporal sizes of the phenomena they represent.

15.4.3   ABSTRACTIONS TO THE RESCUE AND THE ANSWER

It is conceivable to just describe the inferential hierarchy of cortices and make the case that other hierarchies in connected cortices take precedence. The sole focus of prediction processing is to find the environmental factors from which the statistical patterns represented by the lower levels of the inference hierarchy can be most effectively inferred. For instance, many supporters of the hierarchical Bayesian delusion model advocate for the hierarchy of thinking to be arranged in accordance with rising "levels of abstraction." The "lower" levels of the hierarchy correlate to states like perceptions, while the "higher" levels of the hierarchy correspond to states like beliefs. The information processing architecture advanced by predictive processing makes it challenging to coordinate this concept. The different levels of the hierarchy should be arranged according to the corresponding expressive or computational qualities if the inference hierarchy is not ordered by content. They must define inferential hierarchies as still being hierarchies rather than as collections of hierarchically structured perceptual systems interacting with a set of primarily all-inclusive conceptual representations. I feel there is a major risk of violating this constraint—to the "upper" or "top" level of the inference hierarchy—without an explicit statement of the principles by which potentially a unified inference hierarchy is structured. The idea of hierarchy by itself doesn't truly explain anything in this situation. Hierarchy is significant in our model since it is thought that the inference hierarchy is what sets the hierarchical Bayesian model apart from its forerunners.

Second, it could be argued that the distinguishing characteristic of hierarchical Bayesian models of delusion is not their emphasis on an inferential hierarchy as such, but rather their adherence to a single deficit underlying the formation and maintenance of delusional beliefs, as well as their adherence to the importance of bidirectional message passing. In particular, neither the notion that a single disturbance underlies delusions—for instance, an over-confidence in sensory evidence relative to prior beliefs—nor the emphasis on cognitive penetration at the center of hierarchical Bayesian models are directly undermined by my criticism of the idea that all information processing in the brain is subsumed within a single unified inferential hierarchy. These characteristics of hierarchical Bayesian models do, however, exist in the context of an information processing architecture that addresses the idea of an inference hierarchy at its heart and how further contributions of such models can be made by letting go of this hierarchy.

15.5 BAYESIAN BELIEFS AND A BAYESIAN NETWORK MODEL FOR CYBERSECURITY

Evidence exists to refute Bayesian beliefs. Only concentrating on the data supporting a Bayesian theory of belief fixation is insufficient. The Bayesian optimum motion perception framework can be used to explain a variety of motion illusions and their psychophysical effects. The hierarchical Bayesian delusion model looks for failure in the hierarchical Bayesian reasoning process as an explanation for delusions. No one thinks the brain conducts accurate Bayesian inference, as was said in Section 15.2 of this chapter. The mechanisms enabling belief consolidation in healthy populations, according to advocates of hierarchical Bayesian models, are Bayesian in character. We need solid evidence to support the Bayesian interpretation of belief fixation if we want to use delusional beliefs as an example of how the Bayesian reasoning process fails. When making the claim that "human observers act like ideal Bayesian observers," we also strongly reference the work of Ref. [68]. Therefore, it is generally acceptable to deviate from accurate Bayesian inference. To approximate Bayesian inference is the goal of approximation algorithms. The integration of tactile and visual information can be usefully described in terms of Bayesian reasoning, according to a significant showing by Ref. [68]. In certain situations, for instance, the outcomes of such algorithms systematically differ from the outcomes of exact Bayesian inference because they are approximate algorithms [69]. The ability to build a model that characterizes a certain decision, option, or conclusion as being Bayesian optimal does not prove that the particular decision, option, or conclusion is Bayesian optimal. The second, more intriguing response concedes that many of the mechanisms causing belief fixation do not follow a Bayesian model, but maintains that these non-Bayesian features of cognition are unimportant for illogical beliefs. This is the complete antithesis of Bayesian inference.

The hierarchical Bayesian delusion model looks for failure in the hierarchical Bayesian reasoning process as an explanation for delusions. Therefore, a key tenet of such models is that belief fixing in healthy populations follows Bayesian logic. Therefore, it is suggested that the mechanisms underpinning belief fixation in healthy populations are speculatively optimal based on the generally held idea that Bayesian inference is statistically optimal. It is obvious that Cousin Barry's past violation of Bayesian optimality does not call into question the validity of the Bayesian knowledge model. More generally, it is unclear whether this would pose a problem for such Bayesian models if deviations from Bayesian inference were distributed such that, on average, the mechanisms generating belief fixation yielded Bayes optimum judgments.

First, supporters of hierarchical Bayesian models consistently and convincingly assert that Bayesian processes underlie the mechanisms underpinning belief consolidation in healthy populations. Indeed, among supporters of the Bayesian model, it is generally agreed that the low-level, largely unconscious systems involved in cognition and motor control provide the strongest support for the Bayesian model. We have observed them to occur and exist at higher and even the greatest levels of the "hierarchy" of thinking, even among advocates of hierarchical Bayesian models. Therefore, whether or not the mechanisms underpinning belief fixation are Bayesian is not immediately related to showing Bayesian optimality in the area of sensorimotor processing.

It should be noted that:

a) such motivational factors do not currently appear in hierarchical Bayesian models of delusion;

b) the ability to model such phenomena as Bayesian post hoc does not prove that they are Bayesian;

c) it is possible that some cognitive functions positively mandate updating procedures that are specifically intended to produce false beliefs and do not conform to Bayesian updating, although this is not necessarily the case.

d) despite the many assertions made by supporters of the hierarchical Bayesian model of delusion that the mechanism behind belief fixation should be Bayesian, this hypothesis has not been adequately explored in the literature. We therefore anticipate that the processes of belief fixation that act in accordance with Bayesian reasoning have evolved over time, given that this function is best performed using Bayesian reasoning.

A computing-based discipline called cybersecurity uses people, technology, information, and processes to enable assured operations in the face of attackers. To tackle the complexity of network intrusion and detection systems, Ref. [1] created the BN model depicted in Figure 15.4.

A DAG, or BN, is a probabilistic graphical model that depicts a set of variables and their probabilistic dependencies [1]. A random variable is connected to each node of a BN, which is a label that represents an aspect of the problem. These attributes are binary and can take the values TRUE or FALSE. The correspondence between the graphical structure and related probabilistic structure, which enables us to eliminate all the problems of inference problems in graph theory, is one of the many issues with the use of BNs. The operation for transforming the causal graph into a probabilistic representation has another issue. The naïve Bayes, a two-layer BN that presupposes total independency between the nodes, is one method in which BNs have been used in anomaly identification.

The KDDCup 1999 intrusion detection benchmark dataset was utilized in the study by Ref. [1] to create a successful NIDS. A sample of 494,020 instances of primary data with 42 variables from a population of 10 million network traffic data were examined using a combination of open-source SNORT software and other platforms that support Bayesian networks, including NCSS 2019, Pass 2019, GeNIe 2.3, WinBUGS14, BayES, and Analytica 5.1. The BN model underwent structural equation modeling, and the BN structure was created. Discussion was held regarding the effectiveness of the support vector machine (SVM), artificial neural network (ANN), K-nearest neighbour, naive-Bayes, and decision tree algorithms. The employment

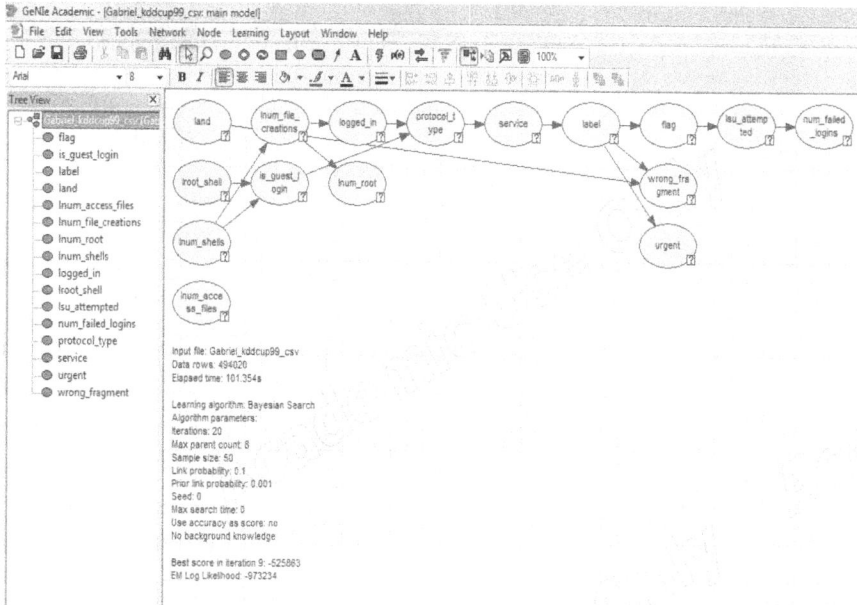

FIGURE 15.4   Bayesian network structure.

of ML methods, notably ANNs, decision tree C4.5, random forests, and SVMs, is addressed as an alternative to the current solutions.

Cybercrimes are steadily on the rise, and there is rising worry about the security and access control of the data that is being held. Host-based intrusion detection systems (HIDSs) and network-based intrusion detection are the two main forms of IDSs (NIDSs). A DAG serving as the visual representation of a set of variables and their probabilistic dependencies is known as a BN. A BN G is a probabilistic graphical model that uses conditional dependencies to describe a joint probability distribution over a set of variables $X = X_1, X_2, \ldots, X_n$. It should be emphasized that the Bayesian network classifier can be trained using training data, with structure learning and conditional probability distribution estimation used in the learning process. Evaluations with a portion of the KDDCUP'99 dataset, which was employed in this study, revealed that the abuse detection module produced a high detection rate with a low false-positive rate, while the anomaly detection component had the ability to discover novel intrusions. From the data, the annealed maximum a posteriori probability (MAP) of the BN was developed, and the computations were displayed on the BN's descriptive statistics and structural equation modeling.

Graphical models come in a variety of forms, each with unique characteristics, structures, and advantages. These forms can be arranged into three groups: undirected graphical models, factor graphical models, and directed graphical models. A random variable is connected to each node of a BN, which is a label that represents an aspect of the problem. These attributes are binary and can take the values TRUE or FALSE. The two main types of hybrid IDSs used today are sequence-based and parallel-based. Sequence-based hybrid IDSs apply either anomaly detection or misuse detection first and then the other one, and parallel-based hybrid IDSs apply multiple detectors simultaneously and base their final decision on multiple output sources. Attacks known as Distributed Denial of Service (DDoS) are a serious threat to Internet security because they aim to render services unreachable by saturating the server's network and end-user systems with falsely produced traffic.

Intrusion attacks can now be detected quite well and efficiently using BN classifiers with strong reasoning capabilities. The best IDS must be able to function constantly without human oversight. The phases of data collection, data preprocessing, intrusion recognition, reporting, and response are frequently included in the intrusion detection process. To combat incredibly intelligent cyber-attacks, effective and efficient IDSs are required to quickly detect and block intrusion. A unique distributed IDS was created utilizing an IDS based on a BN classification modeling technique to detect and stop attacks such denial of service, probes, user to root, and remote to user attacks. Methods for anomaly-based intrusion detection create models from typical activities and locate audited data by calculating the difference between observed and built-in models. Sequential data is present everywhere, for example, in biosequence analysis, text processing, and temporal data. Sequential data is present everywhere. For example, temporal data models a system that is dynamically changing or evolving over time in speech recognition, visual tracking, or financial forecasting. Sequential data also represents changes in the system, such as changes in state, in biosequence analysis, or in text processing. The naive Bayes, a two-layer Bayesian network that presupposes total independency between the nodes, is one method in which BNs have been used in anomaly identification. Each example in the KDD99 dataset reflects the attribute values of a class in the network data flow, and each class is assigned either the label "attack" or "normal."

How to choose important and efficient characteristics from a vast array of potential related features is an issue of significant interest in the training of IDSs. Sequential data modeling is done using DBNs. Very low false-negative and false-positive rates are required for the system. However, BNs only work when there are no cyclic relationships, despite the fact that they are particularly successful for studying microarray data. Nodes can connect to other nodes both inside the same time slice and to nodes in the next slice since DBNs describe their domains as partially observable Markov processes. The application of BN analysis to time is known as a DBN. The directed graph's nodes correspond to the variables that make up the problem, while its edges show the conditional relationships between those variables.

15.6   CONCLUSION

In the developing field of computational psychiatry, hierarchical Bayesian models of delusions have recently been very well-liked and are frequently hailed as one of the most significant success stories. Particularly in the hierarchical and Bayesian aspects, the two theoretical basic elements of such models are thoroughly documented in the literature. Given that there is more to thought than a single information processing inference hierarchy and that the mechanisms underlying belief fixation are not Bayesian, we face significant challenges that have not been addressed. One tries to explain delusional beliefs in terms of the dysfunction of the mechanisms underlying hierarchical Bayesian inference. If this is the case, then not only does the more general hierarchical Bayesian model of the brain have trouble explaining "perception and action and everything in between the mind" but also the hierarchical Bayesian model of delusion. Instead of focusing directly on a hierarchical Bayesian model of neural information processing in general, the research study concentrated on explaining delusional beliefs. Instead of functionally differentiating between bi-hallucinations and delusions, the idea of using Bayesian reasoning to explain delusional beliefs conveys a directional message. Proponents of predictive processing have minimized predicted errors in order to explain everything as a collective effort to model a certain high-level cognitive activity in an effort to explain illusions. Hierarchical Bayesian delusion models thus offer a crucial test case for predictive processing. Predictive processing attempts to explain delusional ideas that encounter significant issues with complexity and clarity, as well as a dearth of supporting data and convincing arguments for central hypotheses. Computational psychiatry must give up universal theories of brain function and cognitive optimality models in favor of other fields, particularly cognitive science and science, if it is to fulfill its promise of illuminating the information processing mechanisms behind mental diseases.

NOTES

1 Technically, a widespread assumption is that such dysfunctions must be *harmful* [47].
2 Ref. [6] argues that positing two factors is "only necessary insofar as there is a clear distinction between perception and inference: a distinction which is not actually compatible with what is known about how the brain deals with the world." This objection to the two-factor framework is confused, however. First, the argument for positing two factors is that it accounts for certain dissociations—cases in which individuals share the same anomalous experience but do not form the delusions—and has nothing to do with whether perception is inferential [52], [64]. Second, the distinction advocated in the two-factor framework is between perception and cognition, not perception and inference. One can think that perception is inferential in a computational sense without abandoning this distinction. In fact, that is the mainstream view in classical cognitive psychology [57], [58]. Take note that hierarchical Bayesian models can embrace two factors.

REFERENCES

[1] Kabanda, G. (2020, May). Performance of machine learning and other artificial intelligence paradigms in cybersecurity. *Oriental Journal of Computer Science and Technology, 13*(1), 1–21, ISSN: 0974–6471, Online ISSN : 2320–848, http://www.computerscijournal.org/vol13no1/performance-of-machine-learning-and-other-artificial-intelligence-paradigms-in-cybersecurity/.

[2] Berman, D.S., Buczak, A.L., Chavis, J.S., and Corbett, C.L. (2019). Survey of deep learning methods for cyber security. *Information 2019, 10*, 122. doi:10.3390/info10040122.

[3] Bringas, P.B., and Santos, I. (2010). *Bayesian Networks for Network Intrusion Detection, Bayesian Network*, Ahmed Rebai (Ed.), ISBN: 978–953–307, 124–4, InTech, http://www. intechopen.com/books/bayesian-network/bayesiannetworks-for-network-intrusion-detection.

[4] Adams, Adams, R., Stephan, K., Brown, H., Frith, C., and Friston, K. (2013). The computational anatomy of psychosis. *Frontiers in P sychiatry*, *4*. http://dx.doi.org/10.3389/fpsyt.2013.00047.

[5] Corlett, P., Taylor, J., Wang, X., Fletcher, P., & Krystal, J. (2010). Toward a neurobiology of delusions. *Progress in Neurobiology*, *92* (3), 345–369. http://dx.doi.org/10.1016/j.pneurobio.2010.06.007.

[6] Fletcher, P., and Frith, C. (2009). Perceiving is believing: A Bayesian approach to explaining the positive symptoms of schizophrenia. *Nature Reviews Neuroscience*, *1 0*(1), 48–58. http://dx.doi.org/10.1038/nrn2536.

[7] Frith, C.D., and Friston, K.J. (2013) False perceptions and false beliefs: Understanding schizophrenia. *Neurosciences and the Human Person: New Perspectives on Human Activities*, *121*, 1–15. [RMR].

[8] Schmack, K., Gomez-Carrillo de Castro, A., Rothkirch, M., Sekutowicz, M., Rossler, H., Haynes, J., et al. (2013). Delusions and the role of beliefs in perceptual inference. *Journal of Neuroscience*, *33*(34), 13701–13712. http://dx.doi.org/10.1523/jneurosci.1778–13.2013.

[9] Clark, A. (2013). Whatever next? Predictive brains, situated agents, and the future of cognitive science. *Behavioral And Brain Sciences*, *36*(0 3), 181–204. http://dx.doi.org/10.1017/s0140525x12000477.

[10] Clark, A. (2016). *Surfing Uncertainty*. Oxford: Oxford University Press.

[11] Friston, K. (2010). The free-energy principle: A unified brain theory? *Nature Reviews Neuroscience*, *11* (2), 127–138. http://dx.doi.org/10.1038/nrn2787.

[12] Friston, K., FitzGerald, T., Rigoli, F., Schwartenbeck, P., and Pezzulo, G. (2017a). Active Inference: A process theory. *Neural Computation*, *29*(1), 1–49. http://dx.doi.org/10.1162/neco_a_00912.

[13] Hohwy, J. (2013). *The Predictive Mind*. Oxford: Oxford University Press.

[14] Griffin, J., and Fletcher, P. (2017). Predictive processing, source monitoring, and psychosis. *Annual Review of Clinical Psychology*, *13* (1), 265–289. http://dx.doi.org/10.1146/annurev-clinpsy-032816-045145.

[15] Margaritis, D. (2003, May). *Learning Bayesian Network Model Structure from Data*. PhD Thesis, Carnegie Mellon University, Pittsburgh, PA.

[16] Boudali, H., and Dugan, J.B. (2006, March). A continuous-time Bayesian network reliability modeling, and analysis framework. *IEEE Transactions on Reliability*, *55*(1).

[17] Xiao, L. (2016). *Intrusion Detection Using Probabilistic Graphical Models*. PhD Dissertation, Iowa State University.

[18] Chater, N., Oaksford, M., Hahn, U., and Heit, E. (2010). Bayesian models of cognition. *Wiley Interdisciplinary Reviews: Cognitive Science*, *1* (6), 811–823. http://dx.doi.org/10.1002/wcs.79.

[19] Tenenbaum, J., Kemp, C., Griffiths, T., and Goodman, N. (2011). How to grow a mind: Statistics, structure, and abstraction. *Science*, *331*(6022) , 1279–1285. http://dx.doi.org/10.1126/science.1192788.

[20] Oaksford, M., and Chater, N. (2007). *Bayesian Rationality*. Oxford: Oxford University Press.

[21] Hahn, U. (2014). The Bayesian boom: Good thing or bad? *Frontiers In P sychology*, *5*. http://dx.doi.org/10.3389/fpsyg.2014.00765.

[22] Knill, D., and Pouget, A. (2004). The Bayesian brain: The role of uncertainty in neural coding and computation. *Trends In Neurosciences*, *27*(1 2), 712–719. http://dx.doi.org/10.1016/j.tins.2004.10.007.

[23] Penny, W. (2012). Bayesian models of brain and behaviour. *ISRN Biomathematics*, *2012*, 1–19. http://dx.doi.org/10.5402/2012/785791.

[24] Marr, D. (1980). *Vision*. New York: Freeman.

[25] Friston, K. (2005). A theory of cortical responses. *Philosophical Transactions of the Royal Society B: Biological Sciences*, *360*(145 6), 815–836. http://dx.doi.org/10.1098/rstb.2005.1622.

[26] Rao, R. P. and Ballard, D. H. (1999). Predictive coding in the visual cortex: A functional interpretation of some extra-classical receptive-field effects. *Nature Neuroscience*, 2(1), 79–87. 10.1038/4580.

[27] Seth, A. K. (2015). The cybernetic Bayesian brain—From interoceptive inference to sensorimotor contingencies. In T. Metzinger and J. M. Windt (Eds.), *Open MIND: 35(T)*. Frankfurt am Main: MIND Group. doi:10.15502/9783958570108.

[28] Jabbari, F., Visweswaran, S., and Cooper, G.F. (2018). Instance-specific Bayesian network structure learning. *Proceedings of Machine Learning Research*, 72, 169–180, PGM 2018.

[29] Tran, T.M., Ko, D.W., Ryul, C., and Dinh, H. (2019). A Bayesian network analysis of reforestation decisions by rural mountain communities in Vietnam. *Forest Science and Technology*. doi:10.1080/21580103.2019.1581665.

[30] Hohwy, J. (2017). Priors in perception: Top-down modulation, Bayesian perceptual learning rate, and prediction error minimization. *Consciousness and Cognition* , 47, 75–85. http://dx.doi.org/10.1016/j.concog.2016.09.004.

[31] Friston, K., Rosch, R., Parr, T., Price, C., and Bowman, H. (2017). Deep temporal models and active inference. *Neuroscience and Biobehavioral Reviews*, 77, 388–402.

[32] Friston, K. (2008). Hierarchical models in the brain. *PLoS Computational Biology*, 4(11), e1000211.

[33] Williams, D. (2017). Predictive processing and the representation wars. *Minds and Machines*. doi:10.1007/s11023-017-9441-6.

[34] Williams, D., and Colling, L. (2017). From symbols to icons: The return of resemblance in the cognitive neuroscience revoluti on. *Synthese*. https://doi.org/10.1007/s11229-017-1578-6.

[35] Lee, T. S., and Mumford, D. (2003). Hierarchical Bayesian inference in the visual cortex. *Journal of the Optical Society of America*, A, 20, 1434–1448.

[36] Mathys, C., Daunizeau, J., Iglesias, S., Diaconescu, A., Weber, L., Friston, K., and Stephan, K. (2012). Computational modeling of perceptual inference: A hierarchical Bayesian approach that allows for individual and contextual differences in weighting of input. *International Journal of Psychophysiology*, 85 (3), 317–318. http://dx.doi.org/10.1016/j.ijpsycho.2012.06.077.

[37] Mathys, C., Lomakina, E., Daunizeau, J., Iglesias, S., Brodersen, K., Friston, K., and Stephan, K. (2014). Uncertainty in perception and the Hierarchical Gaussian Filter. *Frontiers in Human Neu roscience*, 8. http://dx.doi.org/10.3389/fnhum.2014.00825.

[38] Bastos, A., Usrey, W., Adams, R., Mangun, G., Fries, P., and Friston, K. (2012). Canonical microcircuits for predictive coding. *Neuron*, 76 (4), 695–711. http://dx.doi.org/10.1016/j.neuron.2012.10.038.

[39] Hohwy, J. (2016). Attention and conscious perception in the hypothesis testing brain. *Frontiers in Psychology*, 3. doi:10.3389/fpsyg.2012.00096.

[40] Friston, K. J., and Frith, C. (2015). A duet for one. *Consciousness and Cognition*, 36, 390–405.

[41] Murphy, D. (2006). *Psychiatry in the Scientific Image*. Cambridge: MIT Press.

[42] Montague, P., Dolan, R., Friston, K., and Dayan, P. (2012). Computational psychiatry. *Trends In Cognitive Sciences*, 1 6(1), 72–80. http://dx.doi.org/10.1016/j.tics.2011.11.018.

[43] Teufel, C., and Fletcher, P. (2016). The promises and pitfalls of applying computational models to neurological and psychiatric disorders. *Brain*, 139(10) , 2600–2608. http://dx.doi.org/10.1093/brain/aww209.

[44] Friston, K., Stephan, K., Montague, R., and Dolan, R. (2014). Computational psychiatry: The brain as a phantastic organ. *The Lancet Psychiatry*, 1 (2), 148–158. http://dx.doi.org/10.1016/s2215-0366(14)70275-5.

[45] Lawson, R., Rees, G., and Friston, K. (2014). An aberrant precision account of autism. *Frontiers In Human Neu roscience*, 8. http://dx.doi.org/10.3389/fnhum.2014.00302.

[46] Seth, A., Suzuki, K., and Critchley, H. (2012). An interoceptive predictive coding model of conscious presence. *Frontiers In P sychology*, 2. http://dx.doi.org/10.3389/fpsyg.2011.00395.

[47] Chekroud, A. (2015). Unifying treatments for depression: An application of the free energy principle. *Frontiers In P sychology*, 6. http://dx.doi.org/10.3389/fpsyg.2015.00153.

[48] Corlett, P., Honey, G., and Fletcher, P. (2016). Prediction error, ketamine and psychosis: An updated model. *Journal Of Psychopharmacology*, 30(11) , 1145–1155. http://dx.doi.org/10.1177/0269881116650087.

[49] Corlett, P., and Fletcher, P. (2015). Delusions and prediction error: Clarifying the roles of behavioural and brain responses. *Cognitive Neuropsychiatry*, 20 (2), 95–105. http://dx.doi.org/10.1080/13546805.2014.990625.

[50] Firestone, C., and Scholl, B. (2015). Cognition does not affect perception: Evaluating the evidence for "top-down" effects. *Behavioral And Brain Sciences*, 39. http://dx.doi.org/10.1017/s0140525x15000965.

[51] Fodor, J. (1983). *The Modularity of Mind*. Cambridge: The MIT Press.

[52] Maher, B. (1974). Delusional thinking and perceptual disorder. *Journal of Individual Psychology, 30*, 98–113.

[53] Hemsley, D., and Garety, P. (1986). The formation of maintenance of delusions: A Bayesian analysis. *The British Journal of Psychiatry*, 14 9(1), 51–56. http://dx.doi.org/10.1192/bjp.149.1.51.

[54] Bortolotti, L. (2016). Delusion. In *The Stanford Encyclopedia of Philosophy* (Spring 2016 Edition), Edward N. Zalta (E d.). https://plato.stanford.edu/archives/spr2016/entries/delusion/.

[55] Ellis, H., and Young, A. (1990). Accounting for delusional misidentifications. *The British Journal of Psychiatry, 157* (2), 239–248. http://dx.doi.org/10.1192/bjp.157.2.239.

[56] Coltheart, M. (2007). The 33rd Sir Frederick bartlett lecture: Cognitive neuropsychiatry and delusional belief. *The Quarterly Journal of Experimental Psychology*, 60(8), 1041–1062. [RMR].

[57] Colheart, M. (2013). On the distinction between monothematic and polythematic delusions. *Mind & Language, 28* (1), 103–112. http://dx.doi.org/10.1111/mila.12011.

[58] Ross, R., McKay, R., Coltheart, M., and Langdon, R. (2016). Perception, cognition, and delusion. *Behavioral And Brain Sciences*, 39. http://dx.doi.org/10.1017/s0140525x15002691.

[59] Friston, K. J., and Frith, C. D. (2005). Active inference, communication and hermeneutics. cortex. *A Journal Devoted to the Study of the Nervous System and Behavior, 68*, 129–143.

[60] Chadwick, P. (1993). The stepladder to the impossible: A first hand phenomenological account of a schizoaffective psychotic crisis. *Journal of Mental Health*, 2 (3), 239–250. http://dx.doi.org/10.3109/09638239309003769.

[61] Brown, H., Adams, R., Parees, I., Edwards, M., and Friston, K. (2013). Active inference, sensory attenuation and illusions. *Cognitive Processing, 14* (4), 411–427. http://dx.doi.org/10.1007/s10339-013-0571-3.

[62] Teufel, C., Kingdon, A., Ingram, J., Wolpert, D., and Fletcher, P. (2010). Deficits in sensory prediction are related to delusional ideation in healthy individuals. *Neuropsychologia, 48*(14) , 4169–4172. http://dx.doi.org/10.1016/j.neuropsychologia.2010.10.024.

[63] Deneve, S., and Jardri, R. (2016). Circular inference: Mistaken belief, misplaced trust. *Current Opinion in Behavioral Sciences , 11*, 40–48. http://dx.doi.org/10.1016/j.cobeha.2016.04.001.

[64] Bortolotti, L., and Miyazono, K. (2015). Recent work on the nature and development of delusions. *Philosophy Compass, 10*(9), 636–645.

[65] Gadsby, S. (2019). Body Representations and cognitive ontology: Drawing the boundaries of the body image. *Consciousness and Cognition*, *74*, 102772.

[66] Gadsby, S., and Hohwy, J. (2021). Why use predictive processing to explain psychopathology? The case of anorexia nervosa. In S. Gouveia, R. Mendonça, and M. Curado (Eds.), *The Philosophy and Science of Predictive Processing*, London: Bloomsbury.

[67] Notredame, C., Pins, D., Deneve, S., and Jardri, R. (2014). What visual illusions teach us about schizophrenia. *Frontiers In Integrative Neu roscience*, *8*. http://dx.doi.org/10.3389/fnint.2014.00063.

[68] Ernst, M., and Banks, M. (2002). Humans integrate visual and haptic information in a statistically optimal fashion. *Nature*, *415*(687 0), 429–433. http://dx.doi.org/10.1038/415429a.

[69] Sanborn, A., and Chater, N. (2016). Bayesian brains without probabilities. *Trends In Cognitive Sciences*, *20*(1 2), 883–893. http://dx.doi.org/10.1016/j.tics.2016.10.003.

16 Knowledge Representation in AI

Ashwin Kumaar K, Krishna Sai Talupula,
Gogineni Venkata Ashwith, and Mitta
Chaitanya Kumar Reddy

16.1 INTRODUCTION

Artificial intelligence (AI) systems are designed to process and analyze data, make decisions, and perform tasks that would normally require human-level intelligence. In order to do this effectively, AI systems must be able to represent and manipulate the knowledge and information that they use to make decisions and solve problems. This process of representing knowledge is known as knowledge representation.

There are many ways in which knowledge can be represented in AI systems, and the choice of representation can have a significant impact on the system's performance and effectiveness. Some of the most common methods of knowledge representation in AI include:

1. Rule-based systems
2. Ontologies and semantic networks
3. Decision trees
4. Neural networks
5. Case-based reasoning

In this chapter, we will examine each of these methods in more detail and discuss their strengths and limitations [1].

16.1.1 RULE-BASED SYSTEMS

One of the simplest and most widely used methods of knowledge representation in AI is the rule-based system. In a rule-based system, knowledge is represented as a set of rules that specify the conditions under which a particular action should be taken.

For example, consider a simple rule-based system for diagnosing medical conditions. The system might contain rules such as:

- If the patient develops a fever and a cough, then they may have some flu.
- If the patient develops a rash and joint pain, then they probably have a form of arthritis.

DOI: 10.1201/9781003328414-16

To use the system, the user would input the symptoms that the patient is experiencing, and the system would apply the rules to determine the most likely diagnosis.

Rule-based systems are relatively easy to design and implement, and they can be very effective for solving simple problems with well-defined rules. However, they can be difficult to scale up to more complex problems, and they can be inflexible, as it can be difficult to add or modify rules once the system has been implemented [1].

16.1.2 Ontologies and Semantic Networks

An ontology is a decisional representation of a set of concepts and rules within a domain, along with the relationships between those concepts. Ontologies are often used in AI systems to represent the knowledge that the system uses to reason about a particular domain.

For example, an ontology for a medical diagnosis system might contain concepts such as "disease," "symptom," and "treatment," along with relationships such as "cause," "relieve," and "prevent." This knowledge can be used by the AI system to reason about the relationships between different diseases, symptoms, and treatments and to make informed decisions about how to diagnose and treat a particular patient [1].

Semantic networks are similar to ontologies, but they represent knowledge using a graph-like structure, with concepts represented as nodes and relationships represented as edges. Semantic networks are often used to represent more complex or nuanced relationships between concepts, and they can be more flexible than ontologies, as it is easier to add or modify relationships in a semantic network.

Both ontologies and semantic networks can be very effective at representing complex domains and facilitating reasoning and decision-making, but they can be difficult to design and maintain, as they require a high level of domain expertise and can be prone to errors and inconsistencies [1].

16.1.3 Decision Trees

Decision trees are a widely used method of knowledge representation in AI, particularly in the field of machine learning. In a decision tree, knowledge is represented as a tree-like structure, with internal nodes representing decision points and leaf nodes representing the final outcomes [1].

16.1.4 Neural Networks

The structure and operation of the human brain served as the inspiration for the machine learning algorithm known as neural networks. They consist of interconnected layers of "neurons," which process and transmit information through the network.

In the context of knowledge representation, neural networks can be used to learn complex relationships and patterns in data and to make decisions and predictions based on that knowledge. For example, a neural network might be trained to recognize patterns in images and classify them as either "cat" or "dog." The network would learn to recognize these patterns by being presented with a large dataset of labelled images and adjusting the connections between neurons to optimize its performance on the task.

16.2 ADVANTAGES

There are several key advantages to using neural networks for knowledge representation in AI:

1. Neural networks are highly flexible and can learn to represent a wide range of knowledge, from simple patterns to complex relationships.
2. They can learn to represent knowledge automatically, without the need for explicit programming or manual feature engineering.
3. They can learn from large and complex datasets, making them well-suited to tasks that involve extracting knowledge from unstructured or noisy data.
4. They can handle incomplete or uncertain data, making them robust in terms of missing or ambiguous information.

However, there are also some limitations to using neural networks for knowledge representation:

1. They can be difficult to interpret, as the internal representation of knowledge is typically not explicit and can be difficult to understand.
2. They can be computationally intensive and may require specialized hardware to train and run efficiently.
3. They can be prone to overfitting, especially if the training dataset is small or not representative of the underlying problem.

Overall, neural networks are an effective method for representing knowledge in AI and have been applied to a variety of tasks to produce cutting-edge outcomes. They can be a vital asset in any AI system that has to be able to learn and adapt, as they are especially well-suited to jobs that involve learning from big and complicated datasets [1].

16.2.1 CASE-BASED REASONING IN KNOWLEDGE REPRESENTATION IN AI

Case-based reasoning (CBR) is a method of knowledge representation in AI that involves using past examples or "cases" to solve new problems or make decisions. In a CBR system, the knowledge is represented as a collection of cases, each of which consists of a problem and a solution. To solve a new problem, the CBR system retrieves and adapts the most similar case from its memory and uses the solution from that case as a starting point.

One of the key advantages of CBR is that it allows the system to learn and adapt over time by adding new cases to its memory as they are encountered. This makes CBR well-suited to tasks that involve complex or dynamic environments, where the knowledge base may need to be updated and refined continuously.

CBR systems are also generally easy to design and implement, as they do not require explicit programming of rules or decision-making logic. This makes them a popular choice for tasks that involve open-ended or unstructured problems, where it may be difficult to define a set of rules or a clear decision-making process.

However, there are also some limitations to using CBR for knowledge representation. One limitation is that CBR systems may be less efficient than other methods, as they may need to search through a large number of cases to find the most similar one. In addition, CBR systems may struggle with tasks that require more abstract or theoretical reasoning, as they are typically based on concrete examples rather than general principles.

Overall, CBR is a useful method of knowledge representation in AI and has been applied to a wide range of tasks, including diagnosis, planning, and problem-solving. It is particularly well-suited to tasks that involve adapting to changing environments or adapting to new problems based on past experience [1].

16.3 TECHNIQUES IN KNOWLEDGE REPRESENTATION

There are several techniques used in knowledge representation in AI:

1. Rule-based systems: These systems store knowledge as a set of rules that can be applied to a given problem.
2. Semantic networks: These are graphical representations of knowledge that show the relationships between concepts.
3. Ontologies: These are formal representations of the concepts and relationships in a particular domain of knowledge.
4. Frames: These are structures that represent knowledge about a particular concept and the relationships between different aspects of that concept.
5. Logic-based systems: These use logical statements to represent knowledge and can infer new conclusions from that knowledge.
6. Concept maps: These are graphical representations of the relationships between concepts, similar to semantic networks.
7. Case-based reasoning: This technique involves using past experiences (cases) to solve new problems.
8. Neural networks: These are machine learning systems that can learn to recognize patterns in data and make decisions based on that knowledge.
9. Decision trees: These are tree-like structures that represent the possible outcomes of a decision and the conditions under which each outcome is reached.

16.4   RULE-BASED SYSTEM

In a rule-based system, knowledge is represented as a set of rules that can be applied to a given problem. These rules are typically written in the form of if-then statements, with the "if" part representing the conditions that must be satisfied and the "then" part representing the action to be taken [2].

For example, a rule-based system might have a rule that says: "If the animal is a cat, then classify it as a domestic animal." This rule would be applied to the input data (in this case, the type of animal) and the system would use it to make a decision (in this case, classifying the animal as domestic) [2].

Rule-based systems are helpful for representing knowledge and reasoning about it in a clear-cut and organized manner. They are frequently utilized in expert systems, which are created to simulate a human expert's decision-making skills in a certain subject. [2].

16.5   SEMANTIC NETWORKS

A semantic network is a graphical representation of knowledge that shows the relationships between concepts. It consists of a set of nodes, which represent the concepts, and edges, which represent the relationships between the concepts [2].

For example, consider a semantic network that represents the concepts "dog," "cat," "mammal," and "pet." In this network, the node for "dog" might be connected to the node for "mammal" with an edge labeled "is a," indicating that a dog is a type of mammal. Similarly, the node for "cat" might be connected to the node for "mammal" with the same edge. The node for "dog" might also be connected to the node for "pet" with an edge labeled "is," indicating that a dog is a type of pet [2].

Semantic networks are useful for representing and reasoning about knowledge because they allow for the representation of complex relationships between concepts in a graphical and easy-to-understand format. They are often used in natural language processing and information retrieval systems to help understand the meaning of words and phrases [2].

16.6   ONTOLOGIES

A formal depiction of the ideas and connections in a certain field of knowledge is called ontology. It is employed to specify the relationships between the terminology and concepts utilized to describe that domain as well as their definitions. [2].

Ontologies are usually written in a formal language and can be represented as a graph structure, with the concepts as nodes and the relationships as edges. They can be used to represent both the structure of the domain (e.g., the hierarchical relationships between concepts) and the attributes and properties of the concepts [2].

Ontologies are useful for knowledge representation in AI because they allow for the representation of complex domain-specific knowledge in a structured and formalized way. They are often used in natural language processing and information retrieval systems to help understand the meaning of words and phrases and to disambiguate them based on their context. They are also used in the development of intelligent agents and other AI systems that need to reason about and interact with a specific domain of knowledge [2].

16.7   FRAMES

A frame is a structure that represents knowledge about a particular concept and the relationships between different aspects of that concept. It consists of a set of slots, which represent the different characteristics or attributes of the concept, and a set of values, which fill those slots and provide information about the concept [2].

For example, consider a frame for the concept of a "car." This frame might have slots for the make and model of the car, the year it was manufactured, its color, and its engine size. Each car would be represented by a separate frame, with the values for the slots providing specific information about that car [2].

Frames are useful for knowledge representation in AI because they allow for the representation of complex concepts and the relationships between their different aspects in a structured and organized way. They are often used in expert systems and other AI applications that need to reason about and manipulate complex concepts [2].

16.8   LOGIC-BASED SYSTEM

Logic-based systems use logical statements to represent knowledge and infer new conclusions from that knowledge. They are based on a formal system of logic, such as propositional logic or first-order logic, which provides a set of rules for constructing and manipulating logical statements [3].

In a logic-based system, knowledge is represented as a set of logical statements, known as axioms. These axioms can be used to make inferences about other statements, using the rules of the logical system. For example, if the axioms state that "all cats are mammals" and "Fluffy is a cat," then the system can infer that "Fluffy is a mammal" [3].

Logic-based systems are useful for knowledge representation because they provide a precise and rigorous way of representing and reasoning about knowledge. They are often used in applications that require precise and accurate reasoning, such as theorem proving and automated planning [3].

16.9   CONCEPT-BASED

A concept map is a graphical representation of the relationships between concepts. It consists of a set of nodes, which represent the concepts, and edges, which represent the relationships between the concepts [3].

Concept maps are similar to semantic networks, but they are typically more focused on showing the relationships between concepts, rather than the attributes of the concepts themselves. They are often used to visualize and organize complex knowledge structures and to help identify the key concepts and relationships within a domain of knowledge [3].

In AI, concept maps can be used to represent and reason about knowledge in a variety of applications, such as natural language processing and information retrieval. They can also be used as a tool for organizing and structuring knowledge in the development of intelligent agents and other AI systems [3].

16.10 APPLICATION KNOWLEDGE REPRESENTATION IN AI

There are many applications for knowledge representation in AI, including:

1. Expert systems: These AI programs are made to resemble the judgement skills of a human expert in a particular field. To store and make sense of the information and skill of the human expert, they employ knowledge representation approaches.
2. Natural language processing: Knowledge representation techniques are often used in natural language processing systems to represent and reason about the meaning and context of words and phrases.
3. Information retrieval: Knowledge representation is used in information retrieval systems to represent and reason about the meaning and relevance of documents and other information sources.
4. Planning and decision-making: AI systems that are used for planning and decision-making, such as autonomous robots and self-driving cars, rely on knowledge representation to store and reason about the options and consequences of different actions.
5. Knowledge-based systems: These are AI systems that use knowledge representation to store and reason about domain-specific knowledge, such as the rules of a game or the properties of a chemical compound.
6. Intelligent tutoring systems: These are AI systems that use knowledge representation to personalize and adapt their teaching and learning approaches to individual students.

16.11 APPROACHES OF KNOWLEDGE REPRESENTATION IN AI

There are several approaches to knowledge representation in AI, including:

1. Symbolic approach: This approach uses symbolic representations, such as logic or rule-based systems, to represent and reason about knowledge.
2. Subsymbolic approach: This approach uses numerical or statistical representations, such as neural networks, to represent and reason about knowledge.
3. Hybrid approach: This approach combines symbolic and subsymbolic techniques to represent and reason about knowledge.
4. Ontological approach: This approach uses formal ontologies to represent and reason about the concepts and relationships in a particular domain of knowledge.

5. Case-based reasoning: This approach uses past experiences (cases) to solve new problems by storing and reusing examples of similar situations.
6. Frame-based systems: This approach uses frames to represent and reason about the attributes and relationships of complex concepts.

Pros of Knowledge Representation in AI [3]:

There are several advantages to using knowledge representation in AI:

1. It allows for the representation of complex and domain-specific knowledge in a structured and organized way.
2. It enables the creation of intelligent systems that can reason about and manipulate this knowledge, leading to more intelligent and human-like behaviour.
3. It allows for the reuse of knowledge across different applications and domains, which can save time and resources.
4. It can help to improve the accuracy and reliability of AI systems by providing a more structured and rigorous way of representing knowledge.
5. It can enable the integration and exchange of knowledge between different AI systems, leading to more collaborative and interoperable systems.
6. It can facilitate the development of more transparent and explainable AI systems by making it easier to understand how the system is using its knowledge to make decisions.

16.12   LIMITATIONS OF KNOWLEDGE REPRESENTATION IN AI

There are also some limitations to knowledge representation in AI:

1. It can be difficult to represent complex and ambiguous knowledge in a structured and formalized way.
2. It requires the knowledge to be encoded in a specific way, which can be time-consuming and error-prone.
3. It may not be able to represent knowledge that is implicit or contextual or that relies on common sense or human-like reasoning [4].
4. It may not be able to adapt to new or changing knowledge, requiring frequent updates and maintenance.
5. It may not be able to handle incomplete or inconsistent knowledge, leading to incorrect inferences or decisions.
6. It may be vulnerable to errors or biases in the knowledge that is encoded, which can affect the performance and reliability of the AI system.

16.13   CONCLUSION

Knowledge representation is a central aspect of AI, as it determines how knowledge is represented, stored, and used by AI systems. There are several techniques and approaches used in knowledge representation, including symbolic representations, such as logic and rule-based systems; subsymbolic representations, such as neural networks; and hybrid approaches that combine both symbolic and subsymbolic techniques [2].

One of the main advantages of knowledge representation is that it allows for the representation of complex and domain-specific knowledge in a structured and organized way, enabling the creation of intelligent systems that can reason about and manipulate this knowledge. It also enables the reuse of knowledge across different applications and domains, which can save time and resources [3].

However, there are also some limitations to knowledge representation. It can be difficult to represent complex and ambiguous knowledge in a structured and formalized way, and it may not be able to represent knowledge that is implicit or contextual or that relies on common sense or human-like reasoning. It may also be vulnerable to errors or biases in the knowledge that is encoded, which can affect the performance and reliability of the AI system [5].

Overall, knowledge representation is important to AI because it affects how knowledge is employed by AI systems to solve issues and make decisions. In order to overcome the difficulties and shortcomings of the current methodologies, new techniques and approaches are constantly being created in this busy area of research and development.

REFERENCES

[1] "Knowledge Representation in Artificial–Intelligence - Javatpoint," *www.javatpoi nt. com*, 2022. https://www.javatpoint.com/knowledge-representation-in-ai (accessed Jan. 05, 2023).
[2] Wikipedia Contributors, "Knowledge Representation and Reasoning," *Wikipedia*, N ov. 14, 2022. https://en.wikipedia.org/wiki/Knowledge_representation_and_reasoning (accessed Jan. 05, 2023).
[3] "What is Knowledge Representation In AI? Usage, Types & Methods | upGrad Blog," *upGrad blog*, S ep. 17, 2020. https://www.upgrad.com/blog/what-is-knowledge-representation-in-ai/ (accessed Jan. 05, 2023).
[4] Kabanda, G., & Kannan, H. (2023). A systematic literature review of reinforcement algorithms in machine learning. In *Handbook of Research on AI and Knowledge Engineering for Real-Time Business Intelligence*, IGI Global, USA, pp. 17–33.
[5] softlogicsys, "Knowledge Representation in AI," *Software Training Institute in Chennai with 10–% Placements - Softlogic*, J ul. 04, 2022. https://www.softlogicsys. in/knowledge-representation-in-ai/ (accessed Jan. 05, 2023).

17 ANN Model for Analytics

Gaddam Venkat Shobika, S. Pavan Siddharth, Student, Vishwa KD, Hemachandran K, and Manjeet Rege

17.1 INTRODUCTION

An artificial neural network (ANN) is an algorithm used in machine learning that is used to construct data trends. ANNs, like the human brain, are made up of a network of interconnected nodes, or neurons, that can recognize patterns in input data.

An ANN is constructed as a function of artificial intelligence to mimic the neural network, which is similar to the human brain so that computers can make decisions comparable to humans. ANNs are created by programming computers to behave like interconnected brain cells. The human brain consists of approximately 1,000 billion neurons with an association point somewhere between 1,000 and 100,000. There is a way that information is distributed in the brain, enabling us to extract more than one piece of information when needed. A neural network is composed of processing nodes that are connected by edges. Each neuron receives input from other neurons and generates output, which is then passed on to other neurons. Synapses are the edges that connect neurons, and synaptic weight is the intensity of the connection between two neurons. A neural network's output is determined by the weights of the connections between neurons and the inputs to neurons.

The weights are adjusted in such a way that the network's output is a near approximation of the desired output. This is referred to as network training. Once trained, the network can be used to make predictions about new data. This is known as inference. A neural network that has been designed to recognize photographs of cats, for example, can be used to identify a cat in a new image. An example of an ANN is a digital logic gate that receives an input and outputs an output. In the case of a "OR" gate with two inputs, the outcome will be "On." When both of the inputs are set to "OFF," the outcome will be configured to "Off." In this instance, the outcome is determined by the input. Because our brain's neurons continuously learn, the outputs-to-inputs connection is constantly changing.

An ANN has the advantage of being able to be trained to recognize patterns that are too complicated for typical machine learning techniques. For example, ANNs have been used to model handwriting recognition, marketing, operations,

DOI: 10.1201/9781003328414-17

the telecom industry, image classification, and even stock market prediction. The downside of using ANNs is that they can be very computationally intensive and therefore require powerful computers to train. Additionally, ANNs can be difficult to understand and interpret, making them less transparent than other machine learning algorithms.[1]

17.2 AN ARTIFICIAL NEURAL NETWORK'S ARCHITECTURE

To understand what neural networks are, we must first define them. They are made up of many layers as well as artificial neurons. ANNs are made up of a huge number of artificial neurons, which are units that can be placed in layers of different sequences. ANNs are made up of three layers, as seen in Figure 17.1.

Input Layer: This type of programming language can be used to create different sorts of programs, as the name suggests.[2]

Hidden Layer: The hidden layer provides a series of in-between calculations to input and output layers. This is where all the work is done in order to find features and patterns you can't see.

Output Layer: The hidden layer computes a weighted sum of each input with a bias coefficient. This output is represented by what the neural network chooses to show at the new calculated layer.

The weighted total is passed as an input to an activation function that decides whether or not a node should "fire." Some activation functions are sensitive to different types of tasks, which is why they exist.[3]

17.3 TRAINING AN ARTIFICIAL NEURAL NETWORK

Neural networks have two primary advantages. First, as compared to a simple collection of formulas, this mathematical model of neural networks is so much more

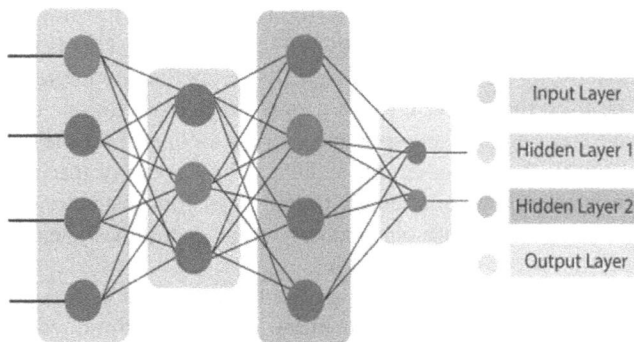

FIGURE 17.1 Architecture of an ANN.

complicated, allowing it to handle a broader range of operational scenarios without fine-tuning. Furthermore, because neural networks are self-learning, they do not require control system knowledge; they only require enough historical data to adequately train themselves.

A network is ready to be trained once it has been established for a particular function. The input values are normally selected at random to start the entire procedure. The next phase is learning or training.

Teaching methods are classified as either supervised or unsupervised. The supervised method uses one of the two methods to provide the network with the necessary output: either by manually "grading" the network's performance or by fusing the needed outputs with the inputs. The network must understand the inputs totally on its own in unsupervised learning. The majority of networks use supervised learning. Inputs are initially defined using an unsupervised method.[4]

17.3.1 SUPERVISED TRAINING

During supervised learning, the sources of data and the findings are delivered. After processing the inputs, the network compares the obtained outputs to the predicted results. Because of the errors propagating back through the framework, the factors that govern the network change. This process is continued as the weights are changed over and over. A "training set" is a collection of data that can be used for training. Since the connection weights improve over time, the same data is analyzed repeatedly while the network is being trained.

Supervised training must save a repository of data that will be utilized to evaluate the system once it has been trained. The developer must clearly assess the inputs, outputs, layers, elements per layer, connections between layers, summation, transfer, training, and initial weights if a network is unable to tackle the task. The modifications necessary to create a strong network are where the "science" of neural networking is found.

Finally, when no further learning is required and the system has been successfully trained, the weights could be "frozen" if required. This network is then translated into hardware in some systems so that it can be quick. When used in production, some systems don't lock themselves in and instead keep learning.[4]

17.3.2 UNSUPERVISED TRAINING

The other kind of training is unsupervised training. Feeding the network inputs but not the anticipated outputs is known as unsupervised training. After that, the system should decide which features to use to categorize the input data. This is frequently described as self-organization or flexibility. The concept of unsupervised learning is continually developing. Because of this ability to adjust to their surroundings, science fiction–style computers would be capable of continuing to learn on their own when confronted with unusual scenarios and new locations. There are many instances in life where precise training sets are lacking. Some of these circumstances entail military action when the use of modern weaponry and battle strategies may be encountered. There is still potential for this subject of study because of the unpredictable nature of life and people's desire to be prepared.[4]

17.4 HOW ARTIFICIAL NEURAL NETWORKS ARE BEING USED

ANN has been undergoing the transformation that happens as an idea is thrust into a harsher world populated by individuals who simply want to get the job done. Considering the fact that the vast majority of networks being created today are statistically extremely accurate, customers who want computers to completely cure their problems still find fault with them. These networks could have an accuracy of 85% to 90%. However, only a few applications could tolerate such inaccuracy. This most current method of forecasting the future is by studying its past, which comes with a set of drawbacks. Giving an explanation for the computer-generated response, for example, as to why a specific loan application was rejected, is one of those issues. Neural networks' inner workings are "black boxes," as has been stated throughout this study. Even using neural networks has been referred to as "voodoo engineering" by some people. It has been challenging to describe how a network learns and why it suggests a specific course of action.

17.4.1 LANGUAGE PROCESSING

The applications of language processing are extremely diverse. Text-to-speech technology, secure voice detection locks, auditory input for computers, assistive technology, computerized language translation, computerized transcription, natural language processing, and assistive technology for the disabled and physically impaired who use voice commands are some of these uses.

17.4.2 PATTERN RECOGNITION

The area of quality control sees the most use of neural networks as pattern recognizers. Several automated applications are currently in use; they are developed to isolate a single defective item from hundreds or thousands. Human inspectors lose focus or grow weary. Systems now assess solder connections, welds, and cuts. One manufacturer is currently developing a prototype of a system that assesses paint colour. To check if fresh paint batches are the proper hues, this technology digitizes images of the paint samples.

The use of neural networks as processors for sensors is another significant area where they are being applied to pattern recognition systems. The few useful bits of information that are occasionally provided by sensors may get lost in the sea of data. They are looking at the display, searching for "the needle in the haystack," and people can become bored. There are numerous uses for sensor processing in the defence sector. These neural network systems have demonstrated success in target recognition. Sensor processors collect information using infrared cameras, earthquake recorders, and sonar sensors. Then, potential phenomena are identified using those data.

17.4.3 FINANCE

The financial industries are embracing neural networks in significant ways. Lending institutions, banks, and credit card businesses all deal with imprecise choices. This is where the ANN comes into the picture. Forms must be filled out as part of the loan approval procedure for a loan officer to decide whether to approve the loan. Trained neural networks using the data from previous decisions are now using the information from these forms. In fact, such packages provide info upon which input, or a mixture of inputs, is being used, which is weighted most heavily on the decision to comply with government criteria regarding the reasons why applications are being denied.

17.4.4 IMAGE/DATA COMPRESSION

Numerous experiments have been conducted to demonstrate the real-time data decompression and compression ability of neural networks. These auto-associative networks can break down 8 bits of data into 3, reversing the process, and then breaking 8 eight bits into 3 again. They are not lossless, though. Due to this bit loss, they are unable to effectively compete with conventional techniques.

17.4.5 SERVO CONTROL

One of the most intriguing applications of neural networks is the control of complex systems. The majority of conventional control systems use a single set of formulas to model how each of the system's processes operates. Those formulas need to be manually tweaked to adapt a system for a particular procedure. It is a time-consuming process that requires adjusting parameters until the right mixture is discovered that yields the anticipated outcome.[4]

17.5 TYPES OF ANNS

17.5.1 HOPFIELD NETWORK

An input buffer, a Hopfield layer, as well as an output layer, comprise the Hopfield network. There is the same proportion of processing elements in each layer. Variable connection weights connect the Hopfield layer's inputs to the results of the relevant processing entities inside the input buffer layer. Other than for itself, all of the other processing element's inputs are wired back to the results of the Hopfield layer. These will be additionally linked with the analogous components throughout the output layer.[5] This is shown in Figure 17.2.

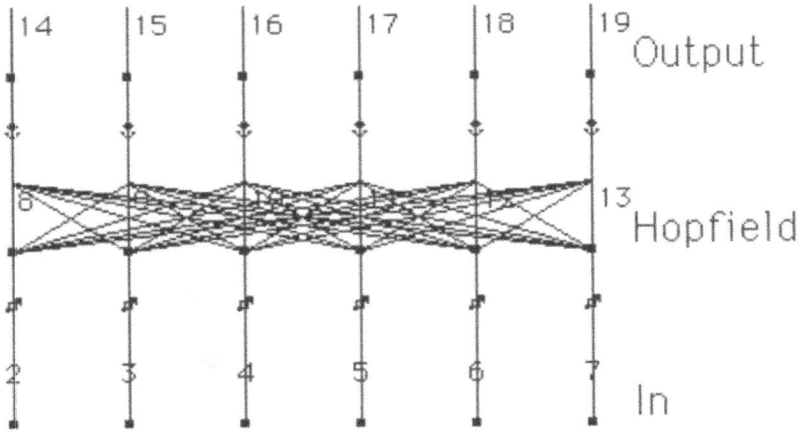

FIGURE 17.2   Hopfield network.[4]

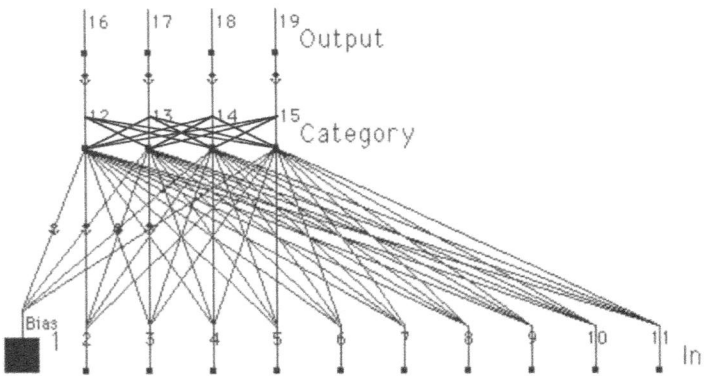

FIGURE 17.3   Hamming network.

17.5.2   HAMMING NETWORK

The front portion of the Hamming network, which is a continuation of the Hopfield network, now includes an expectation-maximization classifier. The Hamming network is composed of three layers. The input layer has an equal quantity of nodes as the number of different binary attributes. It has a Hopfield layer for categories with an equal number of nodes as categories, or classes. The formal Hopfield architecture, which has the same number of intermediate layer nodes as input nodes, is very different from this. In the end, there's an output layer with an equal quantity of nodes as the category layer. See Figure 17.3.

17.5.3   BIDIRECTIONAL ASSOCIATIVE MEMORY

The bidirectional associative storage has the same quantity of output and input processing units. The two hidden layers, which reflect the magnitude of two input vectors, are composed of two distinct associated memories. Although this example uses equal input vector lengths of four each, the two lengths do not necessarily have to match. The middle layers are completely interconnected. For purposes of implementation, the process to enter and fetch data from the network uses the output and input layers.

The middle layers are intended to save linked vector pairs. When an ambiguous pattern vector is used as an input, the intermediary layers iterate unless a stable equilibrium state is achieved. When an ambiguous pattern vector is used as an input, the intermediary layers iterate unless a steady state is achieved. When using the training set's complements as the unknown input vector, the bidirectional associative memory network suffers from the same mistake as the Hopfield network when attempting to locate a taught pattern. See Figure 17.4.

17.5.4   SPATIO-TEMPORAL PATTERN RECOGNITION

Spatio-temporal pattern networks tend to be the most useful at identifying repetitive audio signals. This network has been used by one unit of General Dynamics to categorize different kinds of boats on the basis of the acoustic sounds made by their engine propellers. Due to the attack function's long decay, the framework is able to reliably detect the propeller messages even when the frequency of the input signal fluctuated up to a factor of two. See Figure 17.5.

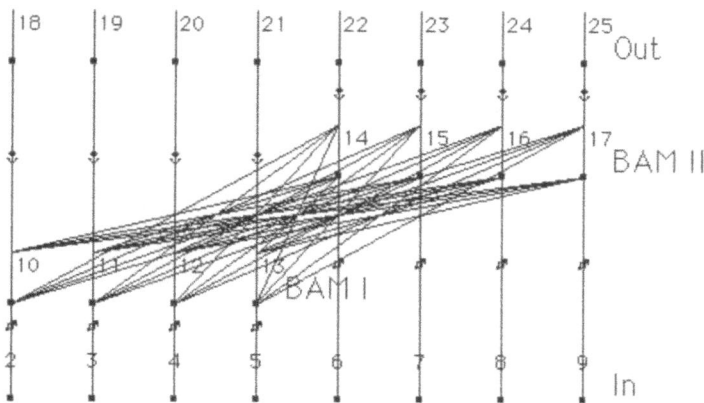

FIGURE 17.4   Bidirectional associative memory.

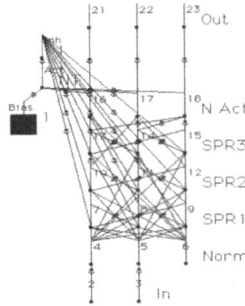

FIGURE 17.5   Spatio-temporal pattern recognition.

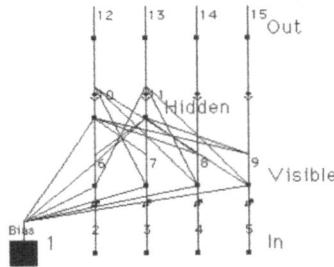

FIGURE 17.6   Recirculation.

17.5.5   RECIRCULATION

Data is solely processed in one way in a recirculation network, and learning is completed entirely with local data. The condition of this processing entity, as well as the input data on the specific connection that needs modification, supply the majority of the information. Because the recirculation network uses the unsupervised method, no favored output vector is required at the final layer. The framework is auto-associative when there are exactly as many outputs as inputs.

The visible and invisible layers of this network are situated between the input and output levels. The learning rule's purpose is to construct an internal depiction of the data shown in the visible layer in the hidden layer. Compressing input data in the hidden layer by employing fewer processing components is a good illustration of this. In this instance, it is possible to think of the hidden representation as a condensed form of the apparent representation. The layers that are both visible and concealed are completely interconnected in both ways. Additionally, every element in both the visible and hidden levels is linked to an unbalanced element. These connections' variable weights adapt in the same way that the network's variable weights do.[4] See Figure 17.6.

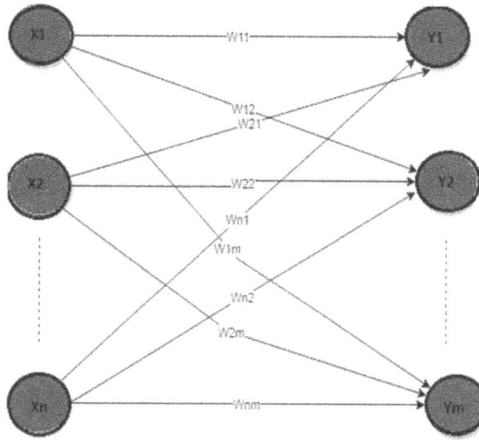

FIGURE 17.7 Single-layer feed-forward network.

17.6 TYPES OF NEURON CONNECTION ARCHITECTURES

17.6.1 SINGLE-LAYER FEED-FORWARD NETWORK

The input layer and the output layer are the fundamental layers in this kind of network; however, the input layer does not count because no calculations are performed in this layer. When constructing the output layer, different weights are added to input nodes to create the combined effect per node.[6] Following all of this, the neurons give an overall signal to the output layer, which determines the output. See Figure 17.7.

17.6.2 MULTILAYER FEED-FORWARD NETWORK

The network also includes a hidden layer that does not directly communicate with the external layer. The inclusion of one or more hidden layers boosts the network's computational capacity and converts it into a "feed-forward" network because data is passed forward through the input function and even the intermediate processes required to arrive at the output Z. The system's outputs are not connected to it by any feedback loops. See Figure 17.8.

17.6.3 SINGLE NODE WITH OWN FEEDBACK

When loops are built in recurrent networks, they're called feedback networks. With recurrent networks, these feedback connections can be chained together to create a looped system. Figure 17.9 shows a simplified representation of one single loop.

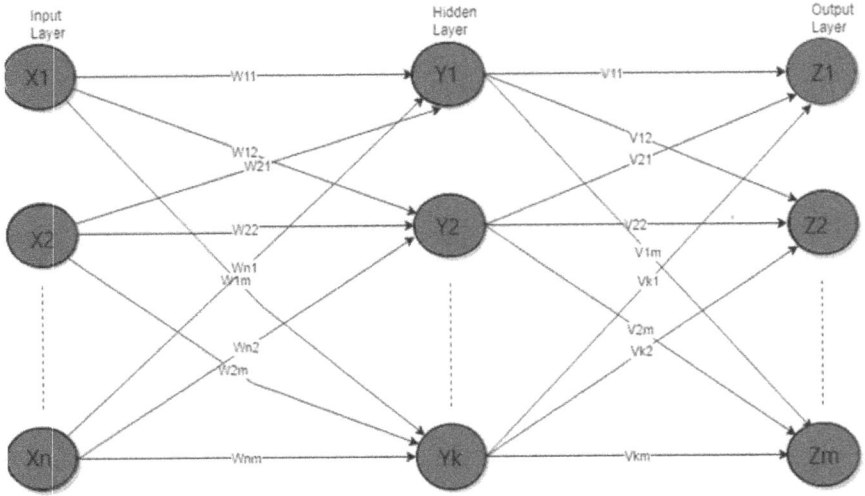

FIGURE 17.8   Multilayer feed-forward network.

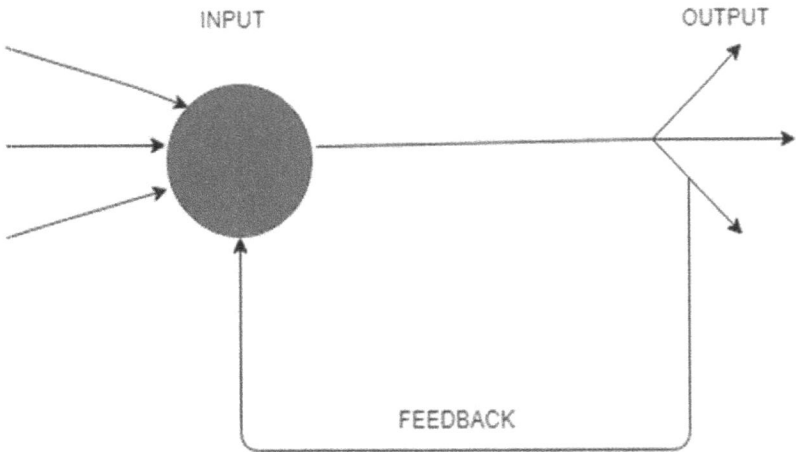

FIGURE 17.9   Single node with own feedback (single loop).

17.6.4 SINGLE-LAYER RECURRENT NETWORK

A recurrent neural network (RNN) is just a branch of artificial intelligence that reconstructs connections to a directed graph across a series. It processes the input sequences using internal state memory, allowing it to demonstrate dynamic temporal behaviors in time sequences. RNNs, unlike feedforward neural networks, cannot be processed by feeding input in only one way. See Figure 17.10.

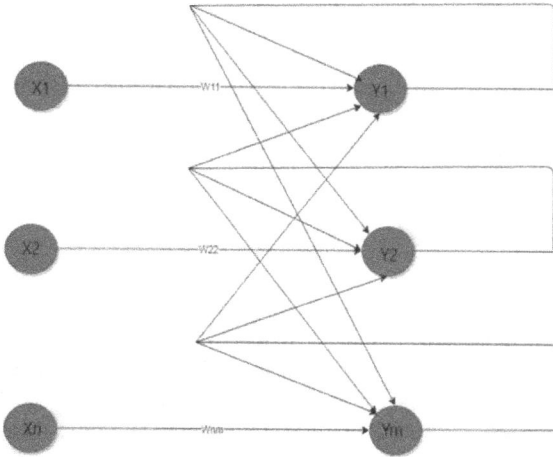

FIGURE 17.10   Single-layer recurrent network.

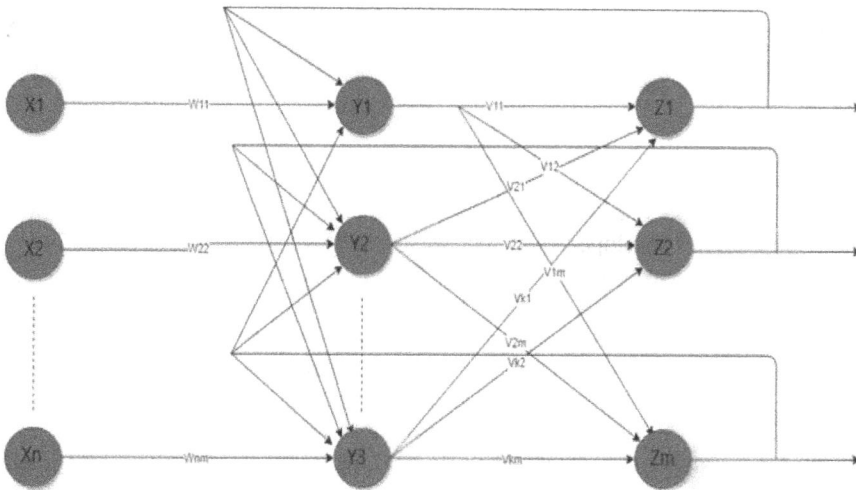

FIGURE 17.11   Multilayer recurrent network.

17.6.5   MULTILAYER RECURRENT NETWORK

The production of a processing element in an RNN can be delivered to an identical processing element in the layer below it as well as the processing element from the previous layer. All of the input data is processed, so no input at a given moment is needed for computation to take place. The main feature of an RNN can be found in its hidden state—the information that it captures about sequences.[7] See Figure 17.11.

17.7 DEEP LEARNING FRAMEWORKS

When it comes to artificial intelligence (AI), any machine learning algorithm is only as good as the framework that builds it. These frameworks are either expansions of existing frameworks or dedicated frameworks built for deep learning algorithm development. Each framework will have its own set of advantages and disadvantages, so let's take a closer look at some of the most powerful and popular deep learning frameworks available today.

17.7.1 TENSORFLOW

TensorFlow is the most prominent deep learning framework in use today, with firms such as NVIDIA and Uber using it in addition to Google. It is also utilized on a regular basis by AI practitioners and data scientists. TensorFlow is a Python framework that may be used to develop deep learning models. However, it requires a lot of code to design the network structure. TensorFlow relies on resources such as a powerful graphics processing unit (GPU) for efficient computations, which means it is an expensive and time-consuming task. The fundamental disadvantage of TensorFlow is that it operates on a static computation graph and must be performed each time any modifications are made. However, the platform itself has been accepted because of its extensibility and capability, implying that the tradeoff may still be worthwhile.

17.7.2 PYTORCH

One of TensorFlow's biggest competitors in the deep learning framework market is PyTorch. Facebook developed this open-source tool that's built with Python. Besides powering most of Facebook's services, other companies, like Johnson & Johnson and Twitter, are also making use of PyTorch. PyTorch is a Python-based library that offers support for debugging tools, as well as an emphasis on machine learning. In contrast to TensorFlow's static graph, PyTorch has a dynamic computation graph and facilitates, which is an easy way to visualize the language that is being used. This makes it easier for developers to see how their code will impact a project without needing to set things up beforehand. PyTorch simplifies neural network training by utilizing modern technologies such as data parallelism and distributed learning. The PyTorch community is also very active, with regularly published pre-trained models. TensorFlow, on the other hand, exceeds PyTorch in terms of cross-platform compatibility, thanks to Google's vertical integration with Android, which makes additional resources available to TensorFlow users.

17.7.3 KERAS

Keras is a deep learning framework that was created on top of well-known frameworks, notably TensorFlow as well as the Microsoft Cognitive Toolkit. Keras isn't quite as programmable as PyTorch or TensorFlow. However, it is the greatest place to start learning neural networks for beginners. Keras simplifies the creation of huge and complicated models with a few commands. While this reduces configurability, it also makes it more accessible as an application programming interface (API), making it more useful in any context.[8]

17.8 CONCLUSION

ANNs have already become an important component of the technology field. They can be used to recognize handwritten text, which can be useful in businesses like banking, as well as many other vital fields such as medicine.[9] ANNs consist of artificial neurons, which are computational models that try to imitate the human brain. They can be used to do complex analyses in a wide range of fields, from medicine to engineering, as well as to design the future generation of computers. Neural networks are useful in the field of medicine because they may be used to design models of the human anatomy that can aid doctors to diagnose ailments more accurately. Because of advances in ANN technology, intricate medical scans may now be evaluated more accurately and efficiently. Many complex problems will be handled by neural network–based devices themselves. They will learn and improve from their mistakes. Perhaps in the future, we will be able to connect humans and machines. This would translate into humans controlling or operating machines and robots. We might be able to engage with our surroundings through our thoughts.

REFERENCES

[1] Kukreja, Harsh, N. Bharath, C. S. Siddesh, and S. Kuldeep. 2016. "An introduction to the artificial neural network." *International Journal of Advance Research and Innovative Ideas in Education* 1: 27–30.

[2] "Architecture of Artificial Neural Network." 2022. *Dot Net Tutorial s*, June 30. https://dotnettutorials.net/lesson/architecture-of-artificial-neural-network/.

[3] "Artificial Neural Networks - Javatpoint. " 2021. *www.jav atpoint.com*. https://www.javatpoint.com/keras-artificial-neural-networks.

[4] Anderson, Dave, and George McNeill. 1992. "Artificial Neural Networks Technology." *Kaman Sciences Corporation* 258 (6): 1–83.

[5] Hopfield, J.J. 1988. "Artificial Neural Networks." *IEEE Circuits and Devices Magazine* 4 (5): 3–10. doi:10.1109/101.8118.

[6] Hemachandran, K., S. Khanra, R. V. Rodriguez, and J. Jaramillo (Eds.), 2022. *Machine Learning for Business Analytics: Real-Time Data Analysis for Decision-Making*. CRC Press, New York.

[7] "Introduction to ANN I Set 4 (Network Architectures) - GeeksforGeeks." 2018. *Geeksfor Geek s*, July 17. https://www.geeksforgeeks.org/introduction-to-ann-set-4-network-architectures/.

[8] Kaushik, Vanshika. 2021. "8 Applications of Neural Networks I Analytics Steps." *An alyticsSteps*. https://www.analyticssteps.com/blogs/8-applications-neural-networks.

[9] Baxt, W.G. 1995. "Application of Artificial Neural Networks to Clinical Medicine." *The Lancet* 346 (8983): 1135–1138. doi:10.1016/s0140-6736(95)91804-3.

18 AI and Real-Time Business Intelligence

J. Bhuvana, M. Balamurugan, Mir Aadil, and Hemachandran K

18.1 INTRODUCTION

The use of artificial intelligence (AI) in the workplace is fundamentally altering how firms operate. There are business applications that use AI to improve customer service and increase sales. Enterprise leaders want to employ AI to enhance their companies' operations and guarantee a return on their investment, but they confront significant obstacles. Expert systems are produced using knowledge engineering technologies. Expert systems combine a rules engine and a sizable, extensible knowledge bank. They are used to support decision-making in a range of areas, including manufacturing, customer service, healthcare, and financial services.

This chapter aims to analyze the origin, evolution, and development of business intelligence (BI) and its relationship with AI. The aim is to define the incidence of BI in business activities and analyze scientific activity and advances of BI to define new research horizons in this field.

This chapter explores the history, development, and link between AI and BI and to establish new research views in this area; thus, it is important to characterize the scope of BI in corporate activities and scientific activity and improvements.

18.2 BACKGROUND

18.2.1 THE BEGINNING OF BUSINESS

In the current industrial age, there is a fusion of AI, robotics, analytics, and other advanced technologies. As AI becomes a massive market, expect business dynamics to evolve. Without a solid AI business model, no one would be able to extract the full value of an AI-based idea or technology.

Prediction has always been an element of human life since the dawn of commerce. Today, organizations must use prediction if they want to keep up with the rapid developments in technology. The Kodak firm bluntly rejected that a digital camera was ever a possibility for the future since it was unable to grasp the good potential of such a device. A BI system is required to complete this task quickly and effectively (Jourdan, Rainer et al. 2008).

DOI: 10.1201/9781003328414-18

18.2.2 ARTIFICIAL INTELLIGENCE AND BUSINESS INTELLIGENCE

Making machines learn from their environment and adjust to it is known as machine learning. There are many different machine learning techniques available, including linear regression, underfitting and overfitting in polynomial estimations, extensions, and more. The creation of AI occurs regardless of the machine learning technique. Neural networks are the most popular machine learning technique supported by multiple complex learning methods and functions (Hastie, Tibshirani et al. 2009). These functions are made up of fixed biases and factors that depend on earlier inputs. This bias, which is a component of the algorithm between the hidden layers in the machine learning process, directs AI along the path that is most likely to be successful and produce the desired results. When thinking about AI ethics, it is essential to evaluate the data that is provided to the computer.

Future projections and decision-making are made easier with the use of business intelligence. A technology called a business intelligence system (BIS) helps managers understand BI. It could be a technique to boost sales, reduce expenses, and improve competition. In the context of an enterprise, tools like BI and AI are becoming more and more important but sometimes misunderstood. These days, add-on features that offer a useful platform are popular with AI and BI. Both are different in terms of some characteristics and abilities (Wixom and Watson 2010).

The study and pattern of human thought serve as the foundation for AI, and the algorithm mimics the workings of the human brain. BI, on the other hand, is actually technology that is utilized to improve business solution decisions. We can highlight a few points as a positive perspective for AI when comparing it to BI of the past. For example, companies are looking for AI solutions to collect, process, analyze, and gain insights from vast amounts of data, also known as big data (BD), in order to help them sell their data as a product. AI software is designed to work long hours and execute jobs more quickly and effectively.

18.2.3 IMPACT OF AI IN BUSINESS

AI is used by businesses for a number of tasks, including data collection and process simplification. A human brain would take much longer to process massive volumes of data than AI is capable of doing. Then, AI software can provide synthesized courses of action to the human user. This will enable us to leverage AI to speed up decision-making by playing out potential outcomes of each action.

18.2.3.1 AI Business Model

Business leaders may create more value for their companies and clients with the assistance of AI. It is essential to comprehend AI's capacity for prediction. Making accurate predictions and practicalizing corporate strategy are both made possible by AI business models. Business people are better able to identify product categories and market segments once their business strategy is clear. To avoid inconsistencies, the key is to be explicit about the business strategy and to ask questions. Use AI to assess whether option A or option B is the superior choice by focusing on certain ideas.

Many businesses seek to use AI technology to boost productivity, reduce operating expenses, boost customer happiness, and enhance revenue. Businesses can gain greatly from using AI. However, enormous advantages also present great obstacles. Renew products and business operations by using intelligent AI technologies like machine learning and natural language processing. One of the best ways to enhance the current business model is through this. AI has a big impact on how businesses operate. By implementing and integrating AI technology, we can automate, improve, and conserve time- and money-consuming traditional procedures. Operational effectiveness and productivity standards will also increase with the support of AI. Making quicker and more logical business decisions is possible with the proper AI business model. AI also aids in preventing human errors (Wixom and Watson 2010; Liang and Liu 2018).

Changes in business operations made possible by AI technology are known as business model innovations. These might consist of novel approaches to client engagement, data management, and analysis, as well as novel approaches to marketing and sales of goods.

18.2.3.2 Innovations in AI Business Model

These cutting-edge technologies not only reduced the quality of the Internet and the software sector but also those of other industries, including construction, the judicial system, the healthcare system, the auto industry, and agriculture (Casadesus-Masanell and Ricart 2011). The majority of top-tier businesses, including Microsoft, Facebook, Apple, Google, Amazon, Myntra, Flipkart, and IBM, are funding applied and AI research and development for their own and their clients' benefit.

A new paradigm for business models is represented by the convergence of blockchain, AI, and Internet of Things (IoT). (Casadesus-Masanell and Ricart 2011). Actually, autonomous automobiles were created with the aid of AI and IoT. Undoubtedly, this model will represent the next development in business, decision theory, and optimization theory, as well as the software industry. Decision-making software is powered by AI models, and real-time data collecting sensors for autos are made possible by IoT sensors. These facts enable the deep learning business model and AI-based algorithms to act quickly and make informed decisions. According to León et al. (2016), the ecosystem phase is described in Figure 18.1.

The most in-demand AI expertise will be deep learning models, Jobs in AI that require deep learning expertise are expanding more quickly. According to León et al. (2016), deep learning is a subset of machine learning that creates artificial neural network algorithms by simulating the structure and operations of the human brain and cognition.

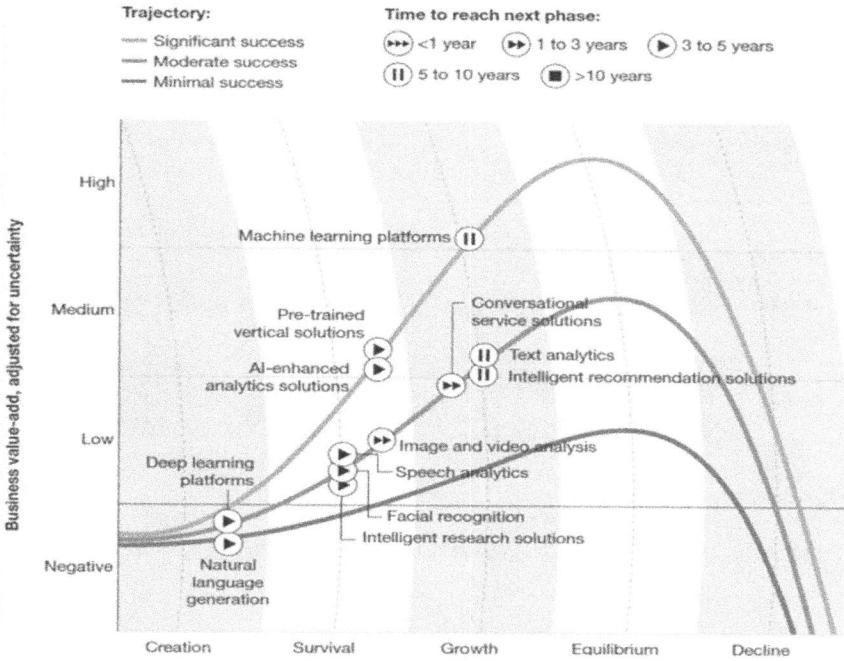

Trajectory:
--- Significant success
--- Moderate success
--- Minimal success

Time to reach next phase:
(►►►) <1 year (►►) 1 to 3 years (►) 3 to 5 years
(II) 5 to 10 years (■) >10 years

FIGURE 18.1   Business ecosystem León et al. (2016).

Voice recognition is accepted by about half of people in Europe, China, the United States, and the majority of people in India, and there are clear signals that these platforms will soon be used in offices. To increase business and workplace productivity, companies including Brooks Brothers, WeWork, Mitsui USA, Capital One, and Vonage have already included Alexa into their business models (Rong et al. 2013).

AI for business is distributing and making inferences from the vast amount of data in real time through instant analytics. As a result, the method ensures that the organization has a strong competitive position by enabling enterprises to make crucial decisions and respond more quickly (Hedman and Kalling 2003). Real-time data analytics will be significantly impacted by the AI and ML business model. In the transportation sector, for instance, drivers may receive information on traffic congestion depending on their location and promptly adjust their routes (Wirtz 2011).

With a better understanding of and satisfaction for customers, AI-based chatbots can respond more quickly and communicate more effectively, which will help businesses win over new clients and keep their existing ones (Johnson et al. 2008).

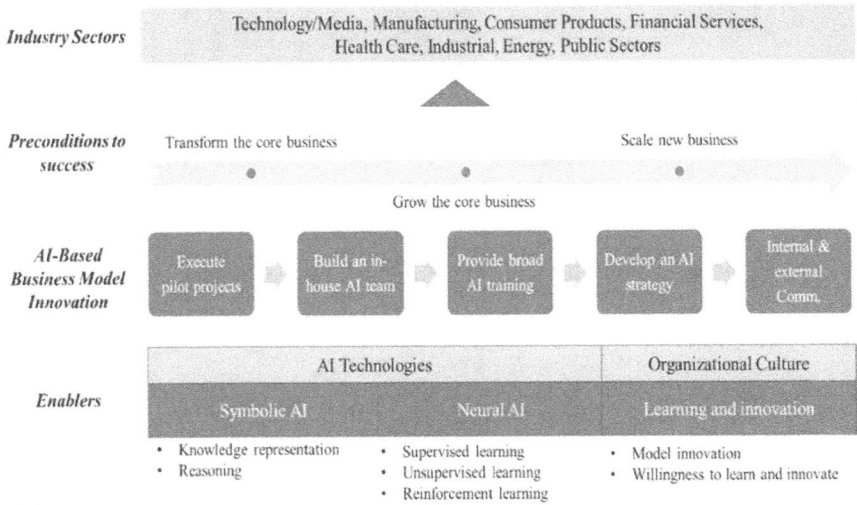

FIGURE 18.2   AI-based business model (Mishra and Tripathi 2021).

The general AI-based business model is shown in Figure 18.2.

Businesses now have the chance to gather more digital data, get crucial perceptions, and change their operations and ideas. The markets will therefore probably progress as a result, which is something that is really essential. This is referred to as an AI cycle. Businesses place a strong emphasis on utilizing AI to improve customer experiences; investigate and analyze data; and anticipate performance to computerize work volume, transactions, trading, and more. The propensity for AI adoption and soliciting falls short of expectations and does little to help intelligence skills improve more quickly.

18.2.4   RESPONSIBLE AI

A governance structure called responsible AI (RAI) outlines how a company is tackling the issues related to AI. It is up to the software engineers and data scientists to create fair, reliable AI standards. Different companies have different requirements for the steps needed to stop prejudice and guarantee transparency. Google and Microsoft have both explicitly urged AI legislation. The word "responsible" is a catch-all phrase that refers to both ethics and democratization. Frequently, the data used to train machine learning models can introduce bias into AI. It stands to reason that when the training data is skewed, the decisions made by the programming would similarly be skewed. It is more and clearer that standards in AI are required as software systems with AI elements proliferate.

18.2.4.1 Principles of RAI

The machine learning models that underpin AI should be thorough, understandable, moral, and effective.

- Completion—thorough to prevent machine learning from being readily exploited, AI has established testing and governance standards.
- Explainable AI is designed to explain its goals, justifications, and decision-making procedures in a way that the average end user can comprehend.
- To identify and remove bias in machine learning models, ethical AI initiatives have procedures in place.
- Continuously operating and able to act promptly to changes in the operational environment are characteristics of effective AI.

One of the main goals of RAI is to avoid situations where a small change in the value of an input can drastically impact the results of a machine learning model. According to corporate governance principles, responsible AI should be a form that cannot be altered by humans, or other programming should be used to meticulously document the model-development procedure. In addition, bias should not be introduced into machine learning models when training on data. See Figure 18.3.

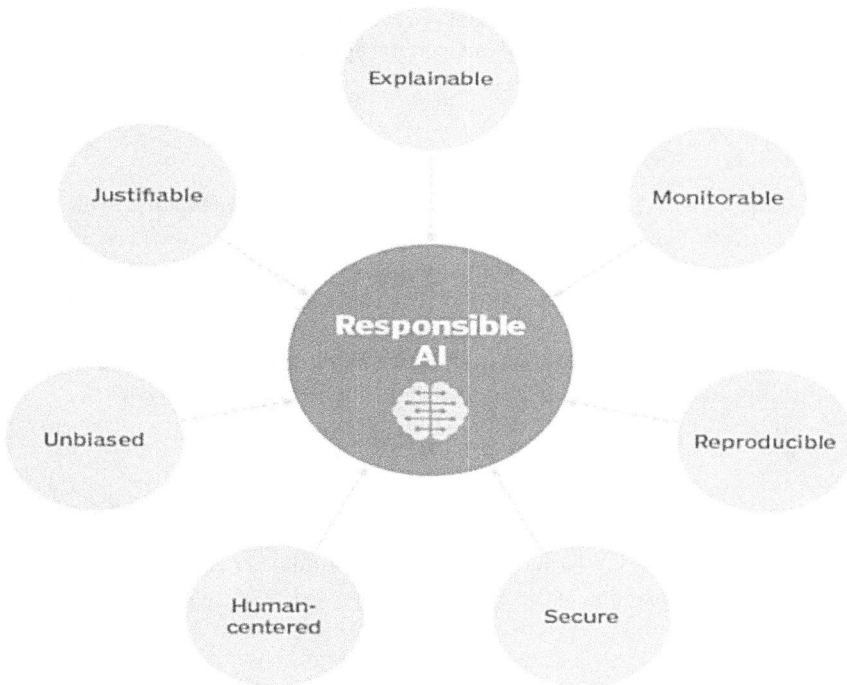

FIGURE 18.3   Principles of RAI (techtarget.com).

18.2.4.2 Traditional vs. Modern BI

Everywhere in the world, business is undergoing constant and quick change as a result of a thriving global economy, changing consumer needs, and the pervasive use of cutting-edge technologies. BI must adapt in order to continue offering the outstanding value that organizations rely on. It must be created to support the type of change that organizations must today manage. The goal of modern BI for everybody is to make it simple for users to share and access real-time data that is pertinent to their work. The differences betweeen traditional BI (TBI) and modern BI (MBI) are listed in Figure 18.4.

When it comes to connecting with various data sources, TBI tools, generally referred to as legacy or on-premise solutions, historically fell short. TBI technologies formerly offered insights that were either preset or manually adjusted by data specialists. All data sources, whether on-premises or in the cloud, are consolidated by MBI solutions into a single analytical cache.

18.3 BENEFITS OF REAL-TIME BUSINESS INTELLIGENCE

For businesses with a lot of assets and a focus on the field, teal-time business information tools are expected to revolutionize industries. To quickly adjust to ever-changing business conditions, they can unearth major events and identify trends with the help of data analytics. The crucial component of real-time BI is providing the appropriate information to the appropriate individuals in the appropriate format and at the appropriate moment.

18.3.1 OPTIMIZING ASSET PERFORMANCE

Utilities must deploy BI systems that are data analytics-driven and provide insight through in-depth visualizations and statistics. Any lag time in addressing maintenance problems and equipment failures might result in inefficient operations. It is possible to avoid unscheduled service outages by identifying potential issues with utility distribution equipment as soon as possible. Monitoring the performance and health of assets will help to avoid equipment failure or decay points. The capacity, dependability, and availability of the assets are all enhanced by BI systems.

18.3.2 ENHANCING CUSTOMER EXPEREINCE

To give customers useful information, utilities might create intelligence systems based on consumption patterns. The best strategies to optimize a customer's consumption can be found by segmenting at the micro level to provide a 360-degree perspective of their usage habits. By offering cost-effective plans that deliver a personalized experience, utilities may increase customer loyalty, increase customer lifetime value, and lower customer turnover. Customers can now use intelligence tools for self-service to view and control their utility usage, consumption patterns, historical usage, billing, payments, and unusual usage conditions.

TRADITIONAL BI	MODERN BI FOR ALL
Forces IT and analysts to restrict business users to specific datasets.	Gives well-governed access to everyone in the organization.
Creates IT bottlenecks due to lack of resources and slows response times by bogging down analysts with routine tasks.	Empowers end-users to make small changes on their own and quickly get what they need to drive the business forward.
Leaves IT pros and data analysts little time to explore new technologies and expand their skillsets.	Frees data experts to tackle bigger challenges and advance the technical acumen of the organization.
Makes business users reluctant to ask questions or submit ad-hoc requests because they end up going nowhere.	Offers the flexibility to add or change items quickly and easily.
Creates a tendency for executives to look at data infrequently, giving them an inexact understanding of their business.	Heightens executives' knowledge of the business by making real-time data easily accessible, anytime and anywhere.
Requires teams to spend hours or even days to compile, format, and deliver useful information to the C-suite.	Allows executives to quickly get what they need on their own, freeing their team to focus on higher value work.
Provides a read-only experience that lets people look at data to make a decision but does little to enable real-time action.	Enables an integrated approach, with insights from reports and analyses automatically feeding back to operational systems to drive near real-time action.
Struggles to grant people in the field with the right access or experience to easily see insights.	Makes it easy for your people to consume data and insights from wherever they are, via any device.
Makes sharing reports with key partners challenging.	Gives key partners access to data that could bolster their business as well, thereby strengthening relationships within the ecosystem, too.

FIGURE 18.4 Differences between TBI and MBI (domo.com).

18.3.3 IMPROVING OPERATIONAL EFFICIENCY

Utility operators can monitor everything from meter reading to uptime for every single piece of data stored in BI dashboards. This aids in gaining great visibility of past asset performance comparisons, access to real-time data, and quick response to assist in making better decisions. In-depth performance reports are also generated by dashboards to assist utility operators in determining the best course of action based on particular key performance metrics.

18.3.4 DYNAMIC PLANNING AND LOAD CONTROL

Utilities have to leverage real-time or historical data of consumption and usage constraints to perform complex analysis. This would help them to identify key performance indicators and patterns for dependable planning of asset failures and performance. Workflow improvements gained by integrating BI solutions with existing automation processes can help avoid asset failures, thus resulting in asset longevity and uptime.

18.3.5 PREVENTION OF UTILITY LOSS AND FRAUD MANAGEMENT

As massive amounts of data streams become available, utilities can utilize real-time and predictive analytics to get insights. Operators may be able to track the status indicators of utility distribution systems, quickly and accurately determine the integrity of the system, and uncover hidden information with this assistance. It can be used to find faults and outages on supply-distribution networks.

18.4 BI IN VARIOUS BUSINESS SECTORS

Terabytes of information are arriving from spreadsheets, payments, diaries, and every other source, and such a large volume of data could seem like a gold mine at first. Extraction of value from the data might be difficult due to privacy issues and a range of data types. Up to 73 percent of a company's data is not used for analytics, according to Inc.com. Real-time BI makes it possible to directly access the operational system and get data in real time. The modern organization can handle and analyze a massive amount of data with a fresh approach to data analytics.

18.4.1 FINANCE SECTOR

By assisting in putting data to use, BI tools help reduce risks. One can readily spot fraudulent behavior by watching client behavior. It is possible to verify that they are adhering to the standards of the profession by watching their behavior. The data can be used to assess credit portfolios and spot possible delinquent cases. BI tools are, in general, a proactive risk management tactic for the financial sector.

18.4.2 RETAIL

BI tools are one of the most important assets, since maintaining current customers is a successful and long-term business plan. If businesspeople have the most recent information on their customers, especially the most lucrative ones, they will be able to market the items and services that are most pertinent to their needs and preferences, including data gathering and analysis to identify which products should be enhanced and which should be phased out.

18.4.3 TRADING

The banking and finance sectors have quickly embraced personalization, which is a hot topic in every market. Therefore, having an edge over the competition is essential. Based on the data availability, they can quickly drive the customer experience using BI technology. Market trends may be used by businesses to plan out new investment opportunities, consumer behavior can be predicted using analytics, and products can be made for the unique needs of each customer.

18.4.4 E-COMMERCE

Most businesses store data across many different systems and formats. Data processing and reporting become challenging and time-consuming as a result. Using a BI solution helps reduce the difficulties associated with data kept in many tools and spreadsheets. BI systems use instantaneous information to provide a more comprehensive view of workplace activity at any given time. The numbers don't lie. With a fully integrated BI solution, it helps to achieve whole corporate success.

18.4.5 MARKETTING

The data from customer relationship management (CRM) can be used to calculate the profitability of marketing initiatives. The effectiveness of a campaign as a whole, the cost of advertising, and email performance can all be utilized to determine where messages are reaching customers and where they need to be improved.

18.5 BI NEED AND USAGE IN THE INDIAN CORPORATE SECTOR

After the entry of foreign banks, the banking industry has become extremely competitive. The banking, finance, and telecom sectors are the early adopters of BI solutions in India. BI tools are helpful in knowing the background profile of the customer, which in turn enables the bank to offer services suited to him.

The BI process includes defining the information needs in an organization [Turban et al. (2010), Technology Advice (2020)], acquiring information, processing it, analyzing it, disseminating it, and using it to make decisions and provide feedback. The BI tools are shown in Table 18.1.

TABLE 18.1

Top 10 Tools Ised in Business Intelligence (Selecthub.com)

S. No.	Name of Tool	Description	Pros	Cons
1	Microsoft Power BI	• Initially created as an additional tool to support Excel, Power BI has since evolved into a self-contained solution. It helps with reporting, data mining, and data visualization to business understanding. Using Power BI, businesses can connect to a multiple data sources and create custom dashboards and reports.	• Seamless Integration with Microsoft • Data Connectivity • Filters and drill Down Functionality	• Learning Curve • Pricing and Licensing • Speed
2	Oracle Analytics Cloud	• A robust analytical and reporting software powered by AI, it provides competence to business of various sizes. Through the convenience of the cloud, it offers a variety of reporting and analysis features. It can be translated into an easy-to-understand format that people can explore and share; data preparation and trend analysis are performed as well.	• User Friendly • Reporting • Augmented Analytics • Interactivity	• Price • Bugs • Support
3	MicroStrategy	• Data analytics platform provides enterprises of all sizes with actionable intelligence. It enables users to create unique real-time dashboards and customize data representations. Users have the option of hybrid, on-premise, or cloud implementation. It distinguishes itself as a pioneer in the enterprise analytics space thanks to its simplicity of use and scalability.	• Data Visualizations and dashboards • Ease of'zuse • Reporting • Mobile Access	• Speed • Cost • Implementation • Resourses and Learning
4	TIBCO Spotfire	• A full business intelligence and data discovery platform, uses artificial intelligence to perform a variety of tasks, including in-depth analysis and powerful visual reporting. It provides data streaming technology that can support big data integration, AI insights, Internet of Things (IoT) integration, and more.	• Easy to use • Interactive Visualizations • Informative Reports • Customizable Dashboards	• Slow Performance • Small Datasets • User Interface • Documenattion

5	Qlik Sense	A self-service data analytics tool that combines artificial intelligence with human intuition to improve data-driven business decisions. Organizations are able to easily study their data and use data insights to generate intuitive representations that are compelling. It improves analytical options to cover the complete insights life cycle and aids businesses in modernizing their approach to intelligence as the next-generation development of QlikView, which was published in 2014. It comes in two editions: Business and Enterprise, both of which require an annual subscription per account. A hosted SaaS public cloud, multicloud, on-premises, or private cloud deployment are all options for businesses.	• Accessibility • Ease of Use • Data Analytics • Speed • Data Visualization • Data management • Training • Functionality	• Securable • Integration • Economy • Alloaction
6	SAS Visual Analytics	Quick solutions to complicated problems derived from datasets of all sizes. It provides users of all technical skill levels supervised analysis, reciprocal control panel, intelligent and agile visualizations, and independent strong analytics, fostering data knowledge and view. By utilizing its adaptable, scalable design, which is focused on data transparency, users may make better business decisions. It encourages intelligent activity that is motivated by insight and is built on a solid in-memory architecture.	• Dta Analysis • Functionality • Data Visualization • Ease of Use	• Cost
7	Domo	Users can integrate structured and unstructured data from all sources, on-premise and cloud-based. Domo's intuitive user interface makes it easy for decision-makers to look at real-time insights. The company's built-in collaboration tool that mimics social media earns bonus points with some users.	• KPI Monitoring • Data Visualization • Data Connectivity • Collaboration • Accessibility	• Cost • Functionality • Customer Support • Speed
8	Dundas BI	Designed with both single-tenant deployments and multitenant deployments, users can access shared, reusable content by the provider. In order to support embedding, an analytics tool for web browsers was developed. Total control over the data stored on an organization's IT system is made possible. Users have the option to dive down into the data as needed and to customizable visualizations.	• Data inquiry and Visualization • Functionality • Ease of Use • Service and Support • Cost	• Trainging • Speed

(Continued)

TABLE 18.1 (Continued)

Top 10 Tools Ised in Business Intelligence (Selecthub.com)

S. No.	Name of Tool	Description	Pros	Cons
9	Sisense	• End-to-end data analytics platform with an embeddable, scalable architecture for the whole customer and provider base of the business; it enables corporate analysts to merge enormous datasets from various sources into a single, unified database. • On the front end, anyone with any level of technical expertise may create visualizations, reports, and dashboards to explore and share insights that advance enterprises. Teams are given the ability to examine important metrics and data insights right where they work thanks to Fusion, its AI-driven, cloud-native analytics service. All sizes of enterprises can use it, and it is available as an on-premises SaaS, dedicated SaaS, and hybrid approach.	• Consumer assitance and guidance • Data synthesis • Data representation • User friendly • Cost	• Data Preparation, Modeling • Training
10	Tableau	• Analytics and data visualization tools that help businesses make decisions based on data. It collects data from several sources to generate timely, useful insights. It allows for simple data analysis methods, including drag-and-drop filtering and natural language searches, regardless of user skill.	• Advisable • Data Accordance • Data Representation • Options for Sharing • Customer Service and Support	• Speed • Price • Need for Training

The various entities under investigation utilize these tools to gather data and base judgments on it. The major objective of these [Selecthub (2020)] tools is to make strategic decisions and present the analytics that could be a deciding element. They have various functionalities and user interfaces.

18.6 CHALLENGES IN BUSINESS INTELLIGENCE

The size of the Indian market is less than 1 percent of the Asia Pacific market. Despite having a high awareness level, Indian businesses have a low level of BI deployment. Many businesses use outdated systems that lack the data needed to apply BI. The accuracy of the data available with the companies is crucial for the successful implementation of BI. The incomplete, unauthenticated, and inaccurate data makes it difficult to apply BI. When it comes to the system's integration, Indian businesses must undergo a major transformation. The BI system should be integrated into the employee's software stack because it affects a much wider variety of workers. Indian businesses are unaware that implementing such applications can negate the cost reductions and higher income the applications provide. Systems for BI will need to handle both analytical and transaction-like questions simultaneously.

18.7 CONCLUSION

BI is used by corporate people for many different reasons, so it helps to get perfect and authentic suggestions in various fields like manufacturing, commercialization, conformity, and employment. Making better business decisions is just one of the many advantages that companies may experience after incorporating BI into their operational frameworks. Additional benefits include greater data quality; rapid, factual reporting and data analysis; better employee satisfaction; and improved economics.

REFERENCES

Casadesus-Masanell, R., & Ricart, J. E. (2011). How to design a winning business model. *Harvard Business Review, 89*(1/2), 100–107.

Hastie, T., Tibshirani, R., Friedman, J. H., & Friedman, J. H. (2009). *The elements of statistical learning: Data mining, inference, and prediction* (Vol. 2, pp. 1–758). New York: Springer.

Hedman, J., & Kalling, T. (2003). The business model concept: Theoretical underpinnings and empirical illustrations. *European Journal of Information Systems, 12*(1), 49–59.

Johnson, M. W., Christensen, C. M., & Kagermann, H. (2008). Reinventing your business model. *Harvard Business Review, 86*(12), 57–68.

Jourdan, Z., Rainer, R. K., & Marshall, T. E. (2008). Business intelligence: An analysis of the literature. *Information Systems Management, 25*(2), 121–131.

León, M. C., Nieto-Hipólito, J. I., Garibaldi-Beltrán, J., Amaya-Parra, G., Luque-Morales, P., Magaña-Espinoza, P., & Aguilar-Velazco, J. (2016). Designing a model of a digital ecosystem for healthcare and wellness using the business model canvas. *Journal of Medical Systems, 40*(6), 1–9.

Liang, T. P., & Liu, Y. H. (2018). Research landscape of business intelligence and big data analytics: A bibliometrics study. *Expert Systems with Applications, 111*, 2–10.

Mishra, S., & Tripathi, A. R. (2021). AI business model: An integrative business approach. *Journal of Innovation and Entrepreneurship, 10*(1), 1–21.

Rong, K., Lin, Y., Shi, Y., & Yu, J. (2013). Linking business ecosystem lifecycle with platform strategy: A triple view of technology, application and organization. *International Journal of Technology Management, 62*(1), 75–94.

Selecthub. (2020). Business intelligence software tools comparison. Acces sed online at https://www.selecthub.com/business-intelligence-tools/, March 2020.

Technology Advice. (2020). Guide to business intelligence software. Acces sed online at https://technologyadvice.com/business-intelligence/, March 2020.

Turban, E., Sharda, R., & Delen, D. (2010). *Decision support and business intelligence systems* (9th ed.). Upper Saddle River, NJ: Prentice Hall Press.

Wirtz, B. W. (2011). Business model management. *Design–Instrumente–Erfolgsfaktoren von Geschäftsmodellen, 2*(1).

Wixom, B., & Watson, H. (2010). The BI-based organization. *International Journal of Business Intelligence Research (IJBIR), 1*(1), 13–28.

19 Introduction to Statistics and Probability

Abha Singh and Reham Alahmadi

19.1 INTRODUCTION

In artificial intelligence (AI), a computer may mimic human behavior, thanks to advancements in technology and science. System learning is a type of AI that lets machines learn on their own from previous data without the need for explicit programming. The objective of AI is to create a computer system that is as intelligent as humans in order to tackle complicated issues. I witnessed multiple instances of misinterpretation of AI in various debates. AI includes machine learning as one of its parts.

The relationship between machine learning and statistics is so close that the lines between the two are blurred at times, and explicit references to statistical study are made when appropriate. Several years before the buzz began, an integrated curriculum between the schools of statistics and computer science taught AI, data science, and machine learning with the goal of training data scientists. We may say that the perspectives of machine learning and statistics are substantially overlapping in terms of content.

Reasoning based on probabilities and statistics has had an impact on a lot of different parts of the theory behind AI. Do the prior interpretations of probability do justice to the fresh applications that have emerged over the course of the last several decades? Gambling and other games of chance were the inspiration for the development of probability theory. What fresh insights into the philosophy of probability can we get from the contemporary use of probability in the field of AI?

19.2 STATISTICS

Statistics is a branch of math that looks at how data is collected, organized, analyzed, interpreted, and shown. It is sometimes known as "numerical analysis." When using statistics to solve a problem in a scientific method, it's common to start with a statistical population or model that will be looked at in more depth later. During the course of their professional and personal lives, engineers and scientists are continuously confronted with collections of information or data, which may be overwhelming. Techniques for organizing and summarizing data, as well as methods for deriving conclusions based on the information contained in the data, are provided by the field of statistics.

When it comes to comprehending the world around us, statistical ideas and approaches are not only valuable, but they are also often required. They give

DOI: 10.1201/9781003328414-19

avenues for acquiring fresh insights into the behavior of a wide range of phenomena that you will face in your chosen area of engineering or scientific concentration. In the face of uncertainty and variance, the field of statistics teaches us how to make educated judgements and informed choices based on available information. There would be no need for statistical procedures or statisticians if there were no uncertainty or variation.

Inferential statistics are used in political polling. Interviewing every Indian of voting age would be costly and impracticable, and this group is known a **population**. Statisticians can afford to survey just a few thousand people to evaluate the opinion of the whole Indian electorate. This group is known a **sample**. Statisticians use data from a random sample of voters to infer (reach a conclusion) about how the whole population wants to vote. Statistical inference helps make such inferences.

Population

The group of all people or things that are being looked at in a statistical study.

Sample

The subset of the population from whom information is gathered and analyzed.

Random sample

A random sample is one that is selected at random. It's more correct to define it as a sample random selection.

Statistical inference

Statistical inference refers to methods for drawing and assessing the trustworthiness of inferences about a population based on information acquired from a sample of the population.

Data

When we talk about data, we're talking about discrete bits of information like numbers or percentages. Data is, in a more technical sense, a collection of numerical or qualitative values pertaining to one or more individuals or things.

19.2.1 COLLECTION OF DATA

There are a number of ways for collecting or obtaining data for use in statistical analysis. Three of the most often used techniques are as follows:

1. Observational Method
This method is depends on the observation of things, people, marketing, and others also.

2. Experimental Method

A coin is flung into the air. The experiment may have two potential results, assuming that the coin does not fall on the edge of the table. The possibilities are heads or tails. When this experiment is performed, no one can predict what will happen as a result of the conclusion. If desired, you may toss the coin as many times as you like.

Many experiments use the experimental method like a blood pressure (BP) measuring experiment, health measuring experiment method, manufacturing product from a machine, etc.

3. Survey Method

A survey is a method used to collect information from people, such as before the polls and marketing surveys. One of the most essential survey metrics is the response rate (i.e., the proportion of those selected who complete the survey). Surveys may be administered in several ways, such as:

- personal interview;
- telephone interview; and
- self-administrated questionnaire

In the world of data, there are two types. The first is qualitative data, and the second is quantitative data. Quantitative data also has two types. One is continuous type data and the second is discrete type data.

Discrete data

Discrete data is a finite or countable number such as 0, 1, 2, 3, . . .
For example, a potato is a finite number.

Continuous data

Continuous range of numbers without gaps, stops, or interruptions or leaps resulting in infinitely many potential values.
For example, the measurement of petrol is 32.6789 liters.

Measurement level

There are four main ways to categorize data: nominal, ordinal, interval, or ratio. We can make better decisions based on the data we collect and which technique we utilize when applying statistics to real-world situations.

Nominal level: The nominal level is presented as a name, a category, and an order such as low or high, but it cannot be arranged in any order.
For example:
1. Person responds like yes/no/cannot say.
2. Is a particular place beautiful or not?

3. It is good for society when some organizations are working as charities.

Ordinal level: The ordinal level is presented as an order like player position, first number, second number, and third number, students' course grade (first, second, third other A, B, C grades), etc.

Interval level: When compared to the ordinal level, the interval level is similar, but it has the extra virtue that the difference of the two data values is significant. This level of information does not have a natural zero beginning point.

For example, year, body temperature, number of things.

Ratio level: The ratio level is defined as larger and smaller, more and less, maximum and minimum measure, such as building number, 0 flour number to 100 flour number, price of clothes is $5 minimum price and $150 maximum price, and 5 stories is better than 16 stories, and so on.

Examine your abilities

1. Explain about the collection of data. How many types of data?
2. Explain about the level of measurements. How many types of level of measurements?
3. Describe the population, sample, and random number.
4. Give four instances of data you may gather in your daily life.

19.2.2   DATA REPRESENTATION VIA TABLES AND GRAPHICAL METHODS

In this part, we looked at how data may be represented in a table. In this case, we take into consideration the final test statistical marks of 20 pupils:

55, 35, 97, 74, 65, 53, 32, 78, 75, 62, 55, 35, 97, 74, 65, 53, 32, 78, 75, 62

In light of the fact that this data is raw data, if we want to look at the maximum (highest scores) or minimum (lowest marks), we must search in the table. If we present the raw data provided to us in an ascending or descending sequence, we can simply determine the highest and lowest grades that students have received.

The data range refers to the difference between the highest and lowest data values.

So, the range = highest mark – lowest mark = 95 – 32 = 63

Example: Consider the following: the average measured BPs of 30 patients at one hospital are as follows:

10	20	36	92	95	40	50	56	60	70	92	88	80	70	72
70	36	40	36	40	92	40	50	50	56	60	70	60	60	88

To refresh your memory, the term "frequency" of BP of patient refers to the practice of measuring a patient's BP several times over the course of a certain period of time.

Take, for example, the BP of four patients is 70. So the frequency of 70 patients is 4. We have presented the information in the form of a table (see Table 19.1).

Example: 100 weekly observations from the city's cost-of-living index study were included in the following list:

96	67	28	32	65	65	69	33	98	96
76	42	32	38	42	40	40	69	95	92
75	83	76	83	85	62	37	65	63	42
89	65	73	81	49	52	64	76	83	92
93	68	52	79	81	83	59	82	75	82
86	90	44	62	31	36	38	42	39	83
87	56	58	23	35	76	83	85	30	68
69	83	86	43	45	39	83	75	66	83
92	75	89	66	91	27	88	89	93	42
53	69	90	55	66	49	52	83	34	36

TABLE 19.1

An Ungrouped Frequency Distribution Table or a Simple Frequency Distribution Table

Blood Pressure (mmHg)	10	10	20	36	40	50	56	60	70	72	80	88	92	95
Number of patients (the frequency)	1	1	1	3	4	3	2	4	4	1	1	2	3	1

TABLE 19.2

Cost of Living Index vs. Number of Weeks Sample 1

Cost of Living Index	Number of Weeks
20–29	3
30–39	14
40–49	12
50–59	8
60–69	18
70–79	10
80–89	23
90–99	12
Total	100

We have condensed the information so that it may be presented to the reader in smaller portions, such as 20–29, 30–39, . . ., 90–99, so that they can more simply understand it. The data we have ranges from 23 to 98; therefore class-size or "class width" refers to the size of these groupings, which are also known as "classes" or "class intervals," and in this example, the size is 10. These classes each have a "lower class limit" and a "higher class limit."

The data presented here is called a grouped frequency distribution table. The lower limit of the classes 90–99 is 90, and upper limit of the classes is 99. Class width is the difference between two consecutive lower class limits or upper class limits: 30 – 20 = 10 or 39 – 29 = 10.

Consider the two class intervals 20–29 and 30–39.

The lower limit of 30–39 = 30, the upper limit of 20–29 = 29; now the difference between the lower limit – upper limit = 30 – 29 = 1 and take half of this ½ = 0.5.

As a result, the new class interval formation from Table 19.1, is 20–29 is (20 – 0.5)—(29 + 0.5), i.e. 19.5–29.5. In this way we can make the continuous classes as follows:

19.5–29.5, 29.5–39.5, 39.5–49.5, 49.5–59.5, 59.5–69.5, 69.5–79.5, 79.5–89.5.

Tables 19.1–19.3 present a variety of graph presentations, each of which may be created by hand or with the assistance of a computer (through a Word document or an Excel sheet, respectively). Figure 19.1 shows the types of graphs can be made.

19.2.2.1 Measure of Central Tendency

We showed the data in a variety of formats, including bar graphs, histograms, and frequency polygons, as well as frequency distribution tables, in the previous sections. These are just some of the ways that the data may be displayed. At this point, we will present the data in a certain format that is known as a *measure of central tendency.*

TABLE 19.3

Cost of Living Index vs. Number of Weeks Sample 2

Cost of Living Index	Number of Weeks
19.5–29.5	3
29.5–39.5	14
39.5–49.5	12
49.5–59.5	8
59.5–69.5	18
69.5–79.5	10
79.5–89.5	23
89.5–99.5	12
Total	100

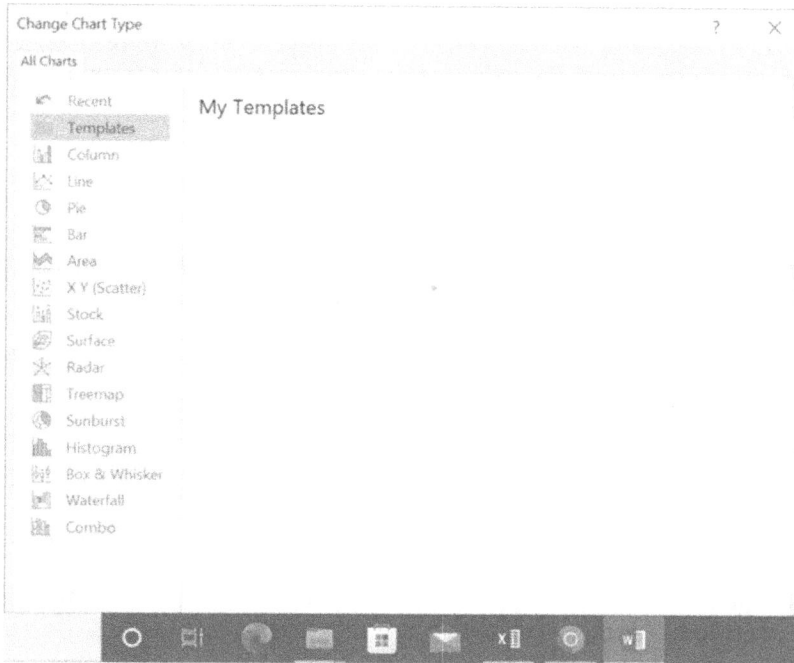

FIGURE 19.1   Types of graphs.

TABLE 19.4

Chocolate Likes Poll

Type of Chocolate	Galaxy	Twix	Dairy Milk	Kit-Kat	Cadbury	Oreos
Men	10	8	9	8	7	6
Women	5	4	7	10	10	10
Children	8	6	3	6	10	9

Example: Take a look at the chocolate made by a few different companies and take a poll to find out which chocolate is the most popular among men, women, and children. See Table 19.4.

Now we found their average response as follows:

Men average response = $(10 + 8 + 9 + 8 + 7 + 6) / 6 = 8$
Women average response = $(5 + 4 + 7 + 10 + 10 + 10) / 6 = 7.67$
Children average response = $(8 + 6 + 3 + 6 + 10 + 9) / 6 = 7$

Our aim is to discuss here the mean (or average), i.e. the mean is the sum of all observations divided by the total number. Mean is denoted by $\bar{x}$, read as "x bar." Therefore, we may refer to the average stated earlier in terms of the **mean**.

Example: Find the mean of the given data 3, 5, 7, 9, 5, 4, 3.

$$\text{Mean}(\bar{x}) = \frac{3+5+7+9+5+4+3}{7} = \frac{36}{7} = 5.143$$

Example: In Table 19.5, we will now determine the mean BP of hospital patients (i.e., mean of the ungrouped frequency distribution).

$$\text{Mean} = \frac{\Sigma fx}{\Sigma f} = \frac{10+20+108+160+150+112+240+280+72+80+176+276+95}{1+1+3+4+3+2+4+4+1+1+2+3+1}$$

$$= \frac{1779}{30} = 59.3$$

In general form of equation of mean $\bar{x} = \frac{\Sigma_{i=1}^{n} f_i x_i}{\Sigma_i^{n} f_i}$ where n is number of observation, i = 1... n.

Example: Now we consider the grouped frequency distribution table. We find the mean of expenses (average of expenses). See Table 19.6.

TABLE 19.5

Determine the Mean BP of Hospital Patients

Blood Pressure (mmHg) (x)	10	20	36	40	50	56	60	70	72	80	88	92	95
Number of patients (the frequency) (f)	1	1	3	4	3	2	4	4	1	1	2	3	1
f x	10	20	108	160	150	112	240	280	72	80	176	276	95

TABLE 19.6

Grouped Frequency Distribution Table

Cost of Living Index	Number of Weeks (f_i)	x_i	$f_i x_i$
20–29	3	(20 + 29) /2 = 24.5	103.5
30–39	14	(30 + 39) / 2 = 34.5	483
40–49	12	(40 + 49) / 2 = 44.5	534
50–59	8	(50 + 59) / 2 = 55.5	444
60–69	18	(60 + 69) / 2 = 66.5	1197
70–79	10	(70 + 79) / 2 = 77.5	775
80–89	23	(80 +8 9) / 2 = 88.5	2035.5
90–99	12	(90 + 99) / 2 = 99.5	1194
Total	$\Sigma_{i=1}^{8} f_i = 100$		$\Sigma_{i=1}^{i=8} f_i x_i = 6766$

$$\text{So, the mean } \overline{x} = \frac{\sum_{i=1}^{i=8} f_i x_i}{\sum_{i=1}^{8} f_i} = \frac{6766}{100} = 67.66$$

19.2.2.2   Median

The median is another often used measure of the center. In essence, the median of a dataset is the value that separates the lowest and highest 50% of the data.

DEFINITION: The median of a set of data, sorting the information in ascending order.

- If there is an odd number of data, the median is the middle observation.
- If there is an even number of observations, the median is the mean of the two middle observations in the ordered list. In both instances, if n represents the number of observations, then the median is located at (n + 1) / 2 in the sorted list.

Example: Consider two instances of shift overtime compensation data for a single business. Determine the median of the datasets.

I dataset salary: $500 500 800 940 300 300 400 300 700 550 900 750 2060

II dataset salary: $500 940 300 400 700 550 900 750 2060 400

Solution: To find the median of dataset I, first we arrange the data in increasing order: 300 300 300 400 500 500 **550** 700 750 800 900 940 2060. The total number of observations is 13, so (n + 1) / 2 = (13 + 1) / 2 = 7. Therefore, the seventh observation in the sorted list, which is 550, represents the median.

To find the median of dataset II, initially, we organize the data in ascending order: 300 400 400 500 **550 700** 750 900 940 2060. The number of observations is 10, so (n + 1) / 2 = (10 + 1) / 2 = 5.5. Therefore, the median is 625, which is the midpoint between the fifth and sixth observations in the sorted list.

19.2.2.3   Mode

The value that comes up most often in a set of data is called the mode.

If no value comes up more than once in the dataset, there is no mode.

Example: I- The mode of set 4 5 3 3 3 4 7 8 8 10 is **3**

II- The mode of set 3 5 6 4 7 2 10 11. There is no mode, because no highest frequency occurs.

Examine your abilities

1. Explain the grouped and ungrouped frequency distribution table.
2. A simple random sample of 21 people's ages at diagnosis and BP measurement were 60 58 52 58 59 58 51 61 54 59 55 53 44 46 47 42 56 57 49 41 43. Make a grouped and ungrouped frequency distribution table using limit grouping with 40–44 and five classes.
3. What is data representation by the graphical method?
4. Establish the mean, median, and mode.
5. Find the mean, median, and mode of the given dataset 2 4 1 2 4 3 3 2 2 1 2 4 2 3.

TABLE 19.7
Skill Measures

Classes	High Skills Measures	Low Skills Measures
5th	12	6
6th	11	0
7th	15	8
8th	20	11
9th	17	12
10th	10	10
11th	14	9

6. Compiles data on student skills. During one year, the numbers of students in different classes of college were tabulated by the high skill and low skill, which resulted in Table 19.7 (this is based on approximation).

Calculate the mean on based on both high skills and low skills measurements.

19.2.2.4 Probability

In the following paragraphs, you will acquire the knowledge necessary to compute the likelihood that a certain result will be obtained from an experiment. Although it was first utilized in gambling, since then, probability has been used a lot in many different fields, like the physical sciences, business, the biological and medical sciences, and even weather forecasting. When we carry out various tests, such as tossing coins, throwing dice, and so on, and saw the results of these endeavors. In the next section, you'll learn how to figure out how likely it is that a certain result will happen in an experiment.

the collection of all the potential results of an experiment is referred to as the **sample space** X. Points y located inside X are referred to as examples of outcomes, realizations, or components. **Events** are a term that refers to subsets of X.

Example. If a coin is flipped twice, then we will get the following results: HH, HT, TH, and TT. The outcome of the first coin toss being heads is denoted by the symbol A, which is read as "HH, HT."

The notation of probability is denoted by "P" and presents the event by symbols "A," "B," and "C." So we can write the probability P(A) = the probability that event A will happen, P(B) = the probability that event B will happen, P(C) = the probability that event C will happen, and so on. The basic rule of probability is to carry out (or watch) a certain process and keep track of the number of times event A really does take place. Assume that a certain process includes n distinct simple events and that each of those simple events has an equal probability of taking place. If event A can occur in any of these different ways, then

$$P(A) = \frac{Number\ of\ Event\ A\ occur}{Total\ number\ of\ different\ event\ occur}$$

$$= \frac{s}{n}$$

Rule Probability:

(i) A probability should always be expressed as a fraction or decimal value that falls between 0 and 1.

(ii) The chance (probability) of something that can never happen is **zero**.

(iii) The chance (probability) of an event that will definitely take place is equal to **one**.

(iv) The chance (probability) of each given occurrence, denoted by the letter A, ranges between 0 and 1. To put it another way, $0 \le P(A) \le 1$.

Complementary Events: The set of all possible outcomes in which the occurrence of the event represented by "A" does not take place is referred to as "A," and it is the complement of the event signified by $\bar{A} = 1 - A$.

Example: There are 8 red balls, 7 blue balls, and 6 green balls in a box. One ball is chosen by chance. How likely is it that the ball is neither red nor green?

Solution: let E = neither red nor green = 7, total sum of ball (S) = 21

$$P(E) = E / (S) = 7/21 = 1/3$$

The probability that it is neither red nor green is 1/3.

Example: Tickets with the numbers 1 through 20 are mixed together, and then one is picked at random. How likely is it that the number on the winning ticket is a multiple of 3 or 5?

Solution: Here, S = {1, 2, 3, 4 . . . 19, 20}.

Let E = the event of getting a multiple of 3 or 5 = {3, 6, 9, 12, 15, 18, 5, 10, 20}.

$$\therefore P(E) = \frac{n(E)}{n(S)} = \frac{9}{20}$$

Example: Table 19.8 contains data from a study of two airlines which fly to Riyadh, in the kingdom of Saudi Arabia.

Find the probability that the flight selected is Flynus Airlines, which was on time.

Solution: The probability of P (Flynus Airlines on time) = 43/87.

TABLE 19.8

Study of Two Airlines

	No. of Flights on Time	No. of Late Flights	Totals
Saudi Airlines	33	6	0.6
Flynus Airlines	43	5	0.4
Totals	76	11	87

Addition Rule: The probability that either event A or event B occurs as the only result of the operation is calculated using the addition rule, which may be expressed as P (A or B) (or that both events occur). The word "or" is the central focus of this paragraph. This is known as the inclusive or, because it may indicate either one or the other, or even both.

Formal Addition Rule:

$$P (A \text{ or } B) = P (A) + P (B) - P (A \text{ and } B)$$

where P (A and B) signifies the chance that A and B both occur as a result of a procedure trial at the same time.

Disjoint or Mutually Exclusive: If events A and B cannot occur concurrently, they are discontinuous (or mutually exclusive). (In other words, discontinuous events don't happen at the same time.)

Complementary events $P (A)$ and $P (B)$ are disjointed, so it is not possible for two things to happen at the same time.

Venn diagrams for events that are not disjointed and that are disjointed are shown in Figure 19.2 and Figure 19.3.

Example: There are 20 boys and 45 girls in a class. The likelihood that a pupil will not be a boy if they are chosen at random is:

$$\text{Prob}(\overline{boy}) = 1 - \text{Prob}(boy)$$

$$= 1 - \frac{20}{65} = 0.692$$

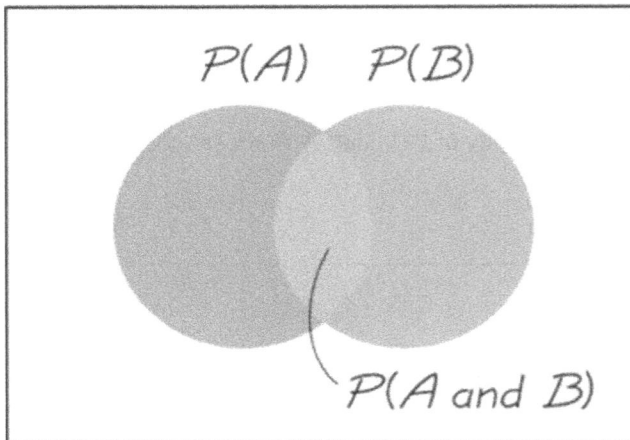

FIGURE 19.2   Venn diagram for events that are not disjointed.

$$Total\ Area = 1$$

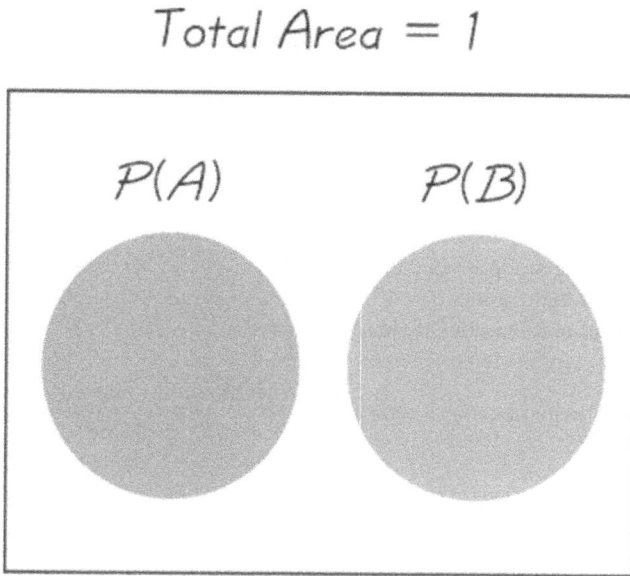

FIGURE 19.3   Venn diagram for disjointed events.

Example: A six-sided die is rolled, then the P(2 or 6) is:

$$P (A\ or\ B) = P (A) + P (B) = P (2) + P (6) = 1/6 + 1/6 = 1/3.$$

Example: What is the probability of getting a number higher than three on a 6-sided die?
  Solution: The die numbers that are greater than 3 are 4, 5, and 6.

$$P (4) = 1/6,\ P (5) = 1/6,\ P (6) = 1/6$$
$$P (4) + P (5) + P (6) = 3/6.$$

19.2.2.5   Multiplication Rule

The basic multiplication rule is used to find P (A and B), which is the chance that event A will happen in the first trial and event B will happen in the second trial.

If the outcome of event A affects the likelihood of event B in some way, it is important to change the probability of event B to reflect what happened in event A.

P (A and B) = P (In the first trial, event A takes place, and in the second trial, event B takes place) = P (AB) = P (A) P (B).

For example, in tossing a fair die, let A = {3, 9, 21} and let B = {5, 7, 9, 12}. Then, A $\cap$ B = {9}, P (AB) = 2/6 = P (A) P (B) = (1/2) $\times$ (2/3).

Now we discuss conditional probability. We have to change the odds of the second event based on what happened with the first. With the notation of conditional probabilities, assuming event A has already happened, P (B|A) is the probability of event B. Similarly, P (A|B) is the probability of event A assuming event B has already occurred.

A and B are independent events if none affects the other's likelihood. (Events are independent if one doesn't alter the others' probability.)

$$P(A \text{ and } B) = P(A) \cdot P(B/A) = P(B) \cdot P(A/B).$$

For example, the results of a survey of undergraduate students at a university's faculty of science disclose the following information (see Table 19.9):

A student is randomly selected. What is the probability of selecting a mathematics major, given that the student selected is male?

The number of male students = 40

The number of mathematics male students = 8

$$\text{Prob (Mathematics major | Male)} = \frac{P(\text{Mathematics major } AND \text{ Male})}{P(\text{Male})}$$

$$= \frac{\dfrac{8}{50}}{\dfrac{40}{50}} = \frac{8}{40}$$

Examine your abilities

1. If the probability of passing International English Language Testing System is 0.0015, then what is the probability of not passing?
2. When P (A) = 0.6, then what is the value of P $(\overline{A})$?

TABLE 19.9

Survey of Undergraduate Students at a University's Faculty of Science

Gender	Major			Total
	Biology	Physics	Mathematics	
Male	22	10	8	40
Female	3	5	2	10
Total	25	15	10	50

TABLE 19.10

Study of Two Airlines that Fly to Riyadh

	No. of Flights on Time	No. of Late Flights	Totals
Saudi Airlines	33	6	39
Flynas Airlines	43	5	48
Totals	76	11	87

TABLE 19.11
Probabilities

	A	B	Totals
D	0.2	0.4	0.6
E	0.28	0.12	0.4
Totals	0.48	0.52	1

3. How likely is it to answer a question wrong in a random guess on a multiple-choice exam with four potential answers?
4. Table 19.10 contains data from a study of two airlines that fly to Riyadh.

Find the probability that the flight selected is Flynas Airlines, which was on time.
Ans: The probability of P(Flynas Airlines on time) = 43/87

5. A random ticket is picked from a hat from among the tickets with numbers ranging from 1 to 20. What's the chance that the winning ticket has a number that's a multiple of three or five?
6. Suppose events A, B, D, and E have probabilities as given in Table 19.11.

What is the probability of P (B/D)?

Ans: The probability of P (B/D) = P (B and D)/P (D) = 0.4/0.6 = 2/3

REFERENCES

[1] K. Morik, "A note on artificial intelligence and statistics." In *Applications in Statistical Computing*, pp. 127–138. Springer, Cham, 2019.
[2] Z. Ghahramani, "Probabilistic machine learning and artificial intelligence." *Nature* 521, no. 7553 (2015): 452–459.
[3] William A. Gale, and P. Daryl, "Artificial intelligence research in statistics." *AI Magazine* 5, no. 4 (1984): 72–72.
[4] William Mendenhall, Robert J. Beaver, and Barbara M. Beaver, *Introduction to Probability and Statistics*. Cengage Learning, Boston, United States, 2012.
[5] Jiaying Liu, Xiangjie Kong, Feng Xia, Xiaomei Bai, Lei Wang, Qing Qing, and Ivan Lee, "Artificial intelligence in the 21st century." *IEEE Access* 6 (2018): 34403–34421.
[6] K. Hemachandran, S. Khanra, R. V. Rodriguez, and J. Jaramillo (Eds.), *Machine Learning for Business Analytics: Real-Time Data Analysis for Decision-Making*. CRC Press, New york, 2022.
[7] Jay L. Devore, *Probability and Statistics for Engineering and the Sciences*. Cengage Learning, Boston, United States, 2011.

20 Real Impacts of Machine Learning in Business

B. R. S. S. Sowjanya, S. Pavan Siddharth, and Vishwa KD

20.1 INTRODUCTION TO MACHINE LEARNING

A subset of artificial intelligence called machine learning analyzes data to improve and make predictions. It enables machines to construct algorithms that can predict the outcome of data and learn from experience, letting them carry out tasks autonomously. The more data you input into the system, the more the algorithms can learn from it and improve the output. All of this is made possible by machine learning and the rapid advancements in artificial intelligence[1]. Without being instructed where to seek, computers use machine learning to search for pertinent data. They achieve this by utilizing algorithms that iteratively learn data. The idea of automating the application of intricate mathematical computations to massive data has just recently emerged, despite the fact that the concept of machine learning has been around for a while. Many individuals are currently interested in it. As we already know by now, one of the most fascinating developments in artificial intelligence is machine learning. The objective of learning from data using machine-specific inputs is accomplished. Understanding how machine learning functions and, consequently, how it might be applied in the future is crucial. The first step in machine learning is feeding the learning data into an algorithm that has already been "trained" on similar data sets. Known or unknown facts are utilized as the training set to create a computerized decision-making system. The method is influenced by the type of input training data, and this idea will be discussed later. The machine learning algorithm is fed fresh input data to see if it functions properly; predictions and outcomes are then contrasted.

Numerous iterations of the method are performed until the data scientist achieves the desired outcome. As a result, the machine learning system can generate an appropriate response and continuously learn on its own, steadily improving its accuracy over time.

Figure 20.1 shows that collection, cleaning, and labelling are examples of stages that are data-oriented. Other stages are model-oriented (e.g., model requirements, feature engineering, training, evaluation, deployment, and monitoring). The workflow contains a lot of feedback loops. Model review and monitoring may loop back to any of the earlier stages, as shown by the larger feedback arrows. The smaller feedback arrow shows how feature engineering may loop back to model training (e.g., in representation learning).

 DOI: 10.1201/9781003328414-20

FIGURE 20.1   The nine stages of the machine learning workflow.

20.1.1   COMMONLY USED MACHINE LEARNING MODELS FOR BUSINESS

20.1.1.1   Lead, Opportunities, and Conversion Model

The importance of this model lies in its capacity to pinpoint the rational and emotional drivers of clients with higher purchase propensities. These insights serve as the basis for a strategy to create more targeted marketing, which customers crave to the point where they're prepared to trade their data for it.

20.1.1.2   Attrition and Customer Retention Model

A business can learn which customers are likely to churn and when by using the attrition and customer retention model. This enables us to assess how well the product resonates with consumers.

20.1.1.3   Lifetime Value Model

The lifetime value model examines the complete customer journey to determine where the lifetime value of the consumer may be raised, which cohorts provide the most value, and which areas of your business have the greatest impact on lifetime value.

20.1.1.4   Employee Retention Model

To prevent employee turnover and guarantee that their experience is top-notch, businesses can use the employee retention model, which forecasts when employees have a high inclination to leave.

20.2   ALGORITHMS

20.2.1   DECISION TREE ALGORITHM

Popular machine learning methods that can be applied to both classification and regression issues include decision trees. A decision tree employs a tree-like structure to manage decisions and the potential outcomes and repercussions of those actions. Every internal node denotes a test on an attribute, and every branch reflects the outcome of the test. A decision tree's outcome will be more accurate the more nodes it contains. Decision trees have the virtue of being logical and simple to use, but they are not the absolute best when it comes to accuracy. In operations research, decision trees are widely applied, particularly in machine learning, strategic planning, and decision analysis.

20.2.2 Support Vector Machine Algorithm

The well-known machine learning technique known as support vector machine (SVM) is frequently used for classification and regression problems. But specifically, it's employed to address classification issues. Finding the best decision boundaries in an XYZ-dimensional space that can divide data points into classes is the basic goal of the SVM algorithm, and the optimum decision boundary is referred to as a hyperplane. Support vectors are the extreme vectors that SVM chooses to find the hyperplane from[2]. See Figure 20.2.

20.2.3 Clustering Algorithm

Data points are clustered or grouped into several clusters based on similarities and differences; this process is called clustering. The items that have the most commonalities stay in the same group and have little to no overlap with items from other groups.

Numerous activities, including image segmentation, statistical data analysis, market segmentation, etc., can benefit from the usage of clustering techniques.

K-means clustering, hierarchical clustering, density-based spatial clustering of applications with noise, and other popular clustering algorithms are some examples[3].

20.2.4 Association Rule Learning Algorithm

With the help of association rule learning, significant relationships between variables in a big data set are discovered. This learning algorithm's primary objective is to determine the dependencies between data items and then map the variables in a manner that maximizes profit. This algorithm is mostly used in continuous production, web usage mining, market basket analysis, etc. Eclat and FP-growth algorithms are a few of the well-known algorithms for learning association rules[3].

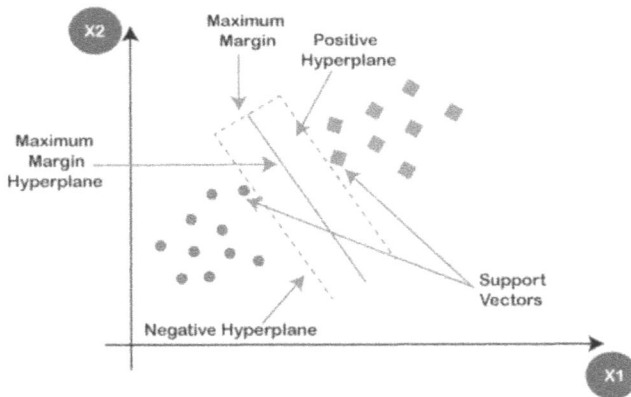

FIGURE 20.2 Support vector machine.

20.2.4.1 Advantages of Machine Learning for Business

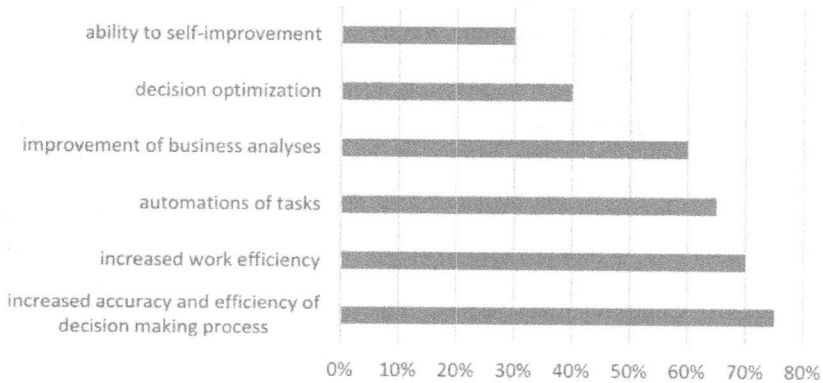

The advantages resulting from machine learning application in business organization

FIGURE 20.3 The advantages resulting from the application of machine learning in a business organization.

BENEFITS OF MACHINE LEARNING FOR BUSINESS

FIGURE 20.4 The benefits of machine learning for a business.

20.2.4.2   Review of Case Studies with Practical Examples

Twenty-two case studies demonstrating the use of automated machine learning in various industries and businesses were included in AI Multiple. A few of which are mentioned next.

The DataRobot automated machine learning solution was employed in the sales and production operations of **Lenovo**, a technology business in Brazil, and it has increased correct predictions from 80% to 87.5% and lowered the model generation time from four weeks to three days.

- **PayPal**, a corporation that offers financial services, found that using the H2O.ai service in the fraud detection domain enhanced the accuracy by 94.8% and cut the model training time to just under two hours.
- In the example of the Canadian company **Imagia**, the use of Google Cloud AutoML in the healthcare sector allowed for a 16-hour to 1-hour reduction in test processing time and an improvement in diagnosis outcomes.
- **Meredith Company**, with the help of the media and entertainment industry, Google Cloud AutoML, was used to classify content and increase customer knowledge about the emerging trends.
- Using DataRobot's email marketing system, a Danish marketing firm, **One Marketing Ltd.**, reduced spam for clients, raised mail open and click rates by 14% and 24%, and increased the ticket sales by 83%.

20.3   REAL-TIME EXAMPLES

The development of artificial intelligence and machine learning is a recent example of significant technological progress shown in figures 20.3 and 20.4. Machine learning is one of the few fields that has the potential to completely alter our way of life, yet many of its uses are now hidden. Here are a few recent machine learning examples.

20.3.1   GOOGLE

Google has been at the forefront of machine learning use since its inception. It offers a helpful and individualized experience to its consumers by using machine learning algorithms. For instance, Google uses machine learning in its picture search and translation capabilities to stop unlawful work or commerce, such as illicit fishing, by utilizing satellite data. Google services already incorporate machine learning, such as Gmail and Search[4].

20.3.1.1   Gmail

Compared to what is in your mailbox, social, promotional, and primary emails may change. Google filters this since it classifies the email appropriately. The computer's threshold is tuned using machine learning, so when a user tags a message consistently, Gmail automatically makes real-time adjustments to its threshold and then learns for subsequent categorization.

20.3.1.2 Google Search and Maps

Further, using machine learning, Google recognizes your search terms as you begin to type them in the search box. Then it offers a list of potential search terms for the same. Due to trends (which everyone is seeking), past searches (recommendations), or your current location, these options are displayed [5].

20.3.1.3 OK Google

This is an intelligent personal assistant app that facilitates task completion, information discovery, and reservation making. When it's pouring outside, you can quickly find nearby restaurants, purchase movie tickets on the go, and locate the theater that is the closest to your location. It also aids in your navigation to the theater. In other words, if you have a smartphone, you don't need to worry about a thing because Google handles everything.

20.3.2 IBM

The platform offered by IBM's Watson artificial intelligence and machine learning tool is made to improve both the intelligence of your company and the performance of your employees. A variety of cutting-edge application programming interfaces (APIs), specialized tools, and software as service applications are available with Watson. Inferring that Watson was created with complex use cases in mind and with experts in mind so that it can effortlessly connect with the platforms they already use for their everyday job, this guarantees easy access to the information you need to make the best decisions.

The cornerstone of your competitive edge, your data, models, learning, and API, are all completely under your control with Watson. Because of its tremendous learning capacity, it can learn more with less. Watson can assist in making judgments that help organizations generate more money by forecasting trends using data.

20.3.3 NASA

The finding of new extraterrestrial objects requires machine learning. Machine learning is essential to find patterns in the massive amount of data produced by NASA satellites and spacecraft in order to make exciting future discoveries. Machine learning can be used to intelligently oversee spacecraft repairs, find undiscovered planets in other galaxies, and uncover other fascinating things.

20.3.3.1 Self-Driving Rovers on Mars

The self-driving Mars rover auto-navigation was launched in 2004 and has been using its machine learning–based navigation and driving system ever since. Curiosity, another 2011-launched rover, employs auto-navigation and is currently scouting the Martian landscape in search of water and other elements[6].

20.3.3.2 Medicine in Space

Exploration Medical Capabilities is a brand-new approach to exploration medicine that NASA is developing. It will employ machine learning to create healthcare options based on the astronauts' expected future medical requirements. These medical solutions will be developed by licensed physicians and surgeons who will grow and change over time in response to the astronauts' experiences.

20.3.4 NETFLIX AND SUCCESS: WHAT DO THE PEOPLE WANT?

What can we learn about successful content from Netflix's United States daily top 10 data?

What the data set includes:

This data set is a gathering of daily top 10 shows on Netflix spanning from March 2020 to March 2022 in the United States.

Rank: A 1–10 scale representing the streaming time of each show for a given date.

Year to Date Rank: A 1–10 scale representing the overall rank relative to all other shows that year (these rankings shift around quite a lot, as this is recalculated by the day).

Last Week Rank: A 1–10 scale showing the overall rank for the prior week.

Title: The name of the show or special in question.

Netflix Exclusive: Whether the show is a Netflix exclusive.

Netflix Release Date: The date which the show debuted on Netflix.

Days in Top 10: How many total days a show has appeared in the top 10 by the as of date.

20.3.4.1 Customer Reviews

The viewership score: This was updated in late 2021 and was not utilized in the research. (Before the redesign, this statistic represented the number of subscribers who watched two minutes or more of a particular program; following the rework, it now represents the number of hours viewed. On November 16, 2021, the *Wall Street Journal* published a report about this.

It is hoped that this analysis would shed light on what being "popular" on Netflix actually means in terms of the numbers.

From the data, Figure 20.5 & Figure 20.6 shows a few of the insights after running a few basic machine learning algorithms and doing sentiment analysis on the reviews[7].

These stand out among the best in both categories. Good benchmarks for the types of Netflix shows American subscribers are most interested in, as well as good prospects for renewal, are provided.

These are also the shows that viewers probably want to see more of, which is noteworthy for other artists. These provide direction on what audiences want for any executives, producers, or even writers searching for a market direction.

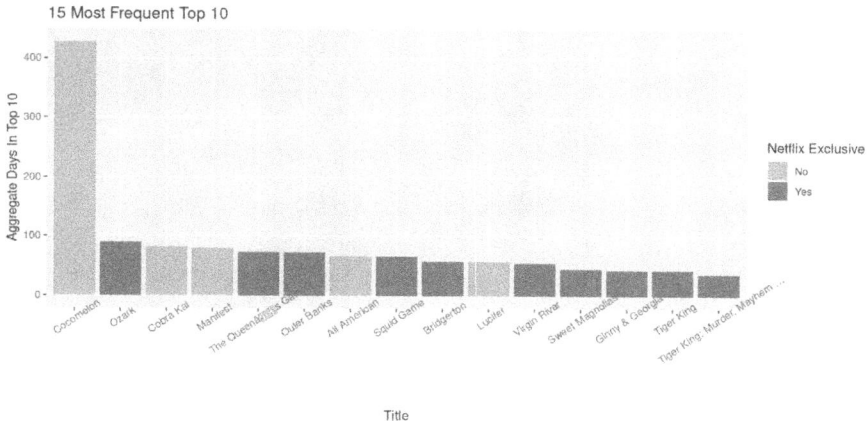

FIGURE 20.5   The 15 most frequent top shows with aggregate days.

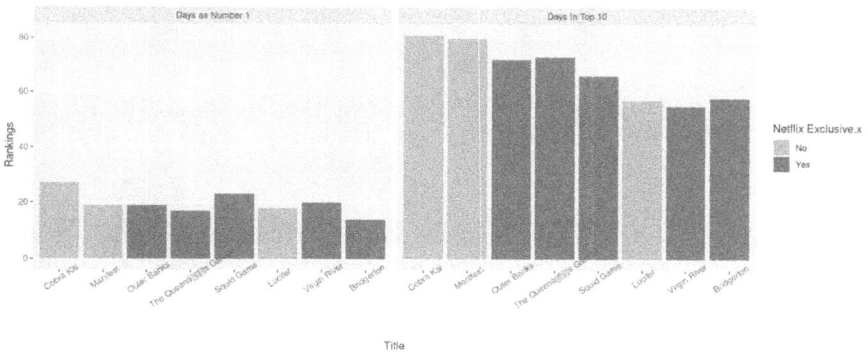

FIGURE 20.6   Plotting of the shows/series as the days on number 1 and days in top 10 simultaneously.

20.3.5   SHAZAM USES SOUND ANALYSIS TO INTELLIGENTLY TAG AND CATEGORIZE SONGS

Another excellent example of employing machine learning to handle a problem that would take a ton of human labor and be inherently tough for human analysis is Shazam.

The company had to deal with the duty of tagging and categorizing the tracks in their library. While this may sound simple, anyone who enjoys a particular musical style can attest to the fact that a genre can encompass a huge variety of musical styles (metal includes black metal, thrash metal, death metal, sludge, and much more).

Even worse, the genre—and particularly the sub-genre—to which a piece of music belongs can be completely arbitrary to the listener. It's uncommon for something to fit into a category without any resistance.

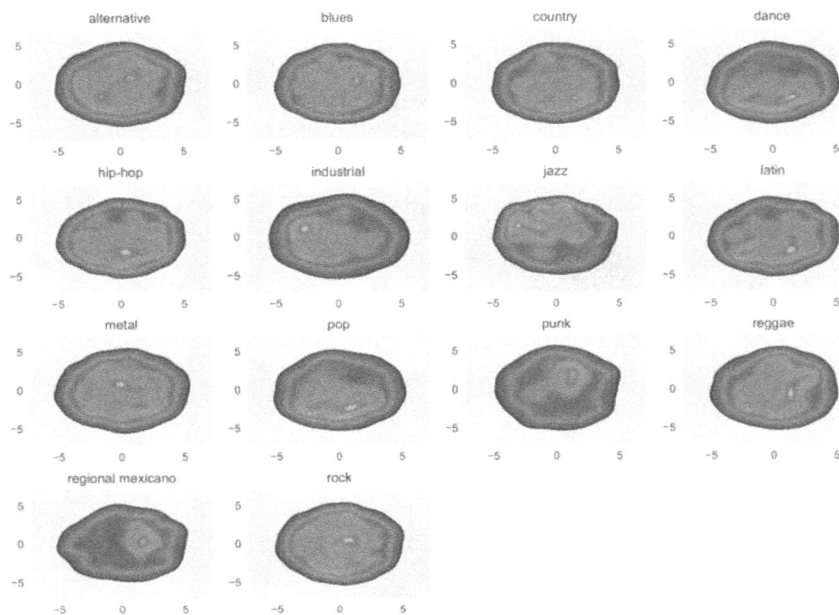

FIGURE 20.7   Amplitude peaks with heat map.

Shazam created a model that examined song excerpts and assigned a "signature" to each in order to address this problem. This functioned by building a spectrogram for the track fragment and then searching for amplitude peaks. See Figure 20.7.

The two "feature" elements for every track were then created from these track signatures so that they could be represented graphically. This provided a heat map of how each musical genre appeared visually after being divided into the human-assigned genres, which could be used to cross-reference with some other tracks to assign a genre automatically.

Even though the specifics are a little complicated, the final result is that Shazam can automatically evaluate songs and assign tags and genres to them while maintaining a high level of accuracy.

20.3.6   STOCK PRICE PREDICTION

A thousand can become a million on the stock market, but millions might also become nothing. The future behavior of the market is impossible to foresee. After the development of social media, it has become especially volatile. A fairly recent example is the "meme" stock Gamestop, which increased by more than 1600% in just 15 days. Another illustration is the company Signal Advance's stock price increase of 1100%. Retail investors began purchasing shares of an unrelated Signal Advance when Elon Musk pushed his Twitter followers to "use Signal," a messaging app. A lot of machine

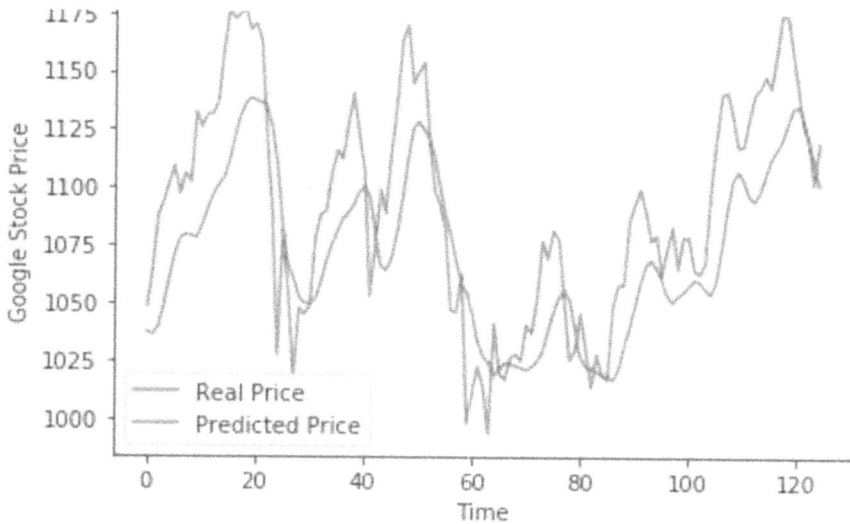

FIGURE 20.8   Google stock price prediction.

learning models speculate if the market is going up or if it going down in the next moment. Here, we discuss a model which predicts the price of the stock based on the fundamentals of the company by using the XGBoost algorithm. See Figure 20.8.

20.4   CONCLUSION

Machine learning is a technology that enables companies to efficiently derive insights from unstructured data. Machine learning algorithms can be used to constantly learn from a set of data to identify patterns and actions, among other things. The machine learning approach is dynamic and constantly evolving, allowing businesses to stay up to date on market and client expectations.

REFERENCES

[1] Osisanwo, F. Y., Akinsola, J. E. T., Awodele, O., Hinmikaiye, J. O., Olakanmi, O., & Akinjobi, J. (2017). Supervised machine learning algorithms: Classification and comparison. *International Journal of Computer Trends and Technology (IJCTT)*, *48*(3), 128–138.
[2] Evgeniou, T., & Pontil, M. (1999, July). Support vector machines: Theory and applications. In *Advanced Course on Artificial Intelligence* (pp. 249–257). Springer.
[3] Mirtaheri, S. L., & Shahbazian, R. (2022). *Machine Learning: Theory to Applications*. CRC Press.
[4] Ahmed, H., Jilani, T. A., Haider, W., Abbasi, M. A., Nand, S., & Kamran, S. (2017). Establishing standard rules for choosing best KPIs for an e-commerce business based on google analytics and machine learning technique. *International Journal of Advanced Computer Science and Applications*, *8*(5).

[5] Taecharungroj, V. (2021). Google Maps amenities and condominium prices: Investigating the effects and relationships using machine learning. *Habitat International, 118,* 102463.

[6] Ono, M., Rothrock, B., Otsu, K., Higa, S., Iwashita, Y., Didier, A., . . . Park, H. (2020, March). MAARS: Machine learning-based analytics for automated rover systems. In *2020 IEEE Aerospace Conference* (pp. 1–17). IEEE.

[7] Hemachandran, K., Khanra, S., Rodriguez, R. V., & Jaramillo, J. (Eds.). (2022). *Machine Learning for Business Analytics: Real-Time Data Analysis for Decision-Making.* CRC Press, New York.

21 A Study on the Application of Natural Language Processing Used in Business Analytics for Better Management Decisions
A Literature Review

Geetha Manoharan, Subhashini Durai,
Gunaseelan Alex Rajesh, and Sunitha
Purushottam Ashtikar

21.1 INTRODUCTION

The field of artificial intelligence and data science known as natural language processing (NLP) is expanding rapidly, and it makes use of sophisticated speech and text processing tools. The goal of this line of research is to provide methods for the automatic analysis and presentation of human language. Automatic summarization, part-of-speech tagging, disambiguation, entity and relation extraction, sentiment analysis, natural language understanding (NLU), and speech recognition are only some of the methods used by NLP to make sense of ambiguities in human language. Numerous NLP-related software tasks, such as morphological and syntactic analysis, have been satisfactorily solved for online use.

So they can be used as software, just like software as a service (SaaS) applications in cloud computing. Users can easily and quickly access a large pool of flexible computing resources over the Internet with minimal overhead thanks to the "cloud computing" approach. Examples of cutting-edge technologies that form the basis of this paradigm include the Internet, virtualization tools, grid computing, and web services. Because of this, cloud computing combines SaaS and utility computing.

DOI: 10.1201/9781003328414-21

The idea of cloud computing is to make flexible, affordable, and reliable computing resources available on-demand.

Researchers and other users with an interest in the topic are currently very interested in cloud-based NLP analysis services. It enables researchers to set up, share, and utilize language processing tools and components in line with the SaaS (data as a service) and PaaS (platform as a service) models. The reviews of cloud-based NLP services, however, are surprisingly scarce. Some of the most well-known cloud-based NLP services and application programming interfaces (APIs) are Amazon Comprehend, Microsoft Azure Cognitive Services, Google Cloud Natural Language, and several third-party alternatives. Machine learning is used by the Amazon Comprehend (AWS) service to identify the language and extract key phrases from a text. Tokenization, sentiment analysis, and text file automation organization are all features of Amazon Comprehend, which integrates with any AWS-supported application.

The portfolio of NLP tools organized into various, more specialized services and uses is called the Microsoft Azure Cognitive Services. By way of illustration, the Azure Text Analytics API can be used by programmers to create tools that analyse the sentiment or determine the language of a given text. The Azure Language Understanding Intelligent Service, on the other hand, is capable of deciphering things like user intent. An invaluable achievement for developers is the creation of chatbots, voice-powered products, and customer care platforms. Additionally, Google Cloud Natural Language can perform entity extraction, sentiment analysis, syntax analysis, and classification. The difference between this API and others is that it is powered by the same comprehensive learning modules that power Google's own language understanding and query comprehension systems for Google Search and Google Assistant.

There are numerous third-party services and APIs for NLP. While Monkey-Learn and other suppliers offer services to automate procedures based on unstructured data, in contrast, as an illustration, businesses like Diffbot provide features via a paid API that enable users to precisely extract data from websites. In contrast to the reviews provided earlier, we investigate the NLP-related methods and technologies that are currently in use, as well as cloud-hosted NLP services, and we talk about NLP and big data methods and technologies, including information extraction through NLP within big data. This chapter's primary goal is to present a review of the literature on various studies carried out in different fields. This is for the purpose of learning and highlighting the essentials of NLP services currently used in various fields for various reasons and purposes.

21.2 NATURAL LANGUAGE PROCESSING

The intersection of linguistics and artificial intelligence seeks to make it possible for computers to understand claims made in human languages. NLP was created to facilitate user tasks and satisfy the demand for a natural language computer interface. Users who lack the time or motivation to learn the machine's language will benefit from NLP because it facilitates communication. One way to define a language is as a system of rules, while another way is to view it as a system of symbols. When two or more symbols are combined, the result is a transmission or broadcast of information.

NLP consists of natural language generation (NLG) concepts and NLU, which are concerned with reading and writing text, respectively. NLU and NLG approaches and developments are related to big data. A few common NLP techniques include probabilistic context-free grammar, part-of-speech tagging, word sense disambiguation, and lexical acquisition. Text mining techniques based on NLP include information extraction, topic modelling, text summarization, classification, clustering, question answering, and opinion mining, to name just a few. In linguistics, "annotation" most often refers to metadata that characterizes words, sentences, or other metadata. Automatically assigning descriptors to input tokens is known as "tagging" in the annotation process. Tagging words as nouns, verbs, and adverbs, among other categories, is called part-of-speech tagging. This method takes a natural language like English and creates a meta-language from it. To begin any NLP task, segmentation must be performed. The unprocessed form of electronic text is merely a string of characters. Because of this, it needs to be broken down into smaller linguistic chunks. It is not a complete list, but it includes things like words, periods, numbers, alphabetic characters, and other symbols.

Tokenization is another name for this process. Linguistic entities (tokens) must be assigned to classes for NLP tasks. The two parts of NLP that concentrate on reading and writing text, respectively, are called NLU and NLG. Big data is relevant to NLU and NLG methodologies and advancements. Common NLP methods include lexical acquisition, probabilistic context-free grammar, part-of-speech tagging, and word sense disambiguation. Methods for text mining that rely on NLP include information extraction, topic modelling, text summarization, classification, clustering, question answering, and opinion mining. An annotation is a type of metadata that describes words, sentences, or other types of metadata in linguistics. "Tagging" refers to the process of automatically assigning descriptors to input tokens during the annotation process. Part-of-speech tagging is the process of translating from a natural language like English into a meta-language made up of different types of words. A crucial step in every NLP process is text segmentation. Electronic text essentially only consists of a string of characters when it is not processed. As a result, it needs to be divided into linguistic units. These units might be anything from words to punctuation to numbers to alphanumeric characters and other symbols or the Apache UIMA project. Statistical, lexical affinity, and keyword spotting techniques all fall under the category of syntax-centred NLP. The most simplistic strategy is keyword spotting, which is also likely the most popular due to its availability and affordability. In comparison to keyword spotting, lexical affinity is a little more sophisticated.

Since the late 1990s, statistical NLP has been the primary focus of researchers in the field. The foundation is in linguistic theory that employs well-known machine learning approaches like support vector machines, expectation maximization, conditional random fields, and maximum likelihood. The semantic features of statistical methods are often inadequate. Web data analysis automatically downloads, extracts, and assesses data from cloud documents and services in order to locate pertinent information. The fields of database management, information retrieval, NLP, and text mining all have connections to web analysis. The study of the Web's data can be broken down into three distinct areas: content mining, structure mining, and usage mining. Text, photos, audio, video, code, metadata, and hyperlinks are just some of the many data sources that can be mined when engaging in web content mining. In this chapter, we provide models for understanding the architecture of hyperlinks on the Internet. However, recovery is a digital problem that needs fixing quickly. Sentiment analysis of data gleaned from social networking sites like Twitter calls for extensive data pre-processing. Due to the data's enormous volume, high degree of unstructuredness, and incredible rate of production, parallel implementations of pre-processing algorithms are required. Frequency mapping, removing any unnecessary words or symbols, converting a string to a vector, and other pre-processing techniques are all examples of pre-processing. According to B. Bharathi and Josephine Varsha (2022), in their study they compared seven transformer model variations on five difficult NLP tasks and seven datasets. They design experiments to focus on their capacity for sustained attention while isolating the effects of pretraining and hyperparameter settings. They also present various approaches for analysing attention behaviours to shed light on model details beyond metric scores. They found that long-range transformers' attention has benefits for content selection and query-guided decoding, but they also have unrecognised drawbacks like failing to pay enough attention to tokens that are far away.

21.3   NATURAL LANGUAGE PROCESSING
AND BUSINESS ANALYTICS

Wang et al. (2022) and others felt that though the interpretability of neural models is a subject of growing concern, there are still no adequate evaluation datasets or metrics available, so this issue is still unresolved. To evaluate the interpretability of neural models and saliency approaches, they proposed a brand-new benchmark in their study. The three typical NLP tasks covered by this benchmark are sentiment analysis, textual similarity, and reading comprehension, all with material that has been annotated in both English and Chinese. To accurately measure the interpretability, they provide token-level explanations that have been meticulously annotated. The consistency between the justifications before and after perturbations is a new metric that they also establish. They experiment on three typical models using three saliency methods based on this benchmark and identify the models' interpretability strengths and weaknesses. Daniel Hershcovich, Frank et al. (2022) has stated that the enormous amount of energy that is increasingly being used for training and running computational models, the climate impact of artificial intelligence, and NLP research in particular, has become a serious issue. As a result, effective NLP is gaining popularity. This significant initiative does not, however, have clear guidelines that would enable systematic climate reporting on NLP research. They contend that this shortcoming is one of the factors contributing to the fact that few NLP publications provide key data that would enable a more in-depth analysis of environmental impact. They suggest a climate performance model card as a workaround whose main goal is to be practically applicable with little knowledge of the experiments and underlying computer hardware. They also explain why taking this action will help raise awareness of how NLP research affects the environment and opens the door to more in-depth discussions. Balkir et al. (2022) and Belz (2022) say that methods in explainable artificial intelligence (XAI) are frequently driven by the desire to identify, measure, and mitigate bias as well as improve the fairness of machine learning models. However, it's frequently not made clear how an XAI method can aid in overcoming biases. They discuss the ways now used to detect and reduce bias using explain ability methodologies, and they briefly explain current ability and fairness research trends. They also highlight the barriers that limit the use of XAI techniques to address fairness issues.

21.4 PROCESS INVOLVED USING NLP IN BIG DATA ANALYTICS

In the big data community, NLP has become increasingly popular, particularly when it comes to creating systems that make decision-making easier. When developing appropriate models, decision-makers frequently use textual sources to locate pertinent data. Although processing a modest quantity of data manually is possible, it is time-consuming. To bridge the gap between efficiency and big data processing, numerous NLP approaches have been created. In order to analyse and extract knowledge from textual sources via linguistic analysis, it employs a set of computational techniques. To better comprehend the informational patterns that arise in human communication, it attempts to widen its methodology to incorporate any language, communication style, or genre that individuals employ.

Although "true" NLU is the NLP's ultimate goal, more research is still required to achieve it. Initially, NLP stood for natural language processing (NLU). It is still being developed and improved to be able to make logical inferences from textual sources while accounting for language complexity in terms of fuzzy knowledge, causality, and ambiguous meaning. At its most elementary, lexical analysis is an NLP system's concentration on the words that are assumed to be the text documents' important components. It relates especially to the process of analysing and arranging the important components of a text into tokens, which are groups of characters with comparable meanings. To put it another way, lexical analysis aids in the comprehension of words by enabling users to relate them to a variety of concepts based on the context in which they are used. Simplified lexical representations are utilized to unify the meanings of words in order to construct sophisticated interpretations at higher meta-levels. For lexical analysis, a lexicon may be needed; lexicons often comprise the method taken by a well-specified NLP system, as well as the nature and extent of the information intrinsic to the lexicon. One of the primary factors in a lexicon's degree of complexity is the amount of information it can store regarding the meaning of words. Accurate and exhaustive sub-categorization lexicons are also required for the advancement of parsing technology and any NLP application that relies on the organization of data tied to a predicate-argument structure.

Several research projects have been created to improve the methods used to analyse words in semantic contexts. Part-of-speech tagging, which aims to describe each word's syntactic role, parsing, and lemmatization are a few of the tasks that comprise lexical analysis in particular. Lemmatization combines a word's inflected forms into a single item that corresponds to its lemma (or dictionary form). Semantic analysis operates on a higher meta-level than parsing, which is the act of grammatically analyzing a sentence in which each word's contribution is considered as a whole with the equivalent in terms of objects related to a lexicon. Semantic processing analyses the relationships between the sentence's word-level meanings to determine potential interpretations. This method may also include the semantic disambiguation of words with diverse meanings, which clarifies the definition of ambiguous phrases so that they can be used in the right semantic representation of the phrase. Any information retrieval and processing system that relies on unclear and insufficient knowledge should take this into account. When addressing statistical properties between ideas in textual sources, NLP has practical applications in inference approaches when extra data from a larger context is utilized.

NLP is the study of the connections between human language and computers, according to Sebastiao Pais et al. (2022). It has recently gained a lot of attention for computationally analysing human language and has broadened its usefulness for a variety of tasks such as machine translation, information extraction, summarization, question answering, and others. Given the fast development of cloud computing services, merging NLP and cloud computing is highly useful. Researchers can conduct NLP-related experiments on large amounts of data handled by big data techniques by utilizing the cloud's huge, on-demand computing capability. Big data, cloud computing, and NLP are all extremely vast topics with a wide range of opportunities and challenges. By removing these barriers and merging these domains, NLP and its applications can reach their full potential. The authors provide a basic review of NLP in cloud computing, with a focus on the comparison of cloud-based NLP services, the challenges of NLP with large datasets, and the need for realistic cloud-based NLP services.

21.5   INFORMATION IS EXTRACTED FROM LARGE AMOUNTS OF DATA USING NLP ALGORITHMS

Data without a predetermined structure, or "unstructured data," is a popular research topic in the field of big data. Textual data falls under this category, and NLP techniques allow for analysis of concepts inside text. Picking the right concepts and types of ties to connect them is the first and most important step in forming relationships between concepts, which is crucial in any decision-making and information-gathering process. Choosing phrases with similar meanings but different lexical levels is one way to ensure that the right amount of information is conveyed (such as "illness" and "sickness"). Large textual databases sometimes contain contradictory information, as well as relationships that seem to be at odds with one another. There are many approaches for assessing the overall behaviour of information included in texts. These include statistically dependent methods that focus on frequency, co-occurrence, and other indicators. Examining the many syntactic and grammatical roles of each textual piece is another common approach. In their study, the authors employ a grammatical approach to the identification of influence relationships between concepts by looking at triples (NP1, VB, NP2), where NP1 and NP2 are noun phrases (e.g., phrases with a noun as their headword) that must contain one or more concepts, and VB is the linking verb that must be connected to an influence type of relationship. Sentiment analysis, which attempts to extract "opinions," or polarity from textual data sources, is one of the more thoroughly studied aspects of NLP. It might help to support the specific information that was extracted. A "positive" general opinion about a given context might also indicate that the information in question is presented favourably.

21.6   BIG DATA DECISION MODELLING USING MACHINE LEARNING

The intelligent automation made possible by machine learning is driving the big data sector forward. It can analyse large amounts of data far more quickly and accurately than humans can, revealing previously unseen patterns, organizational dynamics, and customer preferences. In order to select the most appropriate strategy for processing big data using machine learning techniques, it is essential to be familiar with the benefits and drawbacks of each approach. Supervised learning, supervised learning under supervision, and unsupervised learning under supervision are the three main strategies. There is no perfect solution because each group has unique advantages and disadvantages. Big data analytics can be enhanced by incorporating machine learning methods alongside more conventional procedures. Since it might be prohibitively expensive to hire subject matter experts to manually annotate a large dataset, this issue arises in many real-world contexts.

To solve issues with both labelled and unlabelled data, the semi-supervised learning paradigm merges the two approaches. While supervised learning produces best-guess predictions for unlabelled data, unsupervised learning reveals the underlying data structure. This type of learning is exemplified by the Python-based VADER sentiment analysis tool, which analyses the polarity of each line posted on social media to determine its overall sentiment.

Also popular is the reinforcement learning approach. Based on its current status, it uses complex algorithms to take action. A couple of well-known approaches to reinforcement learning are the Markov decision process (MDP) and a Neuro Evolution of Augmenting Topologies (NEAT) built on neural networks. As a result of its ability to produce insightful analyses, machine learning is ideally suited to addressing many of the challenges associated with dealing with massive amounts of data. The adaptable analytical approach of machine learning allows for rapid evaluations and decisions to be made regardless of the dataset being used.

21.7 NATURAL LANGUAGE PROCESSING IN THE EDUCATION SECTOR

Every day, the market introduces and demands new programming languages and skills, but university curricula are not kept up to date with these developments. Researchers are increasingly interested in developing an intelligence system to aid decision-making, as both students and teachers at technical universities have come to rely on being able to identify in-demand abilities and appropriate courses to take. According to Nhi Ny Voa et al. (2022), the rapid growth of the tech sector necessitates a higher degree of knowledge in its personnel. According to Nhi Ny Voa et al. (2022), the rapid expansion of the IT business has increased the level of competence required by its workers by utilizing recent advances in NLP. Using current advances in NLP, an NLP-based course recommendation system intended specifically for the computer science (CS) and information technology (IT) sectors has been developed. A Named Entity Recognition (CSIT-NER) model in particular has been created to extract entities and tech-related skills. Then, based on these talents, they created a hybrid model that suggests multi-level, customized courses (hybrid CSIT-CRS). Because our CSIT-NER model was trained and improved on a large corpus of text from StackOverflow and GitHub, it outperforms state-of-the-art models in every evaluation criterion. The relevant skills and entities can be accurately extracted. Our combined CSIT-CRS can recommend relevant online courses, job postings, and academic programmes based on a user's interests and professional goals. In our survey of 201 volunteers from tech colleges in Australia and Vietnam, the entire system received positive ratings and comments. Students, academic staff, higher education institutions in the CS/IT field, and participants in tech-related industries will all profit from this research.

According to Garrido-Muñoz et al. (2021), there is a guideline in NLP that should be observed when transferring learning between tasks. According to taxonomy (Houlsby et al., 2019; Rau & Kamps, 2022), there is a structure to visual activities that serves as the foundation for transfer learning for these tasks. The authors offer Cog-Taskonomy, a cognitively inspired framework for learning taxonomy for NLP tasks. The framework is made up of cognitive-neural mapping (CNM) and cognitive representation analytics (CRA). The prior study calculates how similar a job is to a set of language representations for that task using representational similarity analysis, a method commonly employed in computational neuroscience to find a relationship between measurements of brain activity and computational modelling. When neural representations from NLP models are projected onto cognitive signals, the latter learns to recognize task links (i.e., functional magnetic resonance imaging [fMRI] voxels). BERT and TinyBERT are utilized as models for transfer learning in research on 12 NLP tasks. This demonstrates that the proposed Cog-Taskonomy can direct transfer learning and achieve performance comparable to the analytic hierarchy process (Berbatova, 2019) employed in visual Taskonomy (Houlsby et al., 2019) without requiring extensive pair-wise task transferring.

21.8 NLP IN CYBER SECURITY

It was our own responsibility to automatically identify offensive comments and place them in the appropriate categories. Hsu et al. (2022) found a significant rise in cyber bullying, misandry, and other abusive behaviours on social media sites and in numerous other online public forums. Due to the difficulty and length of time involved in locating, reporting, and stopping these offenders, technologies are being developed and used to address these issues. The Tamil text that was code-mixed was present in the datasets made available by the DravidianLangTech@ACL2022 organizers.

21.9 NLP IN LINGUISTIC ANALYSIS

The NLP community has made various attempts to accommodate linguistic variance and support language speakers. However, it is essential to understand that speakers and the content they produce and require vary not only by language but also by culture. Despite the fact that language and culture are inextricably linked, there are still numerous linguistic and cultural variances. Similar to how cross-lingual and multilingual NLP investigate these differences for the benefit of their users, cross-cultural and multicultural NLP investigate these cultural and linguistic variations (Daniel Hershcovich, Webersinke et al., 2022). In addition to examining current and potential approaches, we suggest a moral framework to guide these efforts. According to Lu Cheng et al. (2022), NLP models have been found to discriminate against racial and gender-based groups. In response to the negative effects of these unfavourable biases, researchers have made unprecedented efforts and presented promising techniques for bias mitigation. Despite having a big practical influence, there isn't a lot of information in the existing algorithmic fairness research about how different types of bias relate to one another. It's difficult to eliminate social bias. The term "generalized prejudice," which refers to generalized devaluing sentiments among various groups, is used in a large number of social psychology studies. Detractors of ethnic minorities, for instance, are also likely to be detractors of gay people and women. In order to understand bias correlations in mitigation, this work aims to present the first systematic study in that direction. We explicitly examine bias mitigation in two widespread NLP tasks—toxicity detection and word embeddings—with respect to three social identities, namely race, gender, and religion.

21.10 NLP IN BUSINESS

The most popular websites today all incorporate recommendation systems. Finding the appropriate products and content would be challenging for users without them. Content-based filtering is one of the most widely used techniques for making recommendations. It is dependent on studying product metadata, which is largely composed of textual information. Although they are often utilized, there is currently no method for producing and evaluating content-based recommendations. A study's researchers first looked at the methods currently used for developing, evaluating, and training recommendation systems based on textual data to suggest books to users of the Goodreads website (Chen et al., 2021; Gardner et al., 2020). They took a critical look at current approaches and made recommendations for how to use NLP to enhance content-based recommenders.

Development of software requires the use of the problem-solving method of decomposition. However, it is regarded as the hardest programming ability for beginners to learn. Researchers studied decomposition in basic programming classes using case studies, surveys, and guided experiments. The exponential development of fields like machine learning and NLP has undoubtedly paved the way for more scalable solutions.

According to Nikita Klyuchnikov et al. (2022), neural architecture search (NAS) is a rapidly evolving field of study with promising potential. However, due to its requirement of substantial computational resources for training numerous neural networks, NAS proves impractical for researchers who have limited or no access to high-performance clusters and supercomputers. As a way around this problem and to guarantee repeatable tests, a handful of benchmarks with pre-computed performances of neural architectures have recently been introduced. While useful for computer vision applications in general, these benchmarks were developed using image datasets and convolution-derived architectures. Using the language modelling task, which is fundamental to NLP, has resulted in several important contributions, including the creation of a search space for training recurrent neural networks on text datasets, the development of methods for both intrinsic and extrinsic evaluation of trained models via evaluations of semantic relatedness and language understanding, and the testing of multiple NAS algorithms to show how the pre-computed residuals work. It is believed that the benchmark will aid in the development of new NAS techniques that are well suited for recurrent architectures and give the community access to more trustworthy empirical findings.

21.11 NLP IN SOCIAL MEDIA

As people express their opinions to their family and friends, social networks are crucial sources for learning about people's attitudes and opinions toward various issues. Recent years have seen significant challenges in the fields of NLP and psychology related to the detection of suicidal ideation through online social network analysis. Social media information can be effectively utilized to identify the complex early signs of suicidal ideation, which has the potential to save many lives (Rezaul Haque et al., 2022). The goal of this research is to use a variety of machine learning and deep learning models to identify suicidal ideation on the microblogging service Twitter. The main research aim is to enhance model performance compared to earlier research work in order to precisely identify early warning indicators and stop suicide attempts. They trained various machine learning and deep learning models and employed feature extraction techniques like word embedding and count vectorizer to achieve this.

Unstructured opinionated material, such as free-text comments, on the other hand, is abundant on publicly accessible websites, such as social networking platforms, blogs, forums, and websites that make recommendations (Lewis et al., 2020; Lorenzini et al., 2021; Malamas et al., 2022). In their paper, they describe a novel way for building a knowledge-based recommender system using unstructured (text) data.

The essential components of this method are organizing the text, extracting aspect-based sentiment scores for each text item, and applying an opinion mining algorithm. The sentiment analysis data table is subjected to an action rule mining method. The strategy's proposed application is to the challenge of increasing customer satisfaction ratings. The dataset of consumer reviews of repair services was extensively and appropriately assessed. The research findings were also used to help create a user-friendly recommendation system for the Web that may advise companies on how to improve their services in order to increase their profitability.

21.12   NLP IN THE MEDICAL FIELD

Problems with scanned documents in electronic health records (EHRs) have persisted for decades, and they are unlikely to go away very soon. Methods of processing such as NLP, optical character recognition (OCR), and picture pre-processing are now employed in NLP. However, there hasn't been much research on the interactions between picture pre-processing methods, NLP models, and document layout. In a study, the two key indicators for sleep apnea—oxygen saturation (SaO2) and the apnea hypopnea index (AHI)—were examined using data from 955 scanned sleep study reports. Images were pre-processed using a variety of methods, including greyscale, dilating, erosion, and contrast (Hsu et al., 2022). Tesseract was utilized to implement OCR. Three deep learning models and seven conventional machine learning models were compared. They evaluated combinations of two deep learning architectures and two image pre-processing techniques to achieve the best performance throughout. In order to extract relevant information from scanned reports, a number of linked steps must be followed. Although it would be impossible to test every potential option combination, they did evaluate several of the most significant information extraction procedures, such as image processing and NLP. There is an urgent need to create NLP algorithms to extract useful information from scanned papers because they will likely remain a staple of healthcare for the foreseeable future. Scanned documents may be processed more easily if image pre-treatment and document layout are used properly.

21.13   CONCLUSION

The analysis of publications in Scopus-listed journal articles that employ NLP as its primary analytic method show how textual data may be used to promote management ideas in many fields. The discussion of its use began with the introduction of NLP as an analytical technique, along with the necessary toolkits and procedures, as well as its advantages and disadvantages. This study makes use of this opportunity to draw attention to the technological and managerial limitations associated with the application of NLP in the field of management research; this will help direct future studies. Furthermore, the study discusses the use of NLP in the field of business analytics. Business analytics is an upcoming trend in global business management. By using NLP, performing business analytics becomes more improvised. Like business analytics, big data analytics is also a hot topic to be learned and adopted in various fields of analytics. The study discusses how the information is extracted and the process involved in using NLP in big data analytics. The study also discusses the use of NLP in various fields like education, linguistic analysis, machine learning, cyber security, business, social media, and medicine. Thus, this study helps to learn about NLP, its application techniques, and its current use in various fields.

REFERENCES

Balkir, E., Kiritchenko, S., Nejadgholi, I., & Fraser, K. C. (2022). Challenges in Applying Explainability Methods to Improve the Fairness o f NLP Models. http://arxiv.org/abs/2206.03945.

Belz, A. (2022). A Metrological Perspective on Reproducibility in NLP. Computational Lingu istics, 1–11. https://doi.org/10.1162/coli_a_00448

Berbatova, M. (2019). Overview on NLP Techniques for Content-Based Recommender Systems for Books. Proceedings of the Student Research Workshop Associated with RANLP, 55–61. https://doi.org/10.26615/issn.2603-2821.2019_009

Chen, J., Tam, D., Raffel, C., Bansal, M., & Yang, D. (2021). An Empirical Survey of Data Augmentation for Limited Data Lea rning in NLP. http://arxiv.org/abs/2106.07499.

Cheng, L., Ge, S., & Liu, H. (2022). Toward Understanding Bias Correlations for Mitig ation in NLP. http://arxiv.org/abs/2205.12391.

Gardner, M., Artzi, Y., Basmova, V., Berant, J., Bogin, B., Chen, S., Dasigi, P., Dua, D., Elazar, Y., Gottumukkala, A., Gupta, N., Hajishirzi, H., Ilharco, G., Khashabi, D., Lin, K., Liu, J., Liu, N. F., Mulcaire, P., Ning, Q., . . . Zhou, B. (2020). Evaluating Models' Local Decision Boundaries via C ontrast Sets. http://arxiv.org/abs/2004.02709.

Garrido-Muñoz, I., Montejo-Ráez, A., Martínez-Santiago, F., & Alfonso Ureña-López, L. (2021). A Survey on Bias in Deep NLP. https://doi.org/10.20944/preprints202103.0049.v1

Haque, R., Islam, N., Islam, M., & Ahsan, M. M. (2022). A Comparative Analysis on Suicidal Ideation Detection Using NLP, Machine, and Deep Learning. Technologie s, 10(3), 57. https://doi.org/10.3390/technologies10030057.

Hershcovich, D., Frank, S., Lent, H., de Lhoneux, M., Abdou, M., Brandl, S., Bugliarello, E., Piqueras, L. C., Chalkidis, I., Cui, R., Fierro, C., Margatina, K., Rust, P., & Søgaard, A. (2022). Challenges and Strategies in Cross- Cultural NLP. http://arxiv.org/abs/2203.10020.

Hershcovich, D., Webersinke, N., Kraus, M., Bingler, J. A., & Leippold, M. (2022). Towards Climate Awareness in NLP Research. http://arxiv.org/abs/2205.05071.

Houlsby, N., Giurgiu, A., Jastrzebski, S., Morrone, B., de Laroussilhe, Q., Gesmundo, A., Attariyan, M., & Gelly, S. (2019). Parameter-Efficient Transfer Lear ning for NLP. http://arxiv.org/abs/1902.00751.

Hsu, E., Malagaris, I., Kuo, Y.-F., Sultana, R., & Roberts, K. (2022). Deep Learning-Based NLP Data Pipeline for EHR-Scanned Document Information Extraction. JAMI A Open, 5(2). https://doi.org/10.1093/jamiaopen/ooac045.

Klyuchnikov, N., Trofimov, I., Artemova, E., Salnikov, M., Fedorov, M., Filippov, A., & Burnaev, E. (2022). NAS-Bench-NLP: Neural Architecture Search Benchmark for Natural Language Processing. IEEE Access, 10, 45736–45747. https://doi.org/10.1109/ACCESS.2022.3169897.

Lewis, P., Perez, E., Piktus, A., Petroni, F., Karpukhin, V., Goyal, N., Küttler, H., Lewis, M., Yih, W., Rocktäschel, T., Riedel, S., & Kiela, D. (2020). Retrieval-Augmented Generation for Knowledge-Intensi ve NLP Tasks. http://arxiv.org/abs/2005.11401.

Lorenzini, J., Kriesi, H., Makarov, P., & Wüest, B. (2021). Protest Event Analysis: Developing a Semiautomated NLP Approach. American Behavior al Scientist. https://doi.org/10.1177/00027642211021650

Malamas, N., Papangelou, K., & Symeonidis, A. L. (2022). Upon Improving the Performance of Localized Healthcare Virtual Assistants. Healthcare (Switzer land), 10(1). https://doi.org/10.3390/healthcare10010099

Pais, S., Cordeiro, J., & Jamil, M. L. (2022). NLP-Based Platform as a Service: A Brief Review. Journal of Bi g Data, 9(1). https://doi.org/10.1186/s40537-022-00603-5.

Rau, D., & Kamps, J. (2022). The Role of Complex NLP in Transformers for Text Ranking? https://doi.org/10.1145/3539813.3545144

Varsha, J., & Bharathi, B. (2022). Proceedings of the Second Workshop on Speech and Language Technologies for Dravidian Languages, pages 158-164 SSNCSE NLP@ TamilNLP-ACL2022: Transformer based approach for detection of abusive comment for Tamil language.

Voa, N. N. Y., Vu, Q. T., Vu, N. H., Vu, T. A., Mach, B. D., & Xu, G. (2022). Domain-Specific NLP System to Support Learning Path and Curriculum Design at Tech Universities. Computers and Education: Artificial Int elligence, 3. https://doi.org/10.1016/j.caeai.2021.100042.

Wang, L., Shen, Y., Peng, S., Zhang, S., Xiao, X., Liu, H., Tang, H., Chen, Y., Wu, H., & Wang, H. (2022). A Fine-grained Interpretability Evaluation Benchmark fo r Neural NLP. http://arxiv.org/abs/2205.11097.

22 Detection of Polarity in the Native-Language Comments of Social Media Networks

Sudeshna Sani, Dipra Mitra, and Soumen Mondal

22.1 INTRODUCTION

The quantity of information stored on the Internet grows by the day in our digital age, and the overall to date, there has been a massive amount of data stored. It is no longer possible to determine or study the pattern manually and the behaviors of this massive amount of data from diverse types of people. This data, however, has been made public since it contains extremely useful information on the attitudes of many persons from many categories all across the world. As a result, it has become critical to use automated systems to summarize this massive volume of data.

The analysis of such a vast population's thoughts has become increasingly challenging, demanding the application of new approaches. Several studies on English sentiment analysis have been undertaken, and many of these studies have produced remarkable results. However, because Bangladeshi language sentiment analysis has gotten so little attention, there are a lot of research prospects in this area.

According to this chapter, words' vector representations and sentiment information can both be employed jointly in the analysis of Bengali comments for sentiment. Each comment is rated as excellent or poor based on the opinions of the people who posted it in a Bengali microblogging service. A collection of single and multiline comments of people's viewpoints are collected through surveys, observing that the categorization of feelings is influenced by the sentiment information in the comments as well as the context of the remarks. A new method for combining these two forms of data has been developed, and it has produced impressive results.

22.2 RELATED CONTRIBUTION

The first letter of key terms should be capitalized in major headings, which should be typeset in boldface. English sentiment analysis has been the subject of numerous

DOI: 10.1201/9781003328414-22

studies. Paul Lewis et al. [1] and Cui et al. [2] focused on reviews of products found online. They separated the input into two groups: positive and negative. They examined over 100,000 product reviews from a variety of sources. Jagtap et al. used the support vector machine (SVM) and hidden Markov model (HMM) [3]. Alm and colleagues used a mixed classification algorithm to extract the teacher evaluation emotion, and it worked effectively [4]. Emotions can be classified as positive, negative, or neutral. Phrases were split into three polar categories. Using the winnow parameter modifying approach, they were able to attain 63 percent accuracy. Unigram was used by Agarwal et al. [5]. To extract Twitter sentiments, they used a tree and feature-based model which outperformed the unigram model. They obtained a 61 percent accuracy rate. Zou et al. [6] proposed a method for learning a multilingual unlabeled dataset with word embeddings when it comes to semantic similarity; their model outperformed baselines. Brown clusters, embeddings from Collobert and Weston (2008), as well as hierarchical log- bilinear embeddings were all investigated by Turian et al. [7]. Chen et al. [8] provided a few methods for distinguishing word embedding models, which have been released. They demonstrated that even without possessing the structure, embeddings may detect surprising semantics in texts. Tang et al. [9] proposed a method for obtaining information about words, both contextual and sentiment, by using a methodology they developed. Sentiment-specific word embedding is a technique for obtaining information on both the context and the sentiment about words. They used their model to extract sentiments from Twitter. They acquired an accuracy of roughly 83 percent.

Omer Levy and Yoav Goldberg [10, 11] (2014) and Mikolov et al.'s (2013) skip-gram model with negative sampling was generalized for word representation.

They extracted contexts based on dependency and demonstrated that they produce various forms of similarities. Vocabulary extension, statistical sharing, and embedding structure are considered. According to Andreas and Klein [12] and Hellinger PCA, Lebret and Collobert [13], word embeddings have three potential benefits. They devised a system to determine how words are represented in context, a word co-occurrence matrix. They got an accuracy of roughly 89 percent. To determine word embedding models, Levy et al. [14] used a neural network and word embedding models. Bengali is the subject of a few studies. To determine the emotion of Bengali microblog postings, Chowdhury and Chowdhury [15] used maximum entropy with SVM (MaxEnt). They tried combining the two strategies with different types of attributes. Contextual valency analysis was described by Hasan et al. [16] as a tool for detecting feelings in Bengali literature. They used overall positivity, total negativity, and total neutrality, which can all be determined with a part of speech (POS) tagger, and then calculated the final results. Das [17] proposed an approach for detecting feelings in Bengali and English texts using a computer method. He categorizes feelings into six groups. They are joyful, depressed, angry, disgusted, terrified, and surprised. Hasan et al. [18] introduced an emotion analyzer that can determine people's feelings. Positive and negative sentiment phrase patterns, as well as sentiment orientations, provide sentiment information. Islam et al. used Facebook to detect the emotion of Facebook status in Bengali. In reference [19], the authors employed the naive Bayes model with the naive Bigram and Bayes approach, achieving an F-score of 0.72. We utilized the sentiment of Bengali comments detected using Hellinger PCA and word embedding [11]. A matrix of co-occurrence of words is generated using skip-gram to establish the context. Sliding windows are established to capture pertinent terms in the windows, and information from the comments is collected.

22.3 METHODOLOGY

The Subjectivity Word List and SentiWordNet (Esuli et al., 2006; Wilson et al., 2005) are two extensively used lexical resources in English for subjectivity detection. SentiWordNet is an artificially generated English lexical database that gives each WordNet synset a positivity and negativity score ranging from 0 to 1. SentiWordNet 1.1 for English was released and the same authors have offered English translations. The vocabulary of subjectivity was devised using hand-crafted resources and entries extracted from corpora. The items from the subjective part of the entry's dependability has been categorized as either strong subjective or weak subjective in the subjectivity lexicon.

POS Tagger: A POS tagger scans each word for its assigned parts of speech in a language's text (and other token), such as noun, verb, adjective, and so on; however, most computational applications utilize finer-grained POS tags like 'noun-plural.' Kristina Toutanova created the tagger in the first place. Since then, Dan Klein, Christopher Manning, William Morgan, Anna Rafferty, Michel Galley, and John Bauer have worked to improve language speed, performance, usability, and support.

To tokenize our statement, we use Stanford's POS tagger. Then we choose only those parts of speech that can influence the polarity of a sentence, or polar terms. Table 22.1 lists the abbreviations for several parts of speech.

In the same context, similar phrases occur more frequently. WORD2VEC [20] converts each word into a vector representation. Similar words cluster together in

TABLE 22.1

POS Types

Pos_name	Pos_abbreviation	Sentiwordnet_Abr
Noun	NN	n
Adjective	JJ	a
Verb	VB	v
Adverb	RB	r
Noun	NNS	n
Adjectives	JJS	a

TABLE 22.2

Statistics on Bengali Corpus

	NEWS	BLOG
Number of documents in total	200	–
Number of sentences total	4434	500
The average number of sentences in each paragraph	50	–
Number of different word forms	45,807	8675
A document's average number of word forms	488	–
The total amount of unique word forms	35,176	2478

the WORD2VEC model's vector space. WORD2VEC keeps the words' syntactic meanings and sorts them by syntactic similarity. As a result of their syntactic structure, equivalent words in WORD2VEC vector space stay closer, but opposite emotion polarity words may also stay closer, resulting in poor sentiment classification results. As a result, in sentiment analysis, the polarity score of each word is crucial. Two steps were taken: For polarity detection, we used WORD2VEC word embedding to gather related terms and SentiWordNet to gather lexically similar words. We created a novel strategy that combines WORD2VEC's similarity score of co-occurring words to overcome each word in the query comment's disadvantages of WORD2VEC's emotion polarity score. We used two separate domain corpora, namely NEWS and BLOG, to assess subjectivity. Although sentiment lexicons are often domain agnostic, they are a useful place to start. There are more domains available to provide an adaptation or a fine-tuned approach in the literature. SentiWordNet (Bengali) is used in a subjective classifier that evaluates its coverage using a modest number of rules. Table 22.2 shows the size of both the corpus and the sample.

Then we present our technique for determining the overall polarity of the sentence, which includes terms that could improve, increase, or decrease the polarity of the related word.

22.4   ALGORITHM FOLLOWED FOR POLARITY DETECTION

1) Partition the data into two languages: English and Bengali.
2) Use a file reader object to read the file.
3) Using a POS tagger, parse each text token by token.
4) Each token will be assigned a tag by the POS tagger.
5) Run each token through SentiWordNet to calculate its score and polarity if its tag is 'JJ' or 'JJS' (i.e. the tagged token is an adjective/opinion word).
6) SentiWordNet will yield the word's sentiment type (positive, weak positive, strong positive, negative, strong negative, neutral, etc.) based on its score.
7) For each sentence, count the number of adjectives that are positive (pos count) and negative (neg count).

8) The sentence is regarded as 'negative' as a whole if the negative count is odd. Otherwise, move to step 9. $(-)+(+) = (-)$

9) The sentence is regarded as 'positive' as a whole if neg count is an EVEN integer (consider zero as even). $(-)+(-)$ Equals $(+)$ or $(+)+(+)=(+)$.

For comparison with SentiWordNet (English), the same subjectivity detection methodology was used as applied to the IMDB Movie Review and Multi Perspective Question Answering (NEWS) corpora (English). SentiWordNet (Bengali) has considerable coverage, according to the results of the subjectivity classifier on both corpora. The word list for subjectivity utilized in the subjectivity detection method was developed using the same IMDB corpora as in this study. SentiWordNet is a network of SentiWords (Bengali); on the other hand, it is corpus-independent and has excellent coverage.

The goal of this test is to determine how reliable sentiment lexicon polarity scores are. Beginning with a dictionary words and phrases that are both positive and negative is a frequent approach to sentiment analysis. These lexicons are a collection of lexicons that are used to label the previously out-of-context polarity of entries. In what ways may the present SentiWordNet (Bengali), a previous polarity lexicon, help with text polarity identification? To test the reliability of SentiWordNet (Bengali) polarity scores, a classifier for polarity was developed using SentiWordNet (Bengali) and various other linguistic characteristics. According to the feature ablation approach, the produced SentiWordNet (Bengali) is reliable in terms of the scores related to it. The findings of a SentiWordNet-based polarity classifier are shown in Table 22.3.

The SentiWordNet (Bengali) polarity scores must now be taken seriously. Unfortunately, there isn't a single paper in the literature that discusses the polarity classification accuracy that just uses prior polarity lexicon. A comparable investigation will be necessary in the future, but our findings suggest SentiWordNet is a network of words and could provide a solid foundation (approximately 50 percent accuracy).

22.5 EXPERIMENTAL APPROACH AND RESULTS

In order to calculate the accuracy through the WORD2VEC model we selected 20,000 Bengali comments from News and Blogs and 20,000 English comments from both MPQA and IMDB from the corpus mentioned in Table 22.3. We divided our entire dataset based on their responses; there are two subsets of favorable and

TABLE 22.3

Using SentiWordNet for Polarity-Wise Performance (Bengali)

Polarity	Precision	Recall
Positive	66.59%	62.89%
Negative	85.57%	75.87%

negative remarks. Because the positive and negative training datasets are founded on the views of a variety of persons, the training datasets have a high level of clarity and accurately reflect the actual situation. Though this sort of labelling reflects the actual situation, uncertainty may develop due to the variety of tags. This ambiguity can be eliminated by considering the opinions of a vast number of people, which we have done.

22.6 DISCUSSION AND RESULTS

Our model performed the execution in five steps. We received better accuracy in each stage. After each step 4000 additional observations from the training data were added. Our entire 20,000 comments in a database were thus trained in five steps. After each stage we determined the precision and recall and found that it rises as the dataset grows larger. Table 22.4 shows the average classification accuracy at each step for the Bengali corpus, and Table 22.5 shows a summary of precision and recall for both English and Bengali comments. The progress of classification accuracy for Bengali comments is shown in figure 22.1. The Bengali Blog confusion matrix in clearly shown in Figure 22.2 and Table 22.6.

TABLE 22.4
Average Classification Accuracy at Each Step for Bengali Corpus

Steps of Execution	Data	Accuracy in Terms of TPR and TNR Obtained (%)
1	4000	90%
2	8000	91%
3	12,000	91.5%
4	16,000	92.5%
5	20,000	93%
Average	20,000 sample data	93.20%

TABLE 22.5
Precision and Recall for Highest Number of Dataset

Languages	Domain	Precision	Recall
English	MPQA	86.08%	93.33%
	IMDB	89.90%	96.55%
Average of English		87.99%	94.94%
Bengali	NEWS	82.16%	96.00%
	BLOG	84.6%	90.4%
Average of Bengali		83.38%	93.20%

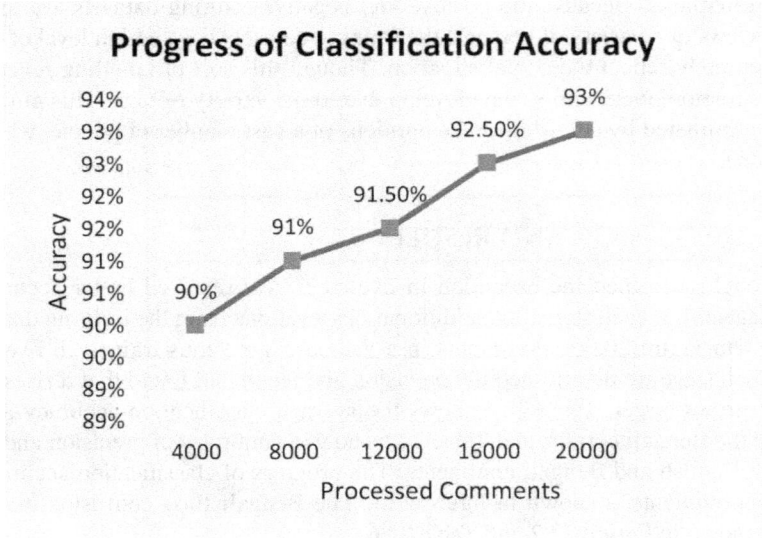

FIGURE 22.1 Classification accuracy for Bengali comments.

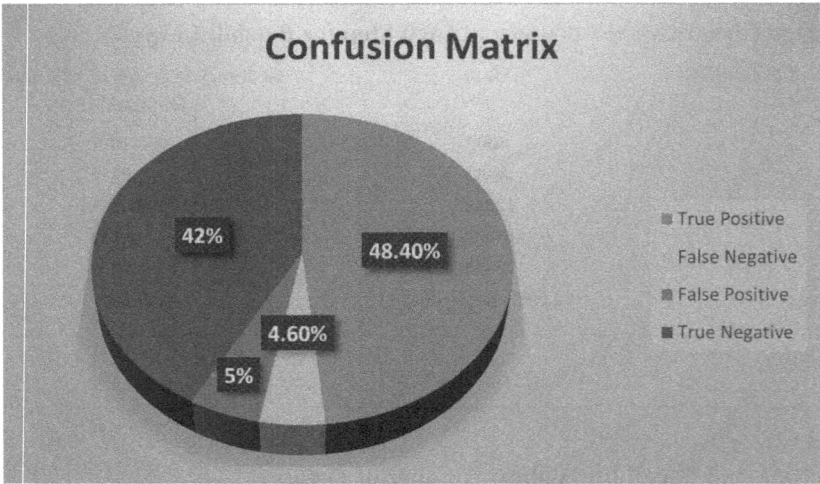

FIGURE 22.2 Testing confusion matrix with 20,000 comments.

TABLE 22.6

Confusion Matrix Bengali Blog

	Model Predicted as Positive	Model Predicted as Negative
Really Positive Comments	48.4%	4.6%
Really Negative Comments	5%	42%

22.7 CONCLUSION

The way words are represented in a sentence can influence their qualities. The context determines the meaning of the phrases. The context and word properties of a number of sentences is determined by word embedding. Other statistical techniques, including Bengali sentiment analysis, are heavily reliant on sentence structure. However, the outcomes of word embedding are feelings determined as a result of the surroundings aspects of the words, which are independent of the sentence patterns. We used word embedding on our own collection of comments, articles, and blogs in Bengali obtained recently because it is a novel technique for analysis. A collection of extremely both good and negative terms, along with their opposition scores, is created. The accuracy is increased when these results, as well as the neutralization word valence shifter, are coupled.

The result is 93.20 percent, which is highly intriguing and important for future research and tuning for native languages. Because the accuracy produced by our model rises with the size of the dataset, we are optimistic that if it is possible to produce a gold-standard dataset, this strategy can be followed where we will not have a sufficient dataset for native language comments. The graph depicting the level of precision shows that as the dataset grows larger, the accuracy increases; therefore, we're aiming to enhance the outcomes.

REFERENCES

[1] M. Paul Lewis, Gary F. Simons and Charles D. Fennig (eds.), *Ethnologue: Languages of the World*, Nineteenth edition. Dallas, Texas: SIL International, 2016.

[2] Hang Cui, Vibhu Mittal and Mayur Datar, "Comparative Experiments on Sentiment Classification for Online Product Reviews," Proceedings of the 21st National Conference on Artificial Intelligence, AAAI, Boston, MA, 2006.

[3] Balaji Jagtap and Virendrakumar Dhotre, "SVM and HMM Based Hybrid Approach of Sentiment Analysis for Teacher Feedback Assessment," *International Journal of Emerging Trends & Technology in Computer Science (IJETTCS)*, Volume 3, Issue 3, May–June 2014.

[4] C. Alm and D. Roth and R. Sproat, "Emotions from Text: Machine Learning for Text-Based Emotion Prediction," Proceedings of Human Language Technology Conference and Conference on Empirical Methods in Natural Language Processing (HLT/EMNLP), ACM, Pages 579–586, 2005.

[5] Apoorv Agarwal, Boyi Xie, Ilia Vovsha, Owen Rambow and Rebecca Passonneau, "Sentiment Analysis of Twitter Data," LSM'11 Proceedings of the Workshop on Languages in Social Media, Pages 30–38, 2011.

[6] Will Y. Zou, Richard Socher, Daniel Cer and Christopher D. Manning, "Bilingual Word Embeddings for Phrase-Based Machine Translation," Proceeding of Conference on Empirical Methods in Natural Language Processing, Pages 1393–1398, 2013.

[7] Joseph Turian, Lev Ratinov and Yoshua Bengio, "Word Representations: A Simple and General Method for Semi-Supervised Learning," Proceedings of the 48th Annual Meeting of the Association for Computational Linguistics, pages 384–394, Uppsala, Sweden, 11–16 July 2010.

[8] Yanqing Chen, Bryan Perozzi, Rami Al-Rfou, and Steven Skiena, "The Expressive Power of Word Embeddings," ICML 2013 Workshop on Deep Learning for Audio, Speech, and Language Processing, Atlanta, USA, June 2013.

[9] Duyu Tang, Furu Wei, Nan Yang, Ming Zhou, Ting Liu and Bing Qin, "Learning Sentiment-Specific Word Embedding for Twitter Sentiment Classification," Proceedings of the 52nd Annual Meeting of the Association for Computational Linguistics, pages 1555–1565, Baltimore, Maryland, USA, June 23–25, 2014.

[10] Omer Levy and Yoav Goldberg, "Dependency-Based Word Embeddings," Proceedings of the 52nd Annual Meeting of the Association for Computational Linguistics (Short Papers), pages 302–308, Baltimore, Maryland, USA, June 23–25, 2014.

[11] Md. Saiful Islam, Md. Al- Amin and Shapan Das Uzzal, "Word Embedding with Hellinger PCA to Detect the Sentiment of Bengali Text," The 19th International Conference on Computer and Information Technology (ICCIT–2016), December 18–20, North South University, Dhaka, 2016.

[12] Jacob Andreas and Dan Klein, "How Much Do Word Embeddings Encode About Syntax?" Proceedings of the 52nd Annual Meeting of the Association for Computational Linguistics (Volume 2: Short Papers), June 2014.

[13] Remi Lebret and Ronan Collobert, "Word Embeddings through Hellinger PCA," Idiap Research Institute, Rue Marconi 19, CP 592, 1920 Martigny, Switzerland, arXiv preprint arXiv:1312.5542, 2013.

[14] Omer Levy, Yoav Goldberg and Ido Dagan, "Improving Distributional Similarity with Lessons Learned from Word Embeddings," *Transactions of the Association for Computational Linguistics*, Volume 3, pp. 211–225, 2015. Action Editor: Patrick Pantel. Submission batch: 1/2015, Revision batch 3/2015, Published 5/2015.

[15] Shaika Chowdhury and Wasifa Chowdhury, "Sentiment Analysis for Bengali Microblog Posts," International Conference on Informatics, Electronics & Vision (ICIEV), 2014.

[16] K. M. Azharul Hasan, Mosiur Rahman and Badiuzzaman, "Sentiment Detection from Bengali Text using Contextual Valency Analysis," 17th Int'l Conf. on Computer and Information Technology, Daffodil International University, Dhaka, Bangladesh, 22–23 December 2014.

[17] Dipankar Das, "Analysis and Tracking of Emotions in English and Bengali Texts: A Computational Approach," Proceedings of the 20th International Conference on World Wide Web, WWW 2011, Hyderabad, India, March 28–April 1, 2011.

[18] K. M. Azharul Hasan, Sajidul Islam, Mashrur-E-Elahi and Mohammad Navid Izhar, "Sentiment Recognition from Bangla Text," Technical Challenges and Design Issues in Bangla Language Processing, IGI Global, 2013.

[19] Md. Saiful Islam, Md. Afjal Hossain, Md. Ashiqul Islam and Jagoth Jyoti Dey, "Supervised Approach of Sentimentality Extraction from Bengali Facebook Status," The 19th International Conference on Computer and Information Technology (ICCIT-2016), December 18–20, North South University, Dhaka, 2016.

[20] Tomas Mikolov, Kai Chen, Greg Corrado and Jeffrey Dean, "Efficient Estimation of Word Representations in Vector Space," Proceedings of International Conference on Learning Representations, ser. ICLR '13, 2013.

23 Machine Learning Techniques for Detecting and Analyzing Online Fake Reviews

Yashwitha Buchi Reddy, Ch. Prathima, Swetha Jaladi, B. Dinesh, and J.R. Arun Kumar

23.1 INTRODUCTION: BACKGROUND

Reviews have become the primary source of information for clients looking to make a judgment regarding services or items [1]. When a user wants to book a room in a resort, they look at online reviews about previous customers' experiences with the hotel's services [2]. They select whether or not to reserve a room based on the feedback from the reviews. They will most likely book the room if they find favorable reviews. As a result, past evaluations have become incredibly reputable sources of information for most individuals in various Internet services. Since studies are legitimate ways of giving input and any effort to change such ratings by providing false or disingenuous information is considered deceptive behavior, the like evaluations have been withdrawn.

In recent years, both corporations and the academic community have paid close attention to fake review identification. Detecting authentic user experiences and opinions is necessary for evaluations to reflect genuine customer views and experiences. Fake evaluations are a significant issue. Supervised learning [3] has become popular and has been one of the primary techniques to resolve the issue. Reviews of products or businesses are abundant on online e-commerce platforms, and they are critical for buyers making purchasing decisions. Some dishonest people are paid to post bogus evaluations, commonly known as "opinion spamming," promoting or degrading specific products and services [4].

The "false pandemic" is spreading across advertising. The rapid technological improvement that allows the manufacture of synthetic buyer outcomes such as hoaxes, as well as the ecosystem that has developed around these synthetic outcomes pertaining to false production, identification, and abatement, are the main determinants of this. Towards that purpose, assessing false reviews has been highlighted as an important agenda item in Internet and social media consumer research.

DOI: 10.1201/9781003328414-23

23.2 LITERATURE SURVEY

A. The significance of existing customers for platforms like walmart.com and Yelp.com, which deal with literature, resorts, and various other items, cannot be underestimated. However, this has given rise to a concerning trend known as slipper juggling, where publications write biased and overly positive reviews about their products. As a result, consumer trust in these reviews is often misplaced and misguided. People who purchase products online, such as audio books, frequently rely on reviews from other clients instead of those from specialized journals. Regrettably, because of the prevalence of the fake assessments from their own publications, those kinds of evaluations are frequently misguided. Fake reviews are difficult to detect [5].

B. Subscriber Internet assessments are a critical asset for customers to make buying choices; an expanding corpus of inquiry examines the influence of Internet review sites on trading volume and cost. Online reviews, in theory, should raise consumer and producer surplus by income and increased ability to judge unknowable customer satisfaction. Nevertheless, one key impediment to the efficiency of evaluations in revealing customer satisfaction is the likelihood of false or "advertising" customer reviews. Critics with a pecuniary interest in customers' buying habits, for instance, may produce assessments that look to be authored by neutral users but are meant to persuade consumers. Although there is a huge economic theory on influence and selling (discussed later), little attention has been paid to a setting of promotion masquerading as ratings and reviews [6].

23.2.1 METHODOLOGY

Naïve Bayes: A naïve Bayes is a classification machine learning model for classification.

The classifier's crux is built on the Bayesian statistics. $P(A \mid B) = \dfrac{P(B) \mid A)P(A)}{P(B)}$

We can use Bayes' theorem to calculate the odds of A happening if B has already happened. A is the assumption, and B is the confirmation [7]. The determinants of individual in this case are expected to be autonomous. In other words, the presence of one trait has no bearing on the existence of the other. As a result, it is naïve. Naïve Bayesian forecasts the probability of various classes depending on multiple attributes using a similar technique. This method is most typically used for text categorization and multiclass issues. Forecasting the class label data set is simple and quick. It also has the ability to forecast many classes. Whenever the isolation requirement is met, a naïve Bayes classification outperforms other algorithms, especially regressors, and requires a smaller data set. It performs effectively using categorical input variables as opposed to numeric input variables. For numerical variables, a normal distribution is assumed. Naïve Bayes is a popular in-text classification method because of its autonomous concept and strong performance in solving multiclass situations [6]. Hence it is used in various applications that uses smaller data sets. Rather than numerical values, naïve Bayes works better with category input variables.

KNN: This is straightforward and quick way to estimate the class of the testing data set. K-nearest neighbor (KNN) is a simple machine learning technique centered on learning that is supervised. The KNN method saves all current information and classifies new data pieces into similar groupings [1]. This means that using the KNN method, new data may be quickly filtered into the well-categorized sets. The KNN technique is useful for both classification and regression applications; however, it is more typically employed for classification. As seen in Figure 23.1, there are significant modules of data clustered using KNN.

Predicting the class label set of data is simple and quick. KNN is a non-parametric technique involving no decisions based on data [8] because it does not instantly understand from the test data set; it also is known as a slow learner algorithm. Rather, it stores the database and then, when the time comes, performs a categorization function on it.

Logistic Regression: A supervised learning approach used to forecast a dependent categorical predicted value is logistic regression. In essence, logistic regression may be useful if you have a huge quantity of data to categorize. For example, if you were given a cat and an apple and asked whether they were animals or not, you would expect the cat to be classified as a creature and the fruits to be classified as non-animal. Your goal is to correctly label the animal, which is dependent on your data. There are just two viable solutions in this example: animal or not an animal. However, you may configure your regression analysis with more than two potential groups (multinomial logistic regression). As seen in Figure 23.2, there are data points calculated using logistic regression.

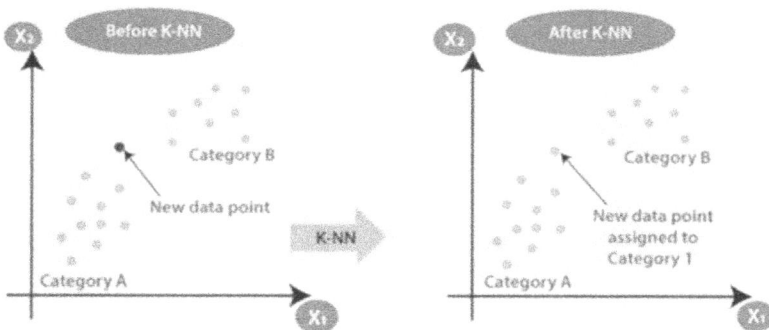

FIGURE 23.1 Data clustered using KNN.

XGBoost: The open-source gradient boosted trees program XGBoost (stochastic gradient enhancing) is famous and effective. Deep learning is an ensemble learning system that involves an aggregate of estimates out of a variety of basic and poorer models in an ability to forecast a target variable properly [1]. The gradient boosted approach thrives in deep learning challenges because of its own strong processing of a variety of information types, connections, patterns, or perfectly acceptable evaluation metrics. XGBoost can help with analysis, categorization (both single and categorical), and sorting problems. As seen in Figure 23.3, accurate predictions can be made using XGBoost.

The XGBoost algorithm distinguishes itself from previous gradient boosting approaches by employing a second estimation of the performance index called the XGBoost Model. Using this approximation, XGBoost can compute the best "if" condition and its influence on performance. XGBoost can then save them in memory for the next decision tree to avoid recalculating it. XGBoost is strong on its own, but it also works well with other tools in your machine learning toolkit. Consider feature engineering, in which the machine learning expert, in order to produce high quality, adds the raw inputs into new input characteristics before handing over control to the model. The XGBoost algorithm takes advantage of designed characteristics to generate a well-interpretable and high-performing model.

CatBoost: CatBoost, often known as categorical boosting, is an accessible enhancing library created by Yandex. CatBoost may be used in ranking, recommender systems, forecasting, and even personal assistants, in addition to

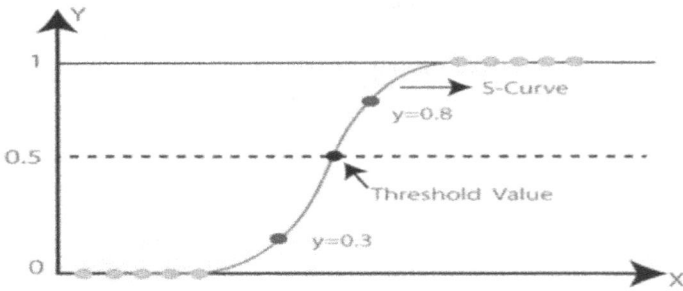

FIGURE 23.2 Logistic regression model.

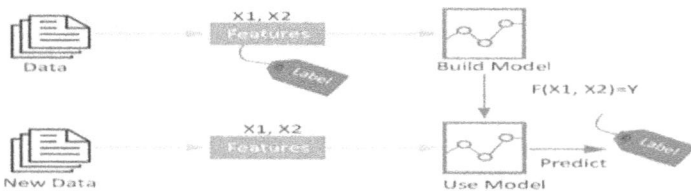

FIGURE 23.3 XGBoost model.

classification and regression problems. It is the first open-source Russian machine learning algorithm. Yandex's machine learning researchers and engineers created the algorithm in 2017 [7]. The goal is to serve several functions, such as personal assistants, self-driving automobiles, weather prediction, and a variety of other jobs. CatBoost is another method in the gradient boosting approach for decision trees. CatBoost has essential advantages, which is one of the major reasons why it was chosen by many boosting algorithms, including Light and the XGBoost method.

23.3   PROPOSED METHOD

The objective of this study is to develop a rapid and accurate method for detecting false reviews. To accomplish this, we utilized a Python-based Django framework that is deeply intertwined with our system. As seen in Figure 23.4 and Figure 23.5, there are significant modules for the assessment of fake reviews.

Advantages:

    i. The precision is excellent.
    ii. Simple to understand.
    iii. Extremely effective.
    iv. There is no requirement for skilled personnel.

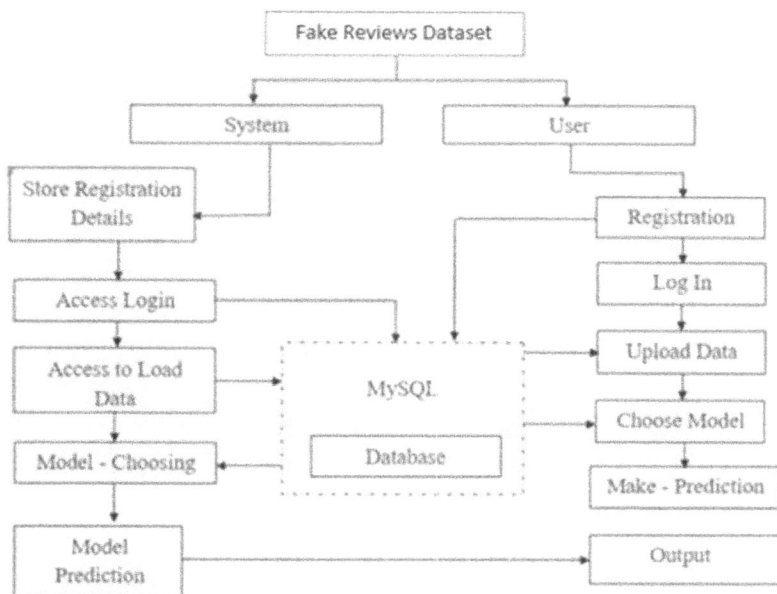

FIGURE 23.4   Significant modules for the assessment of fake reviews.

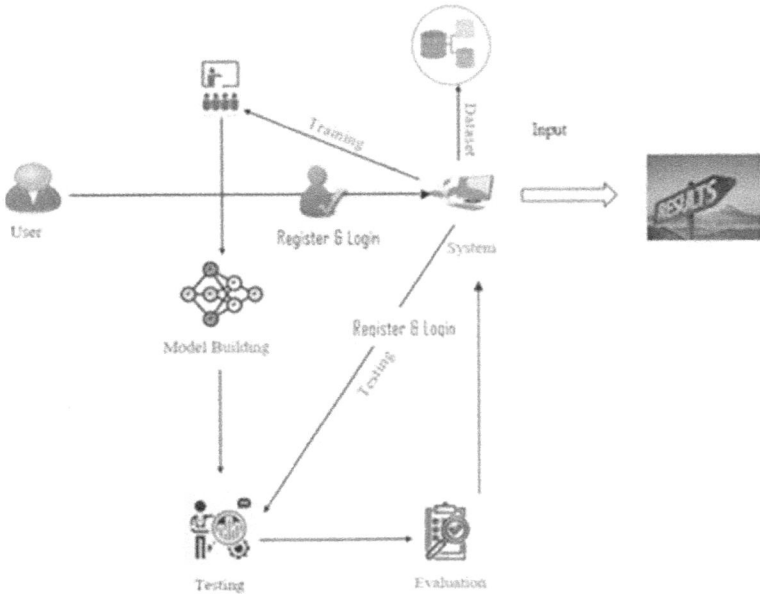

FIGURE 23.5 Model results.

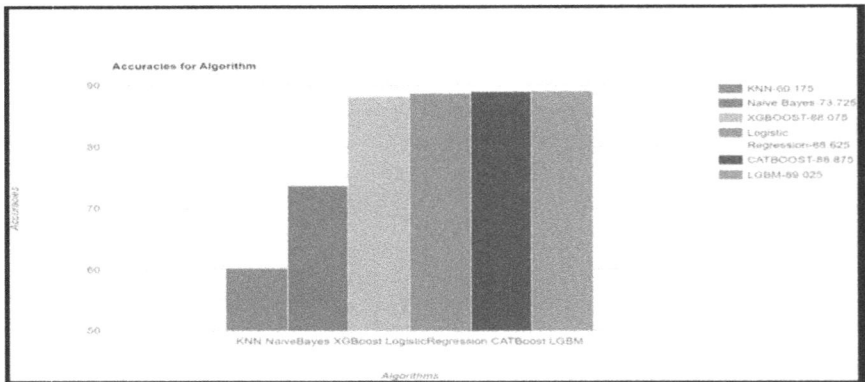

FIGURE 23.6 Graph of accuracies of test cases for the assessment of fake reviews.

23.3.1 RESULTS

Evaluation was done on our proposed system on the Yelp data set. Firstly we get the accuracy of each algorithm. Here are some test cases which are considered on different algorithms: As seen in Table 23.1, there are significant test cases for the assessment of fake reviews. Table 23.2 shows that there are significant values of accuracies of test cases for the assessment of fake reviews. Figure 23.6 shows a graph of accuracies of test cases for the assessment of fake reviews.

TABLE 23.1

Test Cases for the Assessment of Fake Reviews

Test Case	Input	Expected Output	Actual Output	P/F
Examine the test data.	Test- data path.	The data set must be successfully read.	The data set was successfully retrieved.	**Pass**
Preparing the dataset for analysis	Pre-processing starts	Pre-processing should be done on the dataset.	Pre-processing was finished successfully	**Pass**
Model construction	Model construction for the clean data	Models must be created using the necessary algorithms.	The model was successfully created [5]	**Pass**
Fake review estimation	Input provided.	Output should be whether review is fake or not.	Resulted successfully	**Pass**

TABLE 23.2

Accuracy Assessment of Fake Reviews

S. No	Algorithm	x
1	KNN	0.60175
2	Naïve Bayes	0.73725
3	XGBOOST	0.88075
4	Logistic Regression	0.88625
5	CATBOOST	0.88875
6	LGBM	0.89025

Test cases:

Input	Output	Result
Input text	Tested for whether the given review is fake or not	Success

Model construction and test cases:

23.3.2 Conclusion

We have effectively built a detection approach system light gradient boosting model (LGBM) for fake reviews that are found on numerous sites. This is done in a viewer atmosphere using Python programming on the Django framework. More feature

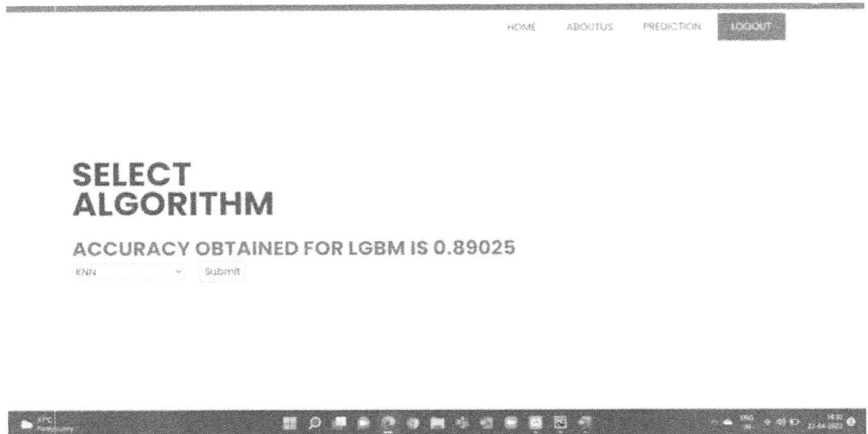

FIGURE 23.7　LGBM accuracy for detecting fake review.

choices and the ability to detect other types of bogus reviews might be implemented in the future. With the improved data set, we want to examine detection approaches and deploy the most effective and valid computational methods for recognition. As seen in Figure 23.7, there is an approach of LGBM accuracy for detecting fake review.

REFERENCES

[1] C. C. Aggarwal, "Opinion mining and sentiment analysis," in: *Machine Learning for Text*. Springer, Cham. pp. 413–434, 2018.
[2] O. A. A. C. A. I. R. Barbado, "A framework for fake review detection in online consumer electronics retailers," *Information Processing & Management*, pp. 1234–1244, 2019.
[3] S. Tadelis, "The economics of reputation and feedback systems in e-commerce," *IEEE Internet Computing*, vol. 20, no. 1, pp. 12–19, 2016.
[4] M. J. H. Mughal, "Data mining: Web data mining techniques, tools and algorithms," *Information Retrieval*, vol. 9, no. 6, pp. 12–15, 2018.
[5] V. V. B. L. A. N. G. A. Mukherjee, "What yelp fake review filter might be doing?" in Seventh International AAAI Conference on Weblogs and Social Media, 2013.
[6] N. J. A. B. Liu, "Review spam detection," in Proceedings of the 16th International Conference on World Wide Web, 2007.
[7] E. E. A. A. Gherbi, "Detecting fake reviews through sentiment analysis using machine learning techniques," *Iaria/Data Analytics*, pp. 564–569, 2017.
[8] R. P. A. U. A. P. W. V. Singh, "Sentiment analysis of movie reviews and blog posts," *Advance Computing Conference (IACC)*, pp. 893–898, 2013.

24 A Study on the Application of Expert Systems as a Support System for Business Decisions
A Literature Review

Geetha Manoharan, Subhashini Durai,
Gunaseelan Alex Rajesh, and
Sunitha Purushottam Ashtikar

24.1 INTRODUCTION

Gabriel Lanzaro and Michelle Andrade (2022) said speed limits balance safety and traffic flow. Establishing a speed limit typically entails choosing a base speed (such as operational speed or design speed) and modifying it in accordance with a number of additional factors. For instance, the typical recommendations in Brazil list a number of factors that influence speed limits but do not outline how to select a speed limit for a particular stretch of highway. So, in accordance with Brazilian practise, the decision-maker must make a decision on a particular issue that primarily depends on an expert opinion. This chapter suggests a fuzzy expert system for determining Brazil's highway speed limits. The system takes into account six input variables. Membership functions and fuzzy rules were generated by expert evaluations of simulated highway scenarios. The experts used linguistic factors and suggested speed limits as they assessed the scenarios. Afterwards, a Mamdani fuzzy controller was created. For the simulated highway scenarios, the expert's responses were compared to the controller's outputs. For additional system validation, some case studies of Brazilian highway segments were used. Results demonstrated that the fuzzy system can generate outputs that concur with professional assessments and current speed limits. This study's fuzzy controller can be used to help professionals set speed limits on Brazilian highways.

24.2 EXPERT SYSTEM IN THE CONSTRUCTION FIELD

24.2.1 REAL-TIME EXPERT SYSTEM FOR SAFETY

According to Han Zhang et al. (2022), real-time safety monitoring at hydroelectric facilities is important for both the facility's safety and its ability to produce power. Due to the vast amount of monitoring data and the complex design of the dam, real-time monitoring of the majority of the dam is not possible. Developing a high-performance monitoring system for safety is advantageous for dam safety management. The paper presents an expert system to manage dam safety that executes real-time dam safety monitoring tasks, including abnormal data imports and modifications. The features of the system are numerous and include remote consultation, visual query, and analysis of dam safety, among others. Ten Chinese dam projects have used the system. The application demonstrates how data anomalies like step changes, multi-point outliers, and single-point jumps can be recognised by reliable data evaluation. Monitoring trends can reveal potential structural alterations. A comprehensive assessment of the safety state of various dam components can be obtained by a comprehensive evaluation. Comprehensive online systems can improve dam safety monitoring and power plant management.

24.3   EXPERT SYSTEM IN AGRICULTURE

24.3.1   Expert System to Diagnose Diseases in Tomato Plants

According to Mr. Mohanad H. Al-Qadi et al. (2022), the destruction of the tomato plant and its crops is without a doubt caused in large part by tomato diseases. These plants suffer obvious harm as a result, making them unusable. Finding out about these illnesses is a necessary first step in getting them treated properly. Using a high degree of accuracy, the diagnosis can be used to determine the course of treatment. Expert systems, used properly, can significantly aid in protecting these plants from harm. According to an accurate diagnosis, the expert system helps farmers find the right treatment for tomato disease. The purpose of this system is to identify tomato plant diseases using an expert system built on C Language Integrated Production System (CLIPS).

24.3.2   Expert System to Diagnose Plant Diseases Using Artificial Intelligence

Among others, Radwan et al. (2022) state that in the world of agriculture, plant diseases are common. Most farmers struggle greatly because of these diseases. We concentrate on seven distinct diseases affecting passion fruit, with symptoms. Expert racially problems in specialised domains, requiring humans to facilitate the technology and making life easier for people today. Joshua Lederberg and Edward Feigenbaum of Stanford University in California, created the first expert system in 1965 and developed Dendral, an expert system for the analysis of chemical compounds used in medical diagnosis, petroleum engineering, and finance. In light of the importance of expert systems to people, an expert system was developed in the agricultural industry that uses the CLIPS expert system language to diagnose plant diseases. The suggested expert system was designed and put into action using the system. Passion-related diseases can be more easily diagnosed thanks to the system. This professional system will aid farmers and others in the agricultural industry in the accurate diagnosis of passion-related diseases and treatment information. Seven diseases, including brown spot, *Septoria* spot, root and crown rot, *Fusarium* wilt, anthracnose, woodiness virus, and scab, are diagnosed by the system's project. Results: Farmers evaluated and applauded the expert system for its successful and useful support.

24.3.3 EXPERT SYSTEM USING CLIPS TO DIAGNOSE DISEASES IN THE MINT PLANT

According to Megdad et al. (2022), mint is a grassy, perennial plant that grows quickly and widely, whose leaves are green, fragrant, tart, and refreshing, with square-shaped legs that are bifurcated and erect, and they can be as tall as 10–201 cm. It calls Europe and Asia home. The significant effects are relieving pain, gallbladder issues, gas release, anti-inflammatory properties, and nerve relaxation. The mint plant also has a number of other advantages. Even though mint is the best plant to use as a starter crop in gardens, it is susceptible to a number of common ailments that stunt its development. Get the proper disease diagnosis and treatment using this expert system's primary objectives. The paper suggests the expert system be designed for farmers and agriculture enthusiasts to diagnose mint diseases such as mint rust, *Verticillium* wilt, anthracnose, powdery mildew, black stem rot, stem and stolon canker, and *Septoria* leaf spot. The study provides a summary of diseases as well as information on their causes and, whenever possible, treatment recommendations. Designing and putting into practise the suggested expert system requires the use of the CLIPS expert system language. Results: Al Azhar University agricultural students and a group of agriculture-interested friends deemed the proposed mint disease expert system satisfactory. The suggested expert method can be extremely helpful to farmers and those who are interested in agriculture.

24.3.4 EXPERT SYSTEM IN IDENTIFYING DISEASES IN BROCCOLI PLANTS

In the study of Ola I. A. Lafi, among others (2022), the large flower head, stalk, and small accompanying leaves of the broccoli plant, a member of the family Brassicaceae and genus *Brassica*, are consumed as vegetables. A broccoli leaf may be affected by one of the diseases described in this paper. When symptoms appear, medical attention is needed in some way. If appropriate survival measures for broccoli disease are not taken promptly, broccoli could perish. Getting the proper disease diagnosis and treatment is the primary objective. This study looks into the development of an expert system to help farmers identify broccoli diseases such as damping off, club root of crucifers or finger and toe disease, *Alternaria* leaf spot, black rot, downy mildew, and white rust.

24.3.5 EXPERT SYSTEM IN CORIANDER PLANTS

Y. I. Aslem and Samy S. Abu-Naser (2022) said that the field that is evolving fastest is artificial intelligence. Today, computer science applications are widely used. Expert systems use artificial intelligence to make predictions and decisions to help solve problems in science, medicine, and even architecture; eventually, they will replace human experts. The user will be able to receive an accurate diagnosis or solution to their problem without having to physically meet with a real-life specialist. This study shows how to use computers in agriculture, like in the coriander plant, and five of its well-known ailments, to build a disease prediction system. The system was created with CLIPS. A thorough summary of the completed work is provided in this article.

24.3.6 Implementation of an Expert System to Diagnose Plant Diseases

Elvis Pawan, among others (2022), state that corn plants are prone to disease and pests and require specialised knowledge to manage. Plant diseases are very upsetting and prevent you from getting the best results. Farmers in the Mura Tami sub-district have found difficulties when they lack the knowledge and skills to identify early corn plant diseases. Through the design of an expert system application, farmers will be able to identify plant diseases as early as possible thanks to the study's goal. Forward chaining, a forward tracing technique, was used to create the expert system. PHP and the MySQL database are used for system development. According to the study's findings, an expert system can successfully implement the forward chaining method. In contrast to user acceptance testing (UAT), which demonstrated 84 percent of respondents strongly agreed with the implementation, black box testing revealed 100 percent of the system's functionality could be done well.

24.4 EXPERT SYSTEM IN PEAR TREES

For Hadeel A. El-Hamrnah et al. (2022), there is no denying that plant diseases are widespread in the agricultural industry. Most farmers, who rely on agriculture for survival, struggle with these diseases. We list the 11 illnesses that affect pear fruit, which exhibit different symptoms for each. Expert systems have been integrated into daily life as they contain various systems and area. For example, using the CLIPS expert system language, an integrated expert system has been developed to diagnose pear diseases. The system simplifies pear-related disease diagnosis. This expert system assists farmers and others working in agriculture in the diagnosis of ailments affecting pears. The purpose of this project is to teach farmers how to properly diagnose and treat diseases that affect pears. The system has a program that diagnoses the following 11 diseases that affect pears: scab, seedling blight, crown gall, white root rot, collar rot, powdery mildew, leaf spot, canker, viral diseases, phytoplasma disease, and bitter rot. Results: Farmers assessed the expert system's assistance to them and gave it high marks.

24.5 USAGE OF EXPERT SYSTEMS AND WEARABLE TECHNOLOGY IN PSYCHOLOGY

24.5.1 To Identify Diseases Connected to the Heart

Mohd Sani et al. (2022), among others, state that the global market is seeing an increase in the popularity of smartphones and wearable technology. These gadgets record information about heart rate, medications, and self-diagnosis of one's health. In addition to collecting personal health data, wearable technology can diagnose hypertension. Within the Malaysian population, hypertension is one of the risk factors for diseases connected to the heart. The ability to diagnose hypertension is not available in any of the numerous mobile applications that are paired with wearable technology to track medical conditions. For this study, we looked at studies that used wearable technology and expert systems to study hypertension. A systematic literature review was conducted based on hypertension risk factors, expert systems, and wearable technology.

After the filtering stage, we located 15 particular research papers. The following three main areas of interest were highlighted by the key findings: the causes of hypertension, expert system techniques, and the different kinds of sensors used in wearable technology. The most frequent cause of hypertension that can be measured by wearable technology is blood pressure. For expert systems, we found that machine learning, neural networks, and fuzzy logic are the three most popular methods. In research on hypertension, the wrist band is the most popular sensor for wearable devices.

24.6 EXPERT SYSTEM IN THE HIGHER EDUCATION FIELD

24.6.1 Designed to Assess Potential

Wulansari et al. (2022), among others, have noted that with technological advancements permeating life and education, this course is designed to help students recognise their true potential and abilities in higher education. This research aims to develop expert system software that students can use to recognise patterns. This expert system assesses higher education potential using software development life cycle (SDLC). This expert system uses multiple intelligences and unified modeling language (UML).

24.6.2 Expert System for Teaching English

In a paper by Jingwei Tangand. To Yi Deng (2022), "As Internet technology evolves, so does the modern educational system," modern education can be improved by integrating technology into the curriculum, a departure from the teacher-centred classroom model. Due to variations in students' knowledge and learning potential, it's difficult to teach students in the traditional educational system. Tiered instruction is one method of addressing this disparity, but there are some drawbacks, for which implicit tiered instruction is a fantastic fix. Artificial intelligence studies human intelligence principles. The current educational system's problems with a shortage of teachers and a single teaching method could be resolved successfully with the help of an expert system for teaching English. It will make it possible for teachers to more easily share their resources, which will enhance the quality of instruction, lighten the load on their shoulders, and help students develop independent learning skills. This study examines current issues in English teaching and introduces a solution for implementing an artificial intelligence–based university English teaching aid system, which employs generative and framework-based knowledge representations of English language experts' subject matter to assist teachers in better understanding their students' mastery of knowledge points. The system assists teachers in understanding students' reasoning and analysis, as well as increasing their interest, learning techniques, and English language proficiency.

24.7 EXPERT SYSTEM DESIGNED AS AN AUXILIARY TEACHING SYSTEM

Huanhuan Chu (2022), College The origin of Japanese professional development can be found in Japanese teaching. Due to increased interactions with Japan in our country's politics, economy, and trade, Japanese is becoming more popular and universal.

Due to the constantly growing trend of using it as a tool for international communication, Japanese instruction in colleges is no longer effective. The study employs an expert system as the theoretical foundation and back propagation (BP) neural network technology as the auxiliary teaching system for Japanese teachers and students to address the needs and growth of society, teacher shortages, and a disregard for students' fundamental knowledge of Japanese language teaching in colleges. This method of test organisation involves categorising and summarising the test questions according to the knowledge domains and level of difficulty. This method helps teachers create tests. By identifying knowledge gaps, this system allows students to learn independently and practise more effectively with half the effort. After extensive data testing and operations, the system has proven to be realistic and useful.

24.8 EXPERT SYSTEM IN THE MEDICAL FIELD

24.8.1 TO IDENTIFY DISEASES IN OBSTETRICS AND GYNAECOLOGY

Mohammed F. El-Habibi and others (2022) state that when a woman has issues with her reproductive organs or during pregnancy, obstetrics and gynaecology specialists are frequently encountered. When she is going through a particularly trying time for herself, such as during her menstrual cycle, pregnancy, or under the influence of a disease affecting any part of her body related to the reproductive organs, women are handled with caution and special care because they often manifest in delicate and difficult-to-treat areas. The female sexual organs are the focus of a gynaecologist's expertise, which includes all illnesses and issues in this area. Regular preventive medical exams are carried out by them, including breast and cervical smear tests. In addition, they offer advice on issues like menopause, infertility, and contraception to women of all ages. This expert system identifies diseases related to women's obstetrics and gynaecology. For the benefit of pregnant women and non-pregnant women, an easy-to-use expert system covering 10 common female illnesses has been developed. Methods: Symptoms include uterine cancer, cervical cancer, infertility, double uterus, ectopic pregnancy, endometrial cancer, female sexual dysfunction, endometriosis, faecal incontinence, and female infertility. The system includes a list of common symptoms, correct diagnoses, and proper treatments. Results: Medical students evaluated the suggested 10 obstetrics and gynaecology diagnostic expert systems, and they were pleased with their performance. Conclusions: Patients with issues related to the reproductive system and recent graduates can all benefit greatly from the proposed expert system.

24.9 EXPERT SYSTEM FOR RELIABLE AND GENERALISED PVC IDENTIFICATION

Among others, Cai et al. (2022) write that premature ventricular contraction (PVC) is one of the frequent ventricular arrhythmias that can result in sudden cardiac death or stroke. Physicians may receive early warnings and even suggestions for diagnoses from automatic long-term electrocardiogram (ECG) analysis algorithms. Regarding robustness, generalisation, and low complexity, however, they are mutually exclusive. The heartbeat classifier and deep learning–based heartbeat template clusterer are combined to create a novel PVC recognition algorithm. For K-means clustering, features from ECG heartbeats were taken out by a long-term memory-based auto-encoder (LSTM-AE) network. So, using the clustering results as a starting point, templates were created and decided upon. Finally, a set of rules, including template matching and rhythm characteristics, was used to determine the PVC heartbeats. Three quantitative parameters, sensitivity (Se), positive predictive value (P+), and accuracy (ACC), were used to evaluate the effectiveness of the proposed method using data from the MIT-BIH Arrhythmia database and the St. Petersburg Institute of Cardiological Technics database. 92.47 percent and 93.18 percent training accuracy on the two test databases. The test accuracy for the two databases was 87.51 percent and 87.92 percent, respectively. The third China Physiological Signal Challenge 2020 training set PVC scores were 36,256 and 46,706. These scores could win open-source competitions. The results showed that an expert system and deep learning can improve PVC identification from single-lead ECG recordings.

24.10 EXPERT SYSTEM DEVELOPED FOR ABLATION STUDIES

According to M. Raihan et al. (2022), the risk of contracting the coronavirus has resulted in many difficulties and barriers for conventional healthcare systems. Through their contributions to various telemedical services, smartphones with IT can play a crucial part in controlling the current pandemic. Because of the widespread availability of smartphones, the focus of this study has been on using them to detect the presence of this virus. The model was created using a public COVID-19 dataset with 33 features from an internal facility. The dataset exhibits a significant class imbalance with 2.82 percent of positive samples and 97.18 percent of negative samples. Since the data in each class is not evenly distributed, the adaptive synthetic (ADASYN) model has been used to address this issue. The model was created using a public COVID-19 dataset with 33 features from an internal facility. The XGB, RF, and SVM classifiers achieved an accuracy of 97.91 percent, 97.81 percent, and 73.37 percent, respectively. Generally, three classification schemes—random forest (RF), extreme gradient boosting (XGB), and support vector machine (SVM)—have been applied for ablation studies. Since the XGB algorithm produces the best results, it was used in the development of web applications and the Android operating system. The created expert system can forecast whether COVID-19 will be found in the main suspect's body by examining the results of 10 users' questionnaires. On the GitHub repository, you can find the preprocessed data and codes.

24.11 EXPERT SYSTEM USING FUZZY NEURAL NETWORKS FOR DIAGNOSING BREAST CANCER

According to Algehyne et al. (2022) and other sources, breast cancer is highly prevalent in Saudi Arabia and ranks as the second-leading cause of death among women. However, due to its complexity and fuzziness, the clinical diagnosis process of any disease, including COVID-19, diabetes, coronary artery disease, and breast cancer, is frequently accompanied by uncertainty. A fuzzy neural network expert system with an enhanced Gini index random forest-based feature importance measure algorithm was proposed as a means of reducing the ambiguity and uncertainty inherent in making a breast cancer diagnosis and relieving the strain placed on the network nodes of the underlying fuzzy neural network system by removing irrelevant features used for prediction or diagnosis. Using an enhanced Gini index RF-based feature importance measure algorithm, the five most appropriate features were selected from the dataset in the Wisconsin breast cancer diagnostic database. Two classification models were created using logistic regression, SVM, K-nearest neighbour, ranaïveforest, and gaussian naive Bayes. Consequently, models with all features (32) and the five best features (31) were used. The comparison's outcome demonstrates that in terms of accuracy, sensitivity, and specificity, the models with the fittest features performed better than those with full features. The five fittest features were used to build an expert system with 99.33% accuracy, 99.41% sensitivity, and 99.24% specificity. The system is more accurate, sensitive, and specific compared to previous studies using fuzzy neural networks or other artificial intelligence techniques on the same dataset to diagnose breast cancer. The z-test result demonstrated improved accuracy for early breast cancer diagnosis.

24.12   EXPERT SYSTEM IN BIOLOGY

24.12.1   Expert System Model for Semi-Automatic Morphology Analysis

Zhou et al. (2022) noticed that electrical status epilepticus in sleep (ESES) is a complex epileptic encephalopathy in children. Electroencephalogram (EEG) slow-wave and continuous spike patterns are present. EEG patterns help diagnose ESES. Most automatic ESES quantification systems ignore signal morphology and subject variability. This study introduces a hybrid expert system that incorporates morphological differences, individual variability, and medical knowledge to model clinical decision-making during ESES quantification. The proposed hybrid system offers a generic framework for moving from a partially automated expert decision model based on morphology analysis to a fully automated ESES quantification using biogeographic optimization (BBO), enabling a precise individualised quantification system to incorporate personalised characteristics by using an individualised parameter-selection framework. Twenty subjects from the Children's Hospital of Fudan University in Shanghai, China, are used to test the method's viability and reliability. The individualised quantitative descriptor ESES estimation error rate is 0.4−4.32 percent and the overall average is 0.9522 percent. The system outperforms others, and the customised system improves ESES quantification accuracy. Positive outcomes imply that ESES can be diagnosed using the hybrid expert system for automatic ESES quantification.

24.13   EXPERT SYSTEM IN THE LINGUISTICS FIELD

24.13.1   Expert System Developed for Single-Valued References

Chao Sun et al. (2022). Referential values evaluate variables with the help of numerical data or linguistic terms and are a key component of the belief rule base (BRB). The rationality and accuracy of these evaluation criteria are crucial to the model's accuracy. Studies on BRB's referential values primarily use single-valued data. Experts' single-valued references don't accurately represent qualitative information due to uncertainty, ambiguity, and vagueness. Using a novel BRB with interval-valued references (BRB-IR), which is proposed in this paper, it is possible to build models by combining qualitative knowledge with quantitative data. To begin with, an optimisation algorithm that is non-linear is used to optimise the interval-valued referential values that experts have provided. The P-CMA-ES algorithm also optimises other model parameters. A pipeline leak detection case study was created to validate the model. Compared to the classic BRB, the proposed BRB-IR demonstrates that the latter is inefficient and fails to adequately capture expert knowledge.

24.14 EXPERT SYSTEM IN MARKETING

24.14.1 Designed to Manage Supply Chain Risk

According to Mehrdokht Soleymani and Maryam OveysiNejad's (2018) study, supply chain risk management (SCRM) is novel in logistical science. The growing global risks in various industries make SCRM more crucial. Using expert systems to manage supply chain risk is the subject of this paper's important literature review. To achieve this goal, we should introduce techniques for qualitatively and quantitatively identifying risks and their effects on performance. Questionnaires are used in qualitative methods, while fuzzy logic and expert systems are the foundations of quantitative methods that use probability theory. According to the findings, expert systems can recognise supply chain risks, which aids leaders in making decisions.

24.15 EXPERT SYSTEM IN SPORTS

24.15.1 To Analyze Somatic Traits in Athletes

Salabun et al. (2022) suggest that if a customised approach to training is used early enough, a swimmer's potential can be fully realised. It's difficult to predict whether a competitor will advance to the international level early in their career. It is influenced by things like swimming technique, the athlete's current skill level, and factors relating to physical and mental characteristics. While some of these factors are outside the scope of swimmers' control, others can be improved through practise. For the study, information was gathered from 30 swimmers. After that, the proposed model was built using the characteristic objects method (COMET) to determine each athlete's predispositions by entering the values of attributes that were specific to them. In addition to being a cutting-edge strategy free of the rank reversal phenomenon, the applied method has been shown to be effective in evaluating the competitors' parameters. Furthermore, the model developed allows for data analysis to see changes in specific criteria affecting the rating. The outcomes are acceptable, but additional work is required to analyse somatic features and produce more precise forecasts.

24.16 CONCLUSION

Traditional management modes require a lot of manpower and time to monitor massive amounts of data, so they can't perform real-time evaluations and miss the best time to solve problems. Therefore, effective computing techniques are needed.

Integration with real-world monitoring systems allows for comprehensive evaluation. It covers abnormal data imports and changes and decision-making consultation. The model uses an intelligent system to make decisions. Traditional systems use streamlined programming. With more mature theories and simulation calculation techniques with more knowledge, we improve inference techniques and knowledge evaluation. The method library, knowledge base, and safety monitoring inference mode are the foundations of the system's expert system development process. The new system also assessed hot spot techniques and unusual monitoring data. The study improved both inference and decision-making procedures. It enabled knowledge expansion and learning. Real-time safety monitoring saves money, improves efficiency, and supports platform management decision-making. The system collects and studies field data to improve inference and knowledge evaluation. The model proved that expert systems can be useful for agricultural purposes, particularly in the diagnosis and management of plant diseases, which helps farmers find treatments.

The proposed expert systems were designed and implemented using CLIPS. System interfaces are smooth and flexible. It helps farmers and farming enthusiasts get faster, more accurate diagnoses. This expert system requires no training and has a user-friendly interface. More tests in all fields were planned for universal application. Future improvements to the algorithm may include deep learning and transfer learning.

REFERENCES

Algehyne, E. A., Jibril, M. L., Algehainy, N. A., Alamri, O. A., & Alzahrani, A. K. (2022). Fuzzy Neural Network Expert System with an Improved Gini Index Random Forest-Based Feature Importance Measure Algorithm for Early Diagnosis of Breast Cancer in Saudi Arabia. Big Data and Cognitive Computing, 6(1). https://doi.org/10.3390/bdcc6010013

Al-Qadi, M. H., El-Habibi, M. F., Megdad, M. M. M., Alqatrawi, M. J. A., Sababa, R. Z., & Abu-Naser, S. S. (2022). Developing an Expert System to Diagnose Tomato Diseases. International Journal of Academic Engineering Research, 6. www.ijeais.org/ijaer.

Aslem, Y. I., & Abu-Naser, S. S. (2022). CLIPS-Expert System to Predict Coriander Diseases. International Journal of Engineering and Informati on Systems (IJEAIS), 6. www.ijeais.org/ijeais.

Cai, Z., Wang, T., Shen, Y., Xing, Y., Yan, R., Li, J., & Liu, C. (2022). Robust PVC Identification by Fusing Expert System and Deep Learning. Biosensors, 12(4): 185. https://doi.org/10.3390/bios12040185

Chu, H. (2022). Research on Expert System of Japanese Auxiliary Teaching Based on BP Neural Network. Mo bile Information Systems. https://doi.org/10.1155/2022/7719392.

El-Habibi, M. F., Megdad, M. M. M., Al-Qadi, M. H., Alqatrawi, M. J. A., Sababa, R. Z., & Abu-Naser, S. S. (2022). A Proposed Expert System for Obstetrics & Gynecology Diseases Diagnosis. International Journal of Academic Multid isciplinary Research, 6. www.ijeais.org/ijamr.

El-Hamarnah, H. A., Lafi, O. I. A., Radwan, H. I. A., Al-Saloul, N. J. H., & Abu-Naser, S. S. (2022). Proposed Expert System for Pear Fruit Diseases. International Journal of Academic and Applied Research, 6. www.ijeais.org/ijaar.

Lafi, O. I. A., El-Hamarnah, H. A., Al-Saloul, N. J. H., Radwan, H. I. A., & Abu-Naser, S. S. (2022). A Proposed Expert System for Broccoli Diseases Diagnosis. International Journal of Engineering and Informat ion Systems (IJEAIS), 6. www.ijeais.org/ijeais.

Lanzaro, G., & Andrade, M. (2022). A Fuzzy Expert System for Setting Brazilian Highway Speed Limits. International Journal of Transpo rtation Science and Technology. https:// doi.org/10.1016/j.ijtst.2022.05.003.

Megdad, M. M. M., Ayyad, M. N., Al-Qadi, M. H., El-Habibi, M. F., Alqatrawi, M. J. A., Sababa, R. Z., & Abu-Naser, S. S. (2022). Mint Expert System Diagnosis and Treatment. International Journal of Academic Informat ion Systems Research, 6. www.ijeais.org/ijaisr.

Mohd Sani, M. I., Abdullah, N. A. S., & Mohd Rosli, M. (2022). Review on Hypertension Diagnosis Using Expert System and Wearable Devices. International Journal of Electrical and Computer Engineering, 12(3), 3166–3175. https://doi.org/10.11591/ijece.v12i3. pp3166-3175.

Pawan, E., Thamrin, Rosiyati M. H., Widodo, Widodo, Bei, Sariaty H. Y., & Luanmasa, Junus J. (2022). Implementation of Forward Chaining Method in Expert System to Detect Diseases in Corn Plants in Muara Tami District. International Journal of Computer and Information System, 3(1).

Radwan, H. I. A., El-Hamarnah, H. A., H Al-Saloul, N. J., A LAfi, O. I., Abu-Naser, S. S., & Edward Feigen Baum, by. (2022). A Proposed Expert System for Passion Fruit Diseases. International Journal of Academic Engineering Research, 6. www.ijeais.org/ijaer

Raihan, M., Hassan, M. M., Hasan, T., Bulbul, A. A. M., Hasan, M. K., Hossain, M. S., Roy, D. S., & Abdul Awal, M. (2022). Development of a Smartphone-Based Expert System for COVID-19 Risk Prediction at Ear ly Stage. Bioengineering, 9(7). https://doi.org/10.3390/ bioengineering9070281.

Salabun, W., Wieckowski, J., & Watrobski, J. (2022). Swimmer Assessment Model (SWAM): Expert System Supporting Sport Potential Measureme nt. IEEE Access, 10, 5051–5068. https://doi.org/10.1109/ACCESS.2022.3141329

Soleymani, M., & Nejad, M. O. (2018). Supply Chain Risk Management using Expert Systems. International Journal of Current Eng ineering and Technology, 8(04). https://doi. org/10.14741/ijcet/v.8.4.12.

Sun, C., Yang, R., He, W., & Zhu, H. (2022). A Novel Belief Rule Base Expert System with Interval-Valued Referen ces. Scientific Reports, 12(1). https://doi.org/10.1038/s41598-022-10636-8.

Tang, J., & Deng, Y. (2022). The Design Model of English Graded Teaching Assistant Expert System Based on Improved B/S Three-Tier Structure System. Mo bile Information Systems. https://doi.org/10.1155/2022/4167760.

Wulansari, R. E., Sakti, R. H., Ambiyar, A., Giatman, M., Syah, N., & Wakhinuddin, W. (2022). Expert System for Career Early Determination Based on Howard Gardner's Multiple Intelligence. Journal of Applied Engineering and Technological Science, 3(2).

Zhou, W., Zhao, X., Wang, X., Zhou, Y., Wang, Y., Meng, L., Fan, J., Shen, N., Zhou, S., Chen, W., & Chen, C. (2022). A Hybrid Expert System for Individualized Quantification of Electrical Status Epilepticus During Sleep Using Biogeography-Based Optimization. IEEE Transactions on Neural Systems and Rehabilitat ion Engineering, 30, 1920–1930. https://doi.org/10.1109/TNSRE.2022.3186942

25 Applications of Artificial Intelligence on Customer Experience and Service Quality of the Banking Sector
An Overview

*Sravani Elaprolu, Channabasava Chola,
Varanasi Chandradhar, and Raul V. Rodriguez*

25.1 INTRODUCTION

There are various stages involved in the banking industry from processing the loan application of the customers to ensuring the safe banking transactions for each customer until they maintain the services with the banks. Customers are looking for better service wherever they collaborate with the products and services; in other words, better customer experience is the demand that the customers put forward. As the technology evolved in the past decades, industries have embarked on using state-of-the-art technology, namely, artificial intelligence, whereby superior quality of service can be delivered to the customers. The importance of the banking industry and its influence on the development of the country is discussed in Section 25.1 and how artificial intelligence and its different applications improves processes involved in the banking industry will be discussed in Section 25.2. And the conclusion part will take place in Section 25.3.

25.1.1 BANKING AND ECONOMY

Banking is a kind of exceptional industry, which deals with its capital to multiply the money regardless of risk (Ghodselahi and Amirmadhi, 2011). Banking institutions have a significant impact on the economy of a country (Park, 2012) as well as financial stability and sustainable development (Gutierrez et al., 2009) of a country, and thus banks should scrutinize loan processing methods to segregate good applications from the total applications. This will be helpful to the banks not only to discourage the sanction of loans to the loss-generating project that will lead to a non-performing

DOI: 10.1201/9781003328414-25

asset at later point in time but also to enhance the process of the loan allocation to the right projects. Approving a loan to a non-profitable project will indicate poor investment of resources, which affects the performance of the banks as well as economic growth of a country. Because providing loans is one of the key functions of the banks, failing in that core activity will severely affect the growth of the banks (Park, 2012).

Besides, banks have to lend the money to the borrowers for making a profit (Ince and Aktan, 2009) that will contribute to the growth of financial activities, economic development activities, and industrial activities (Cetorelli and Gambera, 1999). At the same time, bank loan availability will be dropped significantly if a bank crisis happens, which leads to a reduction in the loan supply offered by the bank (Huber, 2018). In India, public-sector banks possess more than three-fourths of total assets belonging to the entire banking sector wherein the state bank of India itself has 17 percent of the total commercial banking assets (Goldberg, 2009). Banking institutions can perform effectively in providing loans to individuals and firms as per their demand if the market share of the bank is quite large. When banks charge a high interest rate, plan to achieve high margin, and decide to decrease the loan supply, economic growth and creation of jobs will be affected and the unemployment rate will be increased (Feldmann, 2015). In addition to that, if the entry barrier is high in the banking industry, the initial expense will be higher, and as a result the high interest rate will be charged to make a profit and foreign banks will hesitate to invest.

Gross domestic product (GDP) is a measurement of economic development, and is the monetary value of output (finished goods) of production of various industries within a country for a particular time (Atay and Apak, 2013); this GDP estimation is highly affected when the banking industry output is exaggerated (Outlon, 2013). Non-performing assets should be recovered to strengthen the banking industry in a stable manner (Tan and Floros, 2012), which contributes to the economic activities.

25.1.2 THE INTERNATIONAL PRESENCE OF BANKS

With expansion across the globe, banks have a crucial role in stabilizing the economy during the crisis in the hosting country (Goldberg, 2009). Less competition in the banking industry will make it operate at a high cost and deliver poor quality of services, which reduce the need for outside finance and decline the industrial growth. Because banks have international branches, financial systems are integrated across the globe as a result of globalization. This international presence will make the transactions from one country to another country effortless and fill the gap that is not served by local banks, making the banks provide better customer service (Goldberg, 2009). Industries which depend on more outside financing will have faster growth rates due to the competitive nature present in the bank and finance institutions. The banking institution is one of the big sources of funds for companies to gain external funds from which the company can run the business smoothly (Campiglio, 2015).

25.1.3 Banking Crisis

When the depositors withdraw money from the bank due to the perception that the bank is untrustworthy, the bank system will fail. Without deposits from customers, it is difficult for banks to run the business irrespective of whether the situation is normal or in crisis (Kunt et al., 2000). Depositors always look for good-functioning banks and higher interest rate to invest their money (Goldberg, 2009). During the crisis, the bank loan issue will be drastically reduced and bank assets will drop, which paves the way for a reduction in the growth of output and investment. The crisis affects the growth of output volume not only in the year of crisis but also in the following year. After few years of crisis happened, the growth of output can be recovered, but the recovery may not possible for credit of the banks within that time frame with respect to growth. The banking interest rate for deposits will be higher during the crisis and the subsequent year in order to gain more deposits and maintain the existing deposits (Kunt et al., 2000). The inflation rate is inversely proportional to the output of the country, and it is found that that both inflation rate and output growth are correlated in a negative way (Haslag, 1995). Despite an increase in the interest rate, the interest rate should be higher than the inflation rate to lure the customer for gaining deposits, but there is no significant difference in interest rates before and after the crisis and no proof to show banks have given a higher interest rate than the inflation rate. Banks that have maintained liquidity even after depositors withdraw most of the deposits need help from the central bank and its authorities (Kunt et al., 2000).

25.1.4 Customer Experience and Service Quality

Customers are intrigued in availing themselves of newly launched products and services which are created by banks to execute the banking operations quickly (Laketa et al., 2015). Based on the quality of service delivered to the customers, the success of the banks is determined and the differentiation of the bank from its competitors is identified. Customer satisfaction, which decides the survival and the success of the organization in the competitive environment, is an important indicator for evaluating the performance of the organization, especially retail banking that is dependent on customer loyalty to run the business profitably by luring new and maintaining existing customers (Dahari et al., 2015). In spite of the big efforts taken by the banks, the majority of the customers are not satisfied with the banking services provided by the banks. Due to the growing competition in the banking sector, banks have taken steps to enhance the service quality as per customer demand and to intensify the service in reliable way (Johnston, 1997).

Nowadays, businesses should segment the customers to deliver the best service to them as per their various needs, which helps to treat each customer effectively. The customer relationship management methodology that coalesces the marketing strategy with processes and functions performed within the company and with network connections outside the company is developed to maintain the existing customer in the highly competitive market by ascertaining and understanding the needs of the customers. Banking institutions can efficaciously capitalize on customer relationship management to serve the customers better if they focus on these four major elements: preserving the existing customer, enticing the new customer, motivating the customers to have profound collaboration with the bank, and updating customers with the banks' new services (Laketa et al., 2015). Additionally, the banking industry can gain more deposits from the retail depositor if it treats them properly (Puri and Rocholl, 2008).

25.1.5   BANKING PROCESS AND TECHNOLOGY

Banking services, namely,- automated teller machine (ATMs), Internet banking, credit cards, banking apps, etc., have been used by millions of customers on a daily basis, and interestingly, the number keeps on increasing (SundarKumar and Ravi, 2015). These services need a huge manpower, have a high cost, and are time-consuming. Conventional banking that needs many workers to perform the tasks such as receiving deposits, approving loans, and conducting money transfers is changed completely by Internet banking which is smoother and faster with the help of information technology (IT). Internet banking that revolutionizes the banking sector has an important role in serving the individual customer better in order to maintain long-term relationships. The Internet that makes financial and banking services more competitive has helped to have many e-banking products, from ATM to credit and debit cards. This e-banking that loosens entry barriers gives access to each customer anywhere in the world and saves time for customers as well as bank managers and has changed from manual paper work to paperless work with the help of technology and communication (Atay and Apak, 2013).

25.2   ARTIFICIAL INTELLIGENCE AND ITS APPLICATION TO THE BANKING INDUSTRY

25.2.1   CREDIT SCORE

Banks should lend money after critically evaluating credit scores of those customers who are applying for loans (Eletter et al., 2010). The banking industry makes a profit irrespective of the risk because the banks have been controlling and managing the risk. Among various risks, credit risk is one of the prominent risks which should be given more attention to get away from total system failure, due to the fact that it is not easy to compensate for (Ghodselahi and Amirmadhi, 2011) this kind of classification between a good score that has a high probability not to default and a bad score that has a high chance of defaulting. Classification and regression tree models are made with a decision tree algorithm that is one of the artificial intelligence techniques used for classification problems. This technique gives better outcomes in evaluating credit scores than other techniques, namely, logistic regression and discriminant analysis (Ince and Aktan, 2009). Approving loans is a crucial decision for the bank's profit and marketing strategy; this is not easy when different lending approaches are followed by competing banks, and frequent changes in the borrowing behavior of consumers occurs. Loan applications should be classified into two categories: positive credit risk and negative credit risk of the applicants. The positive credit risk indicates high probability of the applicants to fail to repay the loan, whereas the negative credit risk shows low probability of the applicants to fail to repay the loan. The bank managers who are overwhelmed with customer data should take the right decision in order to approve the application with negative credit risk and deny the application with positive credit risk. Now, artificial intelligence has assisted managers to take better decisions (Eletter et al., 2010).

As mentioned earlier, classification is the technique used to classify between good credit and bad credit of the applicant. This credit score is applicable for the company, municipality, state, financial institution, and so on. It is the value obtained from credit score processing, which is used by debt givers, bond buyers, and government officers; the risk involved is inversely proportional to the credit score value, which is based on the various indicators such as economic condition of the applicant, capital involved, collateral offered by the applicant, the capacity of the applicant, and the behavior history of the applicant. Though the most commonly used models are logistic regression and linear discriminant analysis, the former is performing prediction of dichotomous outcomes and linear relationships of the two variables, which are, in fact, not required for multivariate normality assumption and the latter has also a drawback in the assumption that the variables are linear, but in reality the variables are non-linear. Artificial intelligence techniques such as decision tree, genetic algorithm, artificial neural network, and support vector machine are giving better results than traditional statistical methods. Three models, namely, support vector machine, neural network, and decision tree, are used for classification along with the fuzzy C-means clustering technique. However, hybrid approaches have been outperforming the previously mentioned individual methods with respect to the accuracy of the prediction (Ghodselahi and Amirmadhi, 2011).

25.2.2 Credit Card Fraudulent Activities

Fraudulent activities in credit card transactions are executed because of various reasons, namely, improper deployment method to deal with tens of thousands of transactions and the wrong approach in classifying cost rate to the transaction of different specified amounts. Moreover, the data in which the model is performed are skewed, and the transformation of unlabeled data into labeled data is exorbitant takes a great deal of time. K reverse nearest neighborhood is used to eliminate the outliers which, are considered as noise labels, after applying stratified random sampling to deal with 20 percent of the unbalanced data present in the original dataset. To perform the classification of the dataset, a method that minimizes the dimensionality will be executed first, and then support vector machine (SVM), particularly one class support vector machine (OCSVM), is used. This OCSVM is different from SVM in training the dataset because the former is using only one class to train and its performance is significantly better compared to the group method of data handling (GMDH) and probabilistic neural network (PNN) in the detection rate of fraudulent claims while using hybrid under the sampling approach (SundarKumar and Ravi, 2015).

25.2.3 PHISHING WEBSITES

Phishing websites are the websites which allure the people to disclose their user-name and password that will be used for various illegal transactions. Data mining algorithms, one of the techniques of artificial intelligence, is used to detect those phishing websites, and prediction of these websites is also possible with associative and classification algorithms. Various estimations have revealed that the cost per victim keeps on increasing. In particular, emails are used to lure the banking customers to fall into this trap which is promoted by constantly sending spam mails to many people. Data mining technique will be helpful to get the required information that is most pertinent to the user from the tons of data available. There are 27 major feature vectors, which are a conglomerate of different indicators such as URL and domain identity, security and encryption, source code and JavaScript, page style and content, web address bar, and social human factors. Various approaches, namely, PART, PRISM, JRip, C4.5, MCAR, and CBA, are performed by Aburrous et al. (2010b) to find out the best approach. MCAR outnumbered in terms of accuracy and speed among all other methods. A fuzzy data mining algorithm is used to identify automatically the phishing websites, particularly e-banking websites, but still finding an important feature to achieve this goal is not easy with this technique (Aburrous et al., 2010a).

25.2.4 BANKING FAILURES

Banking failures will happen when the banks are not making a profit; this is due to various reasons, namely, high competition in the market, emerging non-banking institutions, unexpected threat to loan portfolios, and financial distress. The failure of big banks is dangerous as it will lead to a disruption in the whole financial activity. In 1980, big banks failed to secure against non-performing loans, which is one of the reasons for the bank failure and system collapse (Boyd and Gertler, 1994). Predicting risk and making the right decision towards the approval of credit will be helpful to avoid the inevitable situations like bankruptcy and fraud detection (Moro et al., 2015).

Financial soundness indicators (FSI) are used to measure the financial vulnerabilities happening in banks, which is classified into two main indicators such as the encouraged indicators and the core set indicators. It can be abridged by different criteria such as capital adequacy, asset quality, management quality, earning ability, liquidity, and sensitive to the market. Three models, namely, discriminant, logit, and probit analyses, were introduced to reveal the banking failure in advance by three years (Fernando et al., 2011). The adoptive neuro-fuzzy inference system (ANFIS) is one of the technique applied in finance, which is useful to predict the failure of the events in the banking system (Messai and Gallali, 2015). Banks are not only assisting in the economic stability of a country but also reinforcing the financial system of a country. Fuzzy logic and neural network techniques are the apparent techniques used to find the change in efficiency and productivity of the banks (Sharma et al., 2013).

In addition to this, three models were executed to predict the currency crisis, and those models are logit regression, decision tree, and artificial neural network (ANN). The unpleasant situation of a bank can be estimated from the ratio of the non-performing loans to the total gross loans. Because the non-performing loans are the major indicators of the probable financial crisis, ANN is performed with main variables of distress of a bank, which are loan loss reserves of non-performing loans, return on equity average, and loan loss provision to gross loans ratio (Messai and Gallali, 2015). The neural network gives the better percentage in predicting bank failure, as concluded by Messai and Gallali (2015) and Elzamly et al. (2017).

25.2.5 ALARM SYSTEM

Improving the banking system security from robberies in the banks and the ATMs, artificial intelligence gives a better solution than the conventional emergency button alarming system. This system performs in three stages: artificial vision first takes a photo for image processing to get the features, the ANN classifies the event from the obtained pattern and gives the status of the warning messages. Based on the classification of the neural network, the output class is determined. If the output is 1, this means the alarm should be activated and a warning message should be sent using global system for mobile communication (GSM) technology (Ortiz et al., 2016)

25.2.6 MOBILE BANKING

Today 65 people out of 100 are using mobile banking. Most customers have a positive opinion about online payment services which is not only attracting the customer from the conventional card transaction but also assisting in the enhancement of banking services due to maximizing revenue generation. This transformation of user experience helps to collect and analyze the user-generated data for delivering better service to each customer as per the patterns or insights extracted from that data (Dubey, 2019).

Mobile devices facilitate mobile banking services that are preferred by customers due to comfort and convenience and also preferred by banking institutions to maintain good relations with the customers. However, 91 percent of participants who have attended a study conducted by Klynveld Peat Marwick Goerdeler (KMPG) recently responded that they did not use their mobile phones to do banking, not even once. This clearly shows customers should be segmented as per their preference, and the segmented customers should be targeted to understand what makes them use mobile banking services and what kind of expectation they have in their perception. Along with convenient service provided in mobile devices, banks can also utilize from the offered service to serve the customer better because this device that can be used by banks to deploy customer relationship management are operated for customers' personal purposes (Awasthi and Sangle, 2013). Recently, artificial intelligence–enabled mobile banking (Payne et al., 2018) has got some customer attraction, which can be utilized by the banks to gain more customers

25.2.7 CUSTOMER LOYALTY

The relationship between banker and customer is paramount not only to keep hold of existing customers but also to enhance the loyalty of the customers. The customer relationship can be built strongly if the needs and expectations of the customers that change over the period of time are fulfilled by the banking institutions. The loyalty of customers can be improved if the customers are attracted from the good-quality services, which should be provided at low prices. The customer loyalty can be predicted in the banking industry by using ANN that is already used by other industries for the same purpose. After collecting the data, important variables should be taken from all the available variables by using factor analysis which makes the data ready for further modeling. In this prediction model, feed-forward back-propagation is used in the algorithm along with the ANN. K-fold cross validation is used, where K subsets are obtained from categorization of the data during the training of the dataset, and performance of the algorithm can be evaluated from the coefficient of efficiency and root mean square error after the testing of the dataset. The obtained result of predicting customer loyalty from the ANN proves that high accuracy is possible (Kishada et al., 2016).

25.2.8 LIQUIDITY RISK ASSESSMENT

Banking institutions are prone to various risks with respect to technological and financial factors; these risks comprise market, operational, and credit risk. Banks have to increase the profitability, which leads to gain more investment from the shareholders and to ensure the liquidity position at any point in time. There should be an appropriate balance between short-term risk that is liquidity and long-term risk that is profitability. The depositors who invest their money for a short term in the bank will do frequent withdrawals, which makes the bank keep up more attention towards the liquidity. However, there is an unwanted situation on both extremes of liquidity: high liquidity indicates poor utilization of available resources and low liquidity describes a bad impression of the banks, which can lead to low deposits and a drop in market share. In both cases, the bank is moving towards an unpleasant situation such as bankruptcy. Banks have very good databases from which the liquidity risk alert system can be modeled with the help of an ANN and genetic algorithm for evaluating the liquidity risk and a Bayesian network for making predictions of the distribution liquidity risk. The measurement of liquidity risk of a company can be done by bifurcating into three elements: market-related features that consider recession, inference in transaction system and disarrangement in the markets (capital), and bank-related features that contemplate many factors such as credit risk and others. A system risk adjustment model for liquidity is a kind of probability measurement of liquidity risk by combining data from market and balance sheets along with pricing methods. The probability distribution is also used to find the liquidity risk, but this requires big data to get a proper output (Tavana et al., 2018).

Tavana et al. (2018) used an ANN that is applied along with a genetic and Levenberg-Marquardt algorithm for identifying important feature vectors, and a Bayesian network is preferred to assess the probability of occurrence of liquidity risk from the previously mentioned feature vectors. This combination gives a better result, which is consistent after training the model properly.

25.2.9 Intelligence and Augmented Reality

Various industries have started using the latest technology to enhance business, which is a very good advantage for business operations. Augmented reality that is assisting individuals by extending the perception and easing the communications delivers virtual aid to alleviate the real-world complex problems by disclosing more details. Many fields from healthcare to games and media have already deployed augmented reality to optimize the existing process, and industries which have high cost operations and high risk involvement can implement this application for business development (Heng, 2015). Augmented reality is also used by banking and finance institutions to understand and demonstrate the customer performance and to make the best recommendation system to enhance the expenditure pattern of the customers (Dubey, 2019). On the other hand, intelligent systems are also installed for improving the operations and reducing manual work. For example, JP Morgan has executed a contract intelligence system that helps to reduce labor activity by 0.36 million hours (Dubey, 2019).

25.2.10 Chatbots

If there is an issue or enquiry related to the products or services offered by banking institutions, the customers have to contact the officers to get the problem solved, but this process is kind of tedious, repetitive, and time-consuming. Due to the advancement of the technology, many industries have been benefitted from this technology and it is working well in the businesses. Moreover, Watson, developed by IBM, is designed to answer the queries; this is done by applying machine learning algorithm and natural language processing (NLP), which helps to retrieve the information and represent the inbuilt domain knowledge. Implementing these bots are really useful to serve the customer better, which is already done by most of big growing banks (Singh et al., 2018).

25.3 CONCLUSION

Millions of customers undergo multiple transactions in a day as a routine matter. This generates data, which is stored and maintained as big database. Moreover, there are lots of manual work to perform to carry out most of the processes in the banking industry. Now, artificial intelligence has made it easy to carry out this work for both bank employees and customers. This kind of sophisticated work has become a simple task, which has never been seen before due to machine learning techniques.

The banking sector has been improving its service quality by providing various effective tools to ensure the safety and comfort of customers. The technology keeps on improving day by day; it is better to incorporate this technology into the different fields of the business. The state-of-the-art technology is mandatory in maintaining as well as enhancing the security in the banking system, and other sections of the banking industry are ready to implement the latest technology. In this digital era, customers are also expecting their bank to be up-to-date. The technology upgrade-ability will not only uplift the service and security but also improve the reputation of the bank. Nowadays, Internet banking and mobile banking are attractive to customers due to its effectiveness and user-friendliness.

Many studies show that different models have been launched to maximize the accuracy of the process, which is a good thing for the banking industries as well as the customers. This is a win-win situation for both. Due to the competition from the non-banking sectors, banks have to adapt the latest trending technologies used in the digital era to improve the service quality. The technology provides more positive effect in the banking industry. To make the process smooth and spontaneous in the business, artificial intelligence techniques should be utilized in the banking industry. Fortunately, artificial intelligence has been giving a plethora of applications to make the banks reach their greatest efficiency, which also paves the way for a new dimension of the bank.

REFERENCES

Aburrous, M., Hossain, M.A., Dahal, K., & Thabtah, F. (2010a). *Associative Classification Techniques Predicting e-Banking Phishing Web Sites*. MCIT. pp. 9–12.

Aburrous, M., Hossain, M. A., Thabtah, F., & Dahal, K. (2010b). Intelligent phishing detection system for e-banking using fuzzy data mining. *Journal of Expert Systems with Applications*, 37 (12). pp. 7913–7921.

Atay, E., & Apak, S. (2013). An overview of GDP and internet banking relations in the Europe–n Union versus China. *Procedia - Social and Behavioral Sciences*, 99. pp. 36 –45. doi:10.1016/j.sbspro.2013.10.469.

Awasthi, P., & Sangle, P. S. (2013). The importance of value and context for mobile CRM services in banking. *Business Process Management Journal*, 19 (6). pp. 864–891. doi:10.1108/BPMJ-06-2012-0067.

Boyd, J. H., & Gertler, M. (1994). The role of large banks in the recent U.S. Banking crisis. *Federal Reserve Bank of Minneapolis Quarterly Review*, 18 (1). pp. 1–21.

Campiglio, E. (2015). Beyond carbon pricing: The role of banking and monetary policy in financing the transition to a low- carbon economy. *Ecological Economics*, 121. pp. 220–230.

Cetorelli, N., & Gambera, M. (1999). *Banking Market Structure, Financial Dependence and Growth: International Evidence from Industry Data*. Federal Reserve Bank of Chicago. pp. 1–39.

Dahari, Z., Abduh, M., & Fam, K. S. (2015). Measuring service quality in Islamic banking: Importance-performance analysis approach. *Asian Journal of Business Research*, 5 (1). pp. 15–28. DOI 10.14707/ajbr.150008.

Dubey, V. (2019). FinTech innovations in digital banking. *International Journal of Engineering Research & Technology (IJERT)*, 8 (10), pp. 597–601.

Eletter, S. F., Yaseen, S. G., & Elrefae, G.A. (2010). Neuro-based artificial intelligence model for loan decisions. *American Journal of Economics and Business Administration*, 2 (1), pp. 27–34.

Elzamly, A., Hussin, B., Naser, S. S. A., Shibutani, T., & Doheir, M. (2017). Predicting critical cloud computing security issues using artificial neural network (ANNs) algorithms in banking organizations. *International Journal of Information Technology and Electrical Engineering*, 6 (2). pp. 40–45.

Feldmann, H. (2015). Banking system concentration and unemployment in developing countries. *Journal of Economics and Business*, 77. pp. 60–78. https://doi.org/10.1016/j.jeconbus.2014.08.002.

Fernando, C., Chakraborty, A., & Mallick, R. (2011). *The Importance of Being Known: Relationship Banking and Credit Limits.* Accounting and Finance. Faculty Publication Series. Paper 4. pp. 1–28.

Ghodselahi, A., & Amirmadhi, A. (2011). Application of artificial intelligence techniques for credit risk evaluation. *International Journal of Modeling and Optimization,* 1 (3). pp. 243–249.

Goldberg, L. S. (2009). Understanding Banking Sector Globalization. *IMF Staff Papers,* 56, 171–197. doi:10.1057/imfsp.2008.31.

Gutierrez, P. A., Segovia-Vargas, M. J., Salcedo-Sanz, S., Hervas-Martinez, C., Sanchis, A., Portilla-Figueras, J. A., & Fernandez-Navarro, F. (2010). Hybridizing logistic regression with product unit and RBF networks for accurate detection and prediction of banking crises. *Omega,* 38, pp. 333–344. doi:10.1016/j.omega.2009.11.001.

Haslag, J. H. (1995). *Monetary Policy, Banking, and Growth.* Federal Reserve Bank of Dallas. pp. 1–29.

Heng, S. (2015). Augmented reality: Specialised applications are the key to this fast-growing market for Germany. *Deutsche Bank Research, Current Issues Sector Research.* pp. 1–14.

Huber, K. (2018). Disentangling the effects of a banking crisis: Evidence from German firms and counties. *American Economic Revi ew,* 108 (3). pp. 868–898. https://doi.org/10.1257/aer.20161534.

Ince, H., & Aktan, B. (2009). A comparison of data mining techniques for credit scoring in banking: A managerial perspective. *Journal of Business Economics and Management,* 10 (3). pp. 233–240.

Johnston, R. (1997). Identifying the critical determinants of service quality in retail banking: Importance and effect. *International Journal of Bank Marketing,* 5/4. pp. 111–116.

Kishada, Z. M. E., Wahab, N. A., & Mustapha, A. (2016). Customer loyalty assessment in Malaysian Islamic banking using artificial intelligence. *Journal of Theoretical and Applied Information Technology,* 87 (1). pp. 80–91.

Kunt, A. D., Detragiache, E., & Gupta, P. (2000). Inside the crisis: An empirical analysis of banking systems in distress. *Journal of International Money and Finance,* 25 (5). pp. 702–718.

Laketa, M., Dusica, S., Laketa, L., & Misic, Z. (2015). Customer Relationship Management: Concept and Importance for Banking Sector. *UTMS Journal of Economics,* 6 (2). pp. 241–254.

Messai, A. S., & Gallali, M. I. (2015). Financial Leading indicators of banking distress: A micro prudential approach: Evidence from Europe. *Asian Social Science,* 11 (21). pp. 1–13.

Moro, S., Cortez, P., & Rita, P. (2015). Business intelligence in banking: A literature analysis from 2002 to 2013 using text mining and latent Dirichlet allocation. *Expert Systems with Applications,* 42 (3). pp. 1314–1324.

Ortiz, J., Marin, A., & Gualdron, O. (2016). Implementation of a banking system security in embedded systems using artificial intelligence. *Advances in Natural and Applied Sciences,* 10 (17). pp. 95–101.

Oulton, N. (2013). Has the growth of real GDP in the UK been overstated because of mismeasurement of banking output? *Centre for Economic Performance.* pp. 1–12.

Park, J. (2012). Corruption, soundness of the banking sector, and economic growth: A cross-country study. *Journal of International Money and Finance,* 31, pp. 907–929. doi:10.1016/j.jimonfin.2011.07.007.

Payne, E. M., Peltier, J. W., & Barger, V. A. (2018). Mobile banking and AI-enabled mobile banking: The differential effects of technological and non-technological factors on digital natives' perceptions and behavior. *Journal of Research in Interactive Marketing,* 12 (3). pp. 328–346. https://doi.org/10.1108/JRIM-07-2018-0087.

Puri, M., & Rocholl, J. (2008). On the importance of retail banking relationships. *Journal of Financial Economics,* 89. pp. 253–267. doi:10.1016/j.jfineco.2007.07.005.

Sharma, D., Sharma, A. K., & Barua, M. K. (2013). Efficiency and productivity of banking sector: A critical analysis of literature and design of conceptual model. *Qualitative Research in Financial Markets*, 5 (2). pp. 195–224.

Singh, M., Singh, R., Pandey, A., & Kasture, P. (2018). Chat-Bot for Banking Industry. International Conference on Communication, Security and Optimization of Decision Support Systems(IC-CSOD 2018). pp. 247–249. ISBN: 978-0-9994483-1-1.

Sundarkumar, G. G., & Ravi, V. (2015). *Engineering Applications of Artificial Intelligence*. Elsevier Ltd. pp. 368–377.

Tan, Y., & Floros, C. (2012). Bank profitability and GDP growth in China: A note. *Journal of Chinese Economics and Business Studies*, 10 (3). pp. 267–273.

Tavana, M., Abtahi, A.R., Caprio, D, D., & Poortarigh, M. (2018). An artificial neural network and Bayesian network model for liquidity risk assessment in banking. *Neurocomputing*, 275. pp. 2525–2554.

26 Prediction of Terrorist Attacks throughout the Globe Using the Global Terrorism Database
A Comparative Analysis of Machine Learning Prediction Algorithms

Happy and Manoj Yadav

26.1 INTRODUCTION

The terrorist attack is one of the most significant threats to a particular geographical area's government. The Institute for Economics and Peace (IEP) has released its annual study related to the Global Terrorism Index (GTI), 2020 [8] and states that a total of 13,826 people lost their lives with an impact of property damage of around $1,777.60 million due to terrorism all over the world even though it fell by 15.5 percent for the fifth following year after reaching a new peak in 2014.

The Global Terrorism Database (GTD) [5] is one of the major widespread unclassified, open-source online databases. The National Consortium for the Study of Terrorism and Responses to Terrorism (START) manages this database. The same is available for research for both individuals and industries.

As per the necessity of the demand for future terrorist attack prediction systems, most research scholars solve the problem based on historical data analysis of terrorist attacks. The prior research mainly focuses on three aspects of GTD. One is predicting the null values available in the GTD, like the organization responsible for that attack. The second is to predict the future event by numerous machine learning and deep learning models. The third is to predict the behavior and the relationship between the terrorist organizations.

This chapter is divided into the following sections: a literature review of related work available in this context in Section 26.1.1. The research methodology explains the data set, data pre-processing steps, tools used, and purposed system details, which are available in Section 26.2. Section 26.4 contains the results after data

DOI: 10.1201/9781003328414-26

analysis, prediction, and comparison of machine learning prediction algorithms. Section 26.5 provides the conclusion, and ideas related to further work are covered in Section 26.6.

26.1.1 LITERATURE REVIEW

The evolution of terrorism, which is more brutal day by day with the help of typical technology changes, leaves a terrible impact on society with lots of long-term effects. It attracts many researchers to produce better solutions with advanced technology and other facilities. The researchers' main aim is to predict the attack as soon as possible and reduce the casualties to the maximum.

Sattar et al., 2021 [14] use deep learning models to propose and implement a recommender system for a GTD-based database of Iraqi terrorist attacks. Researchers use the proposed system's restricted Boltzmann machine (RBM) neural network to predict terrorist attacks.

Pan, 2021 [13] proposed a framework containing data splitting, pre-processing, and different five prediction models. Researchers use sci-kit learn and select K best for feature selection. Further, they use different five kinds of machine learning (ML) prediction algorithms, which produces a significantly better result than earlier proposed systems. Olabanjo et al., 2021 [10] proposed the assembled ML model, which uses support vector machine (SVM) and K-nearest neighbor (kNN) for the prediction of areas prone to attack by terrorist groups.

Huamaní et al., 2020 [7] describe different ML techniques to predict terrorist attacks over the globe. They represent the many other authors' techniques in their paper. They also explain different artificial intelligence (AI) techniques like symbolic AI, evolutionary AI, and ML techniques.

Zhenkai et al., 2020 [19] designed a system that uses the various ML methods and is used for terrorist attack risk predictions. Bhatia et al., 2020 [3] use Hadoop on GTD. They further use the HiveQL to retrieve essential evidence around the attack. Further, they use Tableau to visualize the data set.

Agarwal et al., 2019 [1] uses the GTD to predict the effectiveness of the violence by using SVM, random forests (RF), and dummy classifiers. Here, researchers used the weather data set to find out the relation between weather conditions and terrorist attack patterns. Researchers merged the GTD with the weather data formed and pre-processed the new combined data set, then performed visualization and analysis on the new collective data set. The output of the previous step is further processed to predict the successfulness and prediction of causalities based on weather conditions.

Singh et al., 2019 [15] proposed a system that analyzes terrorist countries and regions with other knowledge. They try to predict terrorist behavior and use different ML techniques. Gao et al., 2019 [4] also described five different ML techniques for terrorist attack prediction. Spiliotopoulos et al., 2019 [16] proposed a system to predict the terrorist attack risk factor in a particular geographical area using GTD data from 2000 to 2017.

Xia & Gu, 2019 [18] proposed a model called Terrorist Knowledge Graph (TKG). This graph is updated regularly for better prediction by extracting information from the Wikipedia page. The TKG also provides a better understanding of both humans and machines to produce results.

Alhamdani et al., 2018 [2] provide a review of the recommendation system of GTD using deep learning techniques to uncover the social media propaganda for terrorism using the GTD.

Kolajo & Daramola, 2017 [9] proposed a system model which uses different social media sources to identify terrorist activities using Apache Spark technology to achieve the desired results. Zijuan & Shuai, 2017 [20] use big data analysis methods to study terrorist attacks and analyze strategies to review the data related to previous terrorist attacks.

Toure & Gangopadhyay, 2016 [17] solve the various aspects of terrorism with the help of various software and methodologies. Using this system, they aim to calculate risks in different locations and risks in near-future terrorist attacks. Hegde et al., 2016 [6] proposed their work on visual analytics on GTD with the current events of social media, i.e., social network analytics.

Pagan, 2010 [12] analyzed different pre-processing techniques to reduce the noise in the data. Further they used other ML techniques to minimize the error and enhance the accuracy of the system. Ozgul et al., 2009 [11] proposed a model to solve various terrorist assaults which were unsettled in the GTD. Using the clustering method, they used the crime prediction model (CPM) to predict unsolved attacks in Turkey between 1970 and 2005.

26.2   RESEARCH METHODOLOGY

This section is divided further into the following sub-sections that describe the methodology used for carrying out the proposed work.

26.2.1   DATASET

For the proposed work, the GTD is used [5]. GTD is an open-source repository based on terrorist incidences from 1970 to 2020 and is updated periodically. The National Consortium is responsible for regularly updating, maintaining, and studying terrorists and answers to terrorism (START) at Maryland University [8]. This database consists of more than 201,183 terrorist incidence information and is updated every year. These incidents are further labeled by more than 135 attributes or columns to dissipate the possibilities and totality of the event. Various essential attributes of the GTD are as follows:

1. *iyear, imonth, iday:* Year, month, and date, respectively, in the numeric form on which the terrorist attack happened.
2. *country, country_txt, city, latitude, longitude:* Country code, country name, city name in which the attack happened, and coordinates in the form of latitude and longitude for that location.

3. *crit1, crit2, crit3:* crit1, crit2, and crit3 are binary numerical values that identify the purpose of the terrorist attack.
4. *attacktype1_txt:* This attribute contains the details of attack categories. Examples: hijacking, bombing/explosion, and many more.
5. *weaptype1_txt:* Kind of firearm used in the violence. Examples: explosives, firearms, incendiary, and many more.
6. *nkill, nkillter:* nkill counts the total number of people who die in that particular incidence, while nkillter only counts the number of terrorists who die in that particular incidence.
7. *Nwound:* Number of confirmed non-fatal injuries for both victims and terrorists.

   A detailed description of all 135 attributes can be found in the GTD Codebook of studying Terrorists and Responses to Terrorism, ID: 231483. This document was initially uploaded on 7 September 2019 and updated in August 2021, as shown in Figure 26.1.

26.2.2 Data Pre-Processing

Preprocessing of the GTD database contains the following steps.

26.2.2.1 Dropping Irrelevant Columns

The GTD has 2,01,183 records or incidents, and 135 attributes or columns further describe each record or incident. Out of these, most attributes are irrelevant based

	iyear	imonth	iday	country	region	latitude	longitude	crit1	crit2	crit3	doubtterr	alternative	multiple
26985	1986	0	0	186	6	?	?	1	1	1	-9	? 0	
15163	1982	0	0	32	1	43.6667	-79.4167	1	1	1	0	? 0	
7255	1979	0	0	160	5	?	?	1	1	1	0	? 0	
7254	1979	0	0	45	3	4.66713	-74.1061	1	1	1	0	? 0	
7253	1979	0	0	101	4	?	?	1	1	1	-9	? 0	
5728	1978	0	0	61	2	13.6929	-89.1992	1	1	1	-9	? 0	
5727	1978	0	0	30	3	-30.0341	-51.2178	1	1	1	-9	? 0	
4412	1977	0	0	69	8	43.4929	1.47484	1	1	1	0	? 0	
4411	1977	0	0	69	8	43.4929	-1.47484	1	1	1	0	? 0	
4410	1977	0	0	101	4	35.6891	139.748	1	1	1	0	? 0	
4409	1977	0	0	101	4	35.6891	139.748	1	1	1	0	? 0	
4408	1977	0	0	101	4	35.6891	139.748	1	1	1	0	? 0	
3486	1976	0	0	209	10	39.9308	32.7675	1	1	0	1	1.0 0	
3485	1976	0	0	209	10	41.1062	28.6899	1	1	0	1	1.0 0	
2745	1975	0	0	153	6	33.594	73.0691	1	1	1	0	? 0	
2166	1974	0	0	98	8	41.891	12.4901	1	1	1	0	? 0	
2165	1974	0	0	69	8	48.8566	2.34233	1	1	1	-9	? 0	
1691	1973	0	0	45	3	?	?	1	1	1	0	? 0	
1124	1972	0	0	160	5	11.5860	122.754	1	1	1	0	? 0	
2	1970	0	0	130	1	19.3719	-99.0866	1	1	1	0	? 0	
108036	2011	1	0	153	6	32.9811	69.7494	1	1	1	0	? 0	
94896	2010	1	0	153	6	34.9279	72.4152	1	1	1	0	? 1	
94894	2010	1	0	153	6	34.9279	72.4152	1	1	1	0	? 1	
94893	2010	1	0	153	6	34.9279	72.4152	1	1	1	0	? 1	
94892	2010	1	0	153	6	34.9279	72.4152	1	1	1	0	? 1	
94891	2010	1	0	153	6	34.9279	72.4152	1	1	1	0	? 1	
94890	2010	1	0	153	6	34.9279	72.4152	1	1	1	0	? 1	
79362	2006	1	0	95	10	33.3036	44.3718	1	1	1	0	? 0	
79361	2006	1	0	95	10	33.3036	44.3718	1	1	1	0	? 0	
79360	2006	1	0	95	10	35.4521	44.3755	1	1	1	0	? 0	
79359	2006	1	0	95	10	33.3036	44.3718	1	1	1	0	? 0	
79358	2006	1	0	97	10	31.4132	34.6417	1	1	1	0	? 0	
79357	2006	1	0	4	6	32.37	62.1285	1	1	1	0	? 0	
73690	2002	1	0	45	3	2.53593	-75.5277	1	1	1	0	? 0	
71735	2001	1	0	6	10	36.2628	2.21802	1	1	1	0	? 0	
71700	2001	1	0	45	3	7.36005	-73.8531	1	1	1	0	? 0	
71687	2001	1	0	45	3	3.80089	-76.6413	1	0	1	1	3.0 0	

FIGURE 26.1 Sample of GTD before data pre-processing.

on the user's requirement, so some of the attributes or columns were dropped in the pre-processing step. The irrelevant columns were dropped based on the following criteria.

26.2.2.2   Based on Percentage of Null Entries

Remove columns with more than 50% of null values. The null entry percentage can be calculated using the following equation:

Null Enteries(%)=((Total Numbers of Null Entry)/(Total Numbers of Enteries))* 100 (10.1)

Example: If any column or attribute of GTD has more than 100,592 null values, that attribute or column is eternally removed from the dataset.

2.2.3. *Based on the user requirement:* Remove the columns that are not fit as per the user's desires.

2.2.4. *Renaming the columns:* Column or attribute names in the GTD were also renamed for ease of work. These changes help in better understanding and reduce the confusion at the time of data analysis. For example, 'gname' is renamed to 'Group_Name,' 'targtype1_txt' to 'Targtype_Name' and others.

2.2.5. *Replacing unknown values:* We use the most frequent item method to replace null values. We replace unknown cities, attack types, and target types with the most frequent cities, attack types, and target types in that specific country, along with replacing unknown group names and weapon types with the most frequent group names and weapon types in that specific region. Pseudo-code for replacing unknown values is shown in algorithm 1.

2.2.6. *Handling irrelevant values:* Even after removing null values, some of the GTD columns and attributes still have irrelevant values in the form of '−' (negative) values like the 'doubtter' attribute or '0' (zero) in the 'iday,' 'imonth,' and 'iyear.' For the 'doubtter' negative values were directly replaces with '1' because they are most likely to be true for this attribute. The 'iday,' 'imonth,' and 'iyear' unknown values were replaced with the help of most frequently day, month, and year, respectively.

2.2.7. *Dropping the duplicate and missing values rows:* Finally, we drop the duplicate and missing values from the GTD to remove the redundancy and data inconsistency.

2.2.8. *Terrorist attack criteria:* One of the fundamental and most significant problems in any attack for law enforcement agencies is to identify if an attack is a terrorist attack or not. To differentiate between attacks, we use 'Crit1,' 'Crit2,' and 'Crit3' as the different motivations behind the terrorist attack.

ALGORITHM 1. REPLACE UNKNOWN
VALUES WITH MOST FREQUENT

Require: replace (i, j)   # Replace Function

```
1:   create ← [City, AttackType Name, Targtype Name]
2:   df ← gtd.csv
3:   for j = create[0] to create[2] do
4:       for i = 0 to len(df[j]) do
5:           if (df[j][i] ==' Unknown') then
6:                                                   df[j][i] = replace(i, j)
7:                               end if
8:               end for
9:   end for
```

26.3   JUPYTER LAB AND ORANGE TOOL

After executing all the pre-processing steps on the GTD, cleaned data was saved in comma-separated value (CSV) format. The same was passed to the Jupyter lab and Orange tool for further data analysis and machine learning predictions. Jupyter lab is a web-based interface with an updated version of the Jupyter notebook. The Orange tool kit is used for data mining, visualization, and machine learning. Jupyter lab and Jupyter notebook both are part of the Anaconda navigator shown in Figure 26.2.

	Year	Month	Day	Country_Code	Region_Code	Region_Name	Latitude	Longitude	Crit1	Crit2	Crit3
149624	2019	12	1	153	6	South Asia	32.9431	69.955	1	1	0
149625	2019	12	1	33	11	Sub-Saharan Af...	12.7228	1.05029	1	1	0
149626	2019	12	1	37	11	Sub-Saharan Af...	6.03549	10.1213	1	1	1
149627	2019	12	1	214	9	Eastern Europe	50.4408	30.5087	0	1	1
149628	2019	12	1	4	6	South Asia	35.4477	65.8896	1	1	1
149629	2019	12	1	228	10	Middle East & ...	12.8671	44.984	1	1	1
149630	2019	12	1	1004	11	Sub-Saharan Af...	9.81778	33.5456	1	1	1
149631	2019	12	1	95	10	Middle East & ...	34.0356	45.4406	1	1	0
149632	2019	12	1	95	10	Middle East & ...	34.0356	45.4406	1	1	0
149633	2019	12	1	95	10	Middle East & ...	33.3036	44.3718	1	1	1
149634	2019	12	2	4	6	South Asia	34.5230	69.1403	1	1	1
149635	2019	12	2	4	6	South Asia	36.0842	55.7614	1	1	1
149636	2019	12	2	4	6	South Asia	36.1163	65.6884	1	1	1
149637	2019	12	2	200	10	Middle East & ...	36.6865	38.3502	1	1	0
149638	2019	12	2	123	11	Sub-Saharan Af...	14.3095	-2.73751	1	1	0
149639	2019	12	2	141	6	South Asia	28.6512	80.4499	1	1	1
149640	2019	12	3	33	11	Sub-Saharan Af...	14.0863	-2.43389	1	1	0
149641	2019	12	3	33	11	Sub-Saharan Af...	13.379	-3.20308	1	1	0
149642	2019	12	2	4	6	South Asia	33.3153	69.9285	1	1	1
149643	2019	12	2	4	6	South Asia	32.6204	65.4421	1	1	1
149644	2019	12	2	138	5	Southeast Asia	20.8695	92.5272	1	1	1
149645	2019	12	2	4	6	South Asia	34.6569	61.1097	1	1	1
149646	2019	12	4	37	11	Sub-Saharan Af...	11.1615	14.0145	1	1	1
149647	2019	12	3	137	11	Sub-Saharan Af...	-10.8758	40.3695	1	1	1
149648	2019	12	3	95	10	Middle East & ...	33.7875	42.4561	1	1	0
149649	2019	12	3	153	6	South Asia	30.9395	66.5145	1	1	1
149650	2019	12	3	160	5	Southeast Asia	15.9142	120.542	1	1	1

FIGURE 26.2   Sample of GTD after data pre-processing.

26.3.1 Proposed System

The proposed system in this chapter is mainly focused on data analysis and visualization using existing historical data to uncover information, such as year-wise attack count, a prime target, attacking methods, region-wise terrorist activity, most affected countries list, and others, along with the prediction of future terrorist attacks using various machine naïveing algorithms such as kNN, naive Bayes, neural network, logistic regression, RF, tree, Cn2 rule inducer, and SVM. Machine learning algorithms are further used to predict the severity of terrorist attacks based on the consolidated yield of data analysis and visualization.

Further accuracy of the proposed system for the different machine learning algorithms is compared to find the best suitable algorithm for this problem, which is further compared with the existing systems to evaluate enhancement achieved using the proposed approach.

The system architecture shown in Figure 26.3 represents the methodology used to achieve this desired task in diagrammatic representation to carry out the proposed work. The pseudo-code for one of the machine learning algorithms used to achieve the task, i.e., kNN for training and implementing the GTD, is shown in Algorithm 2.

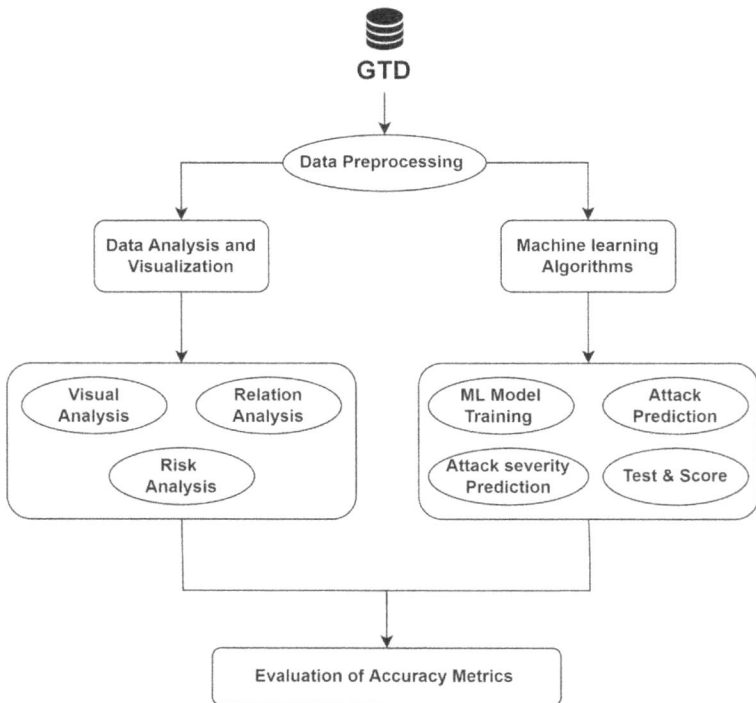

FIGURE 26.3 System architecture.

26.4 RESULT

ALGORITHM 10.2 TRAINING AND IMPLEMENTING KNN FOR GTD

```
1: function TRAIN KNN(D, GTD)
2:          GT D' ← preprocessed(GT D)
3:          k ← select k(D, j)
4:          return GT D', K
5: end function
6: function IMPLEMENTED KNN(D, GTD', k, d)
7:                    Sk ← Calculate Nearest Neighbors (GT D', k, d)
8:                    for every DiɛD do
9:                              pj ← |sk ∩ Dj |/k
10:                   return arg max, pj
11:          end for
12: end function
```

This segment contains numerous observations, including data visualization, prediction of terrorist attacks, and severity of attack, and comparisons between various machine learning prediction algorithms.

In the initial findings, Figure 26.4 shows that the number of terrorist attacks in 2014 reached the new maximum from 1970 to 2019. Along with this, Figure 26.5 revealed that bombing/explosions, armed assault, and assassination are the top three attacking methods used by terrorists.

Figure 26.6 shows that during the 1970 to 2009 period, private citizens and property were prioritized in any attack, followed by the military, police, and government (general). People who belong to these occupations or groups are on the hit list of attackers.

The data set also clearly shows in Figure 26.7 that after 2010, North Africa and the Middle East were heavily impacted by terrorists, followed by Eastern Europe.

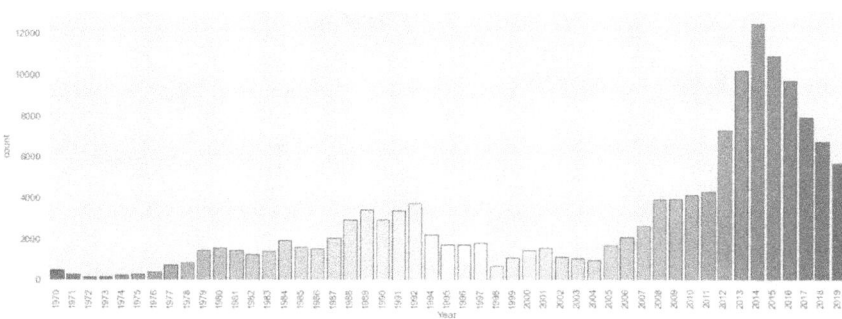

FIGURE 26.4   Year-wise terrorist attacks count.

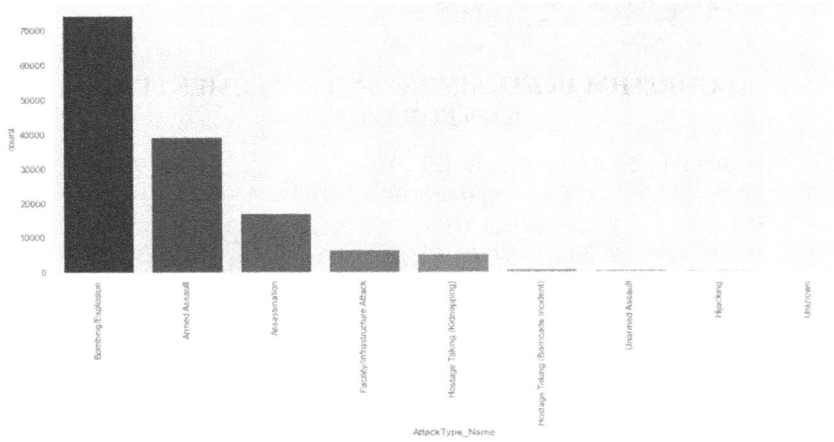

FIGURE 26.5 Attacking methods used by terrorists.

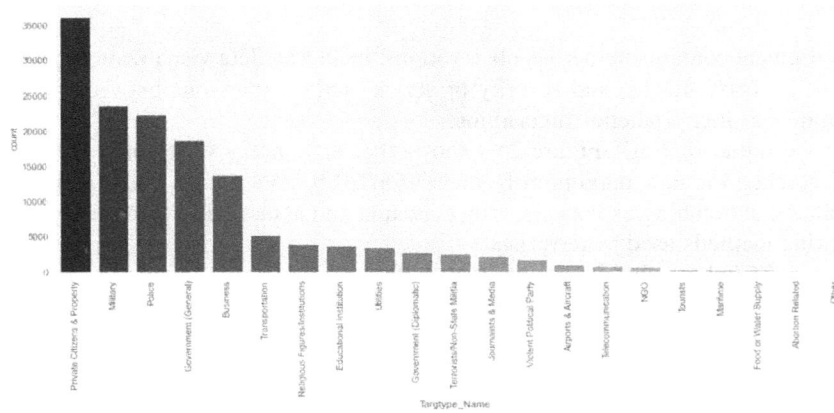

FIGURE 26.6 Main targets of terrorist attacks.

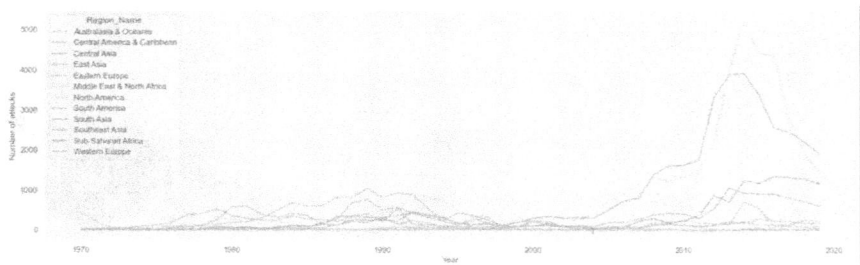

FIGURE 26.7 Year-wise terrorist activities in various regions of the world.

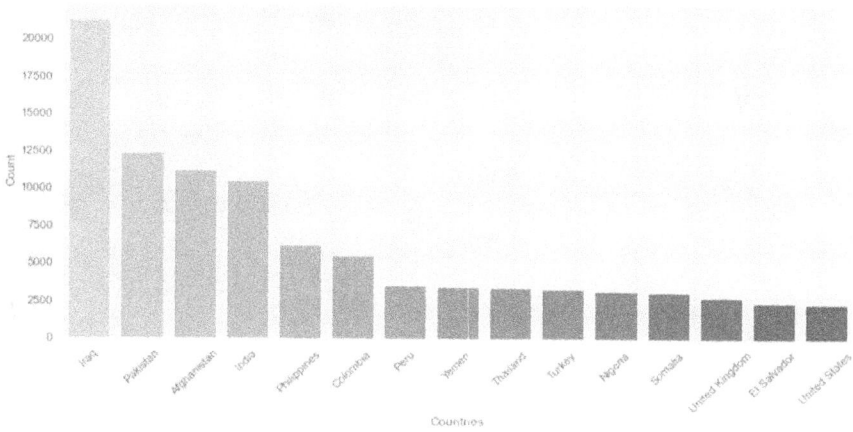

FIGURE 26.8 Most affected countries by terrorism.

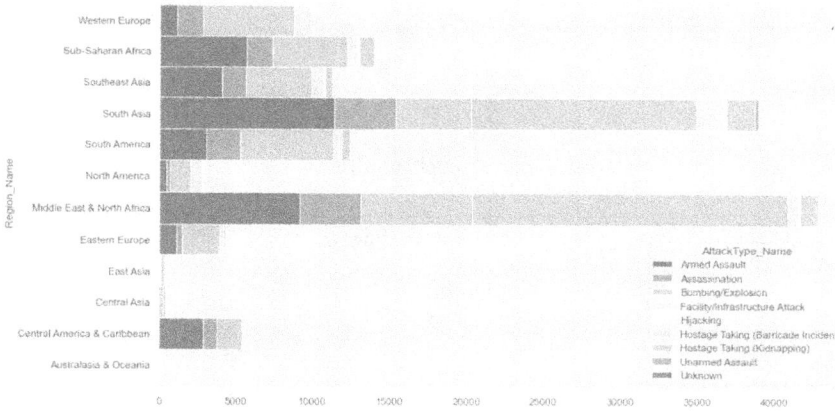

FIGURE 26.9 Country-wise attack types.

In the analysis in Figure 26.8, Iraq is the most impacted nation by terrorism, followed by Pakistan and Afghanistan, whereas India is the fourth-most affected country by terrorism.

The visual representation in Figure 26.9 clearly shows that in South Asia, facility/infrastructure attacks, followed by armed attacks, were the primary weapon used by the terrorists belonging to particular groups.

Figure 26.10 shows that Taliban became the most active terrorist group worldwide after the year 2000.

The visual representation in Figure 26.11 reveals the pattern between attacks vs. those killed in a particular country.

Figure 26.12 shows that the months of May, April, and August are the most exposed months with the highest risk ratio, i.e., 10.05, 9.68, and 9.25, respectively, for terrorist

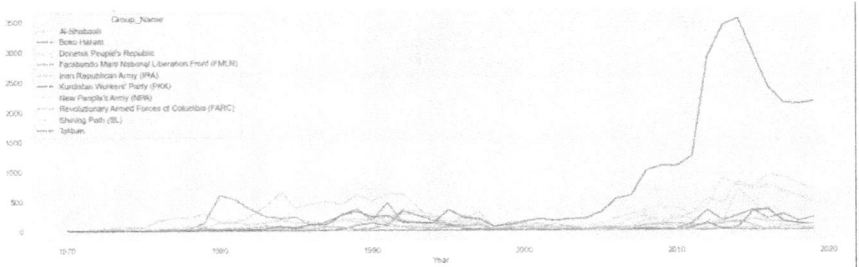

FIGURE 26.10 Year-wise terrorist group activity.

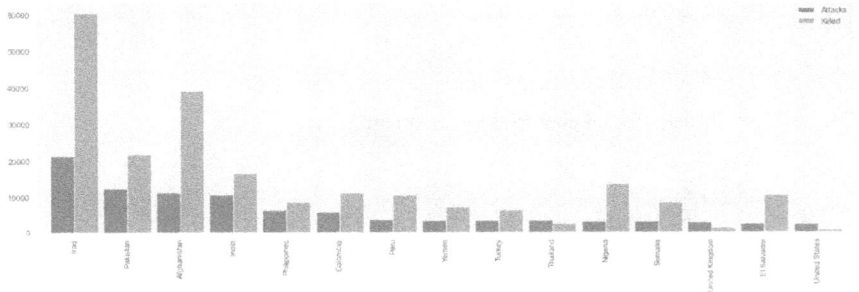

FIGURE 26.11 Country-wise attack vs. number killed.

Month	Percentage
5	10.05%
4	9.68%
8	9.25%
6	9.03%
7	9.00%
11	8.57%
10	8.27%
1	8.14%
3	7.44%
9	7.02%
12	6.96%
2	6.57%

FIGURE 26.12 Month-wise terrorist attack risk.

attacks and the 25th and 15th of every month are the riskiest for a terrorist attack compared to other days of the month. The 25th of the month has the highest risk ratio, i.e., 3.67, followed by the 15th with a 3.61 risk ratio, as shown in Figure 26.13.

Further, for the prediction of whether terrorist attacks are successful or not, the terrorist attack prediction model created in the Orange data mining tool is shown in Figure 26.14

Day	Percentage	Day	Percentage
1	3.48 %	16	3.32%
2	3.36 %	17	3.10%
3	3.11 %	18	3.10%
4	3.28 %	19	3.33%
5	2.87 %	20	3.18%
6	3.02 %	21	3.37%
7	3.52 %	22	3.01%
8	3.52 %	23	3.06%
9	3.16 %	24	3.13%
10	3.53 %	25	3.67%
11	3.24 %	26	3.47%
12	3.30 %	27	3.40%
13	3.42 %	28	3.30%
14	3.12 %	29	2.96%
15	3.61 %	30	3.15%
		31	1.74%

FIGURE 26.13   Day-wise terrorist attack risk.

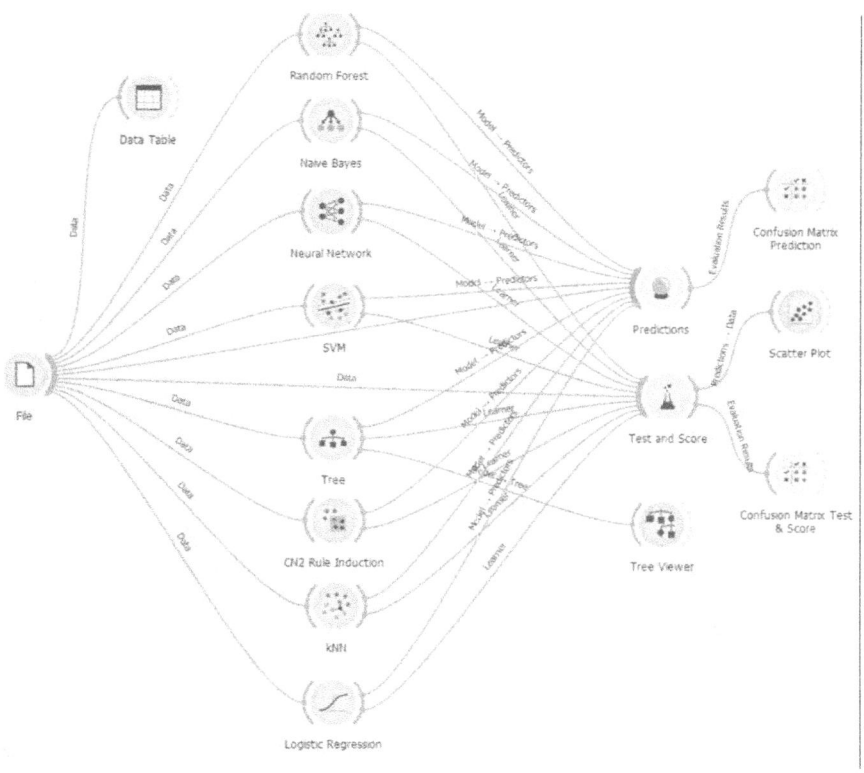

FIGURE 26.14   Attack prediction model.

Model	AUC	CA	F1	Precision	Recall
Random Forest	0.946	0.982	0.982	0.982	0.982
Neural Network	0.946	0.982	0.982	0.982	0.982
Naive Bayes	0.946	0.982	0.982	0.982	0.982
Logistic Regression	0.945	0.982	0.982	0.982	0.982
kNN	0.946	0.982	0.982	0.982	0.982
CN2 rule inducer	0.945	0.982	0.982	0.982	0.982
Tree	0.887	0.963	0.961	0.964	0.963
SVM	0.500	0.079	0.024	0.015	0.079

FIGURE 26.15   Attack prediction accuracy matrix.

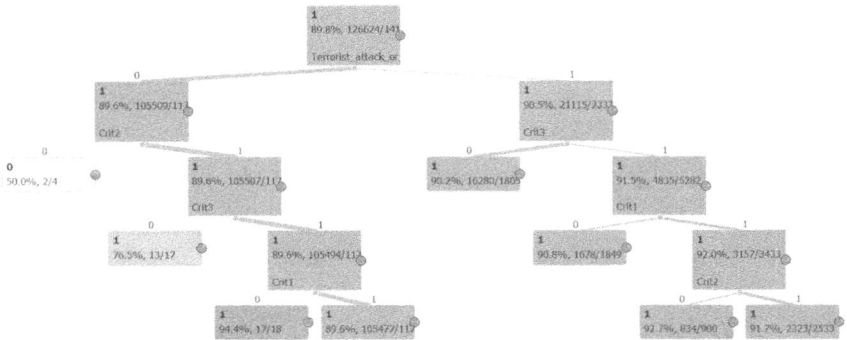

FIGURE 26.16   Tree representation for attack severity prediction.

The terrorist attack prediction model produces the results based on various machine learning prediction algorithms such as RF, neural network, naive Bayes, logistic regression, kNN, CN2 rule inducer, decision tree, and SVM.

For the same area under the ROC curve (AUC), classification accuracy (CA), F1, precision and recall for attack prediction, and if an attack is successful or not are as shown in Figure 26.15.

The tree representation of attack severity prediction of successful terrorist attacks using machine learning tree algorithms is shown in Figure 26.16. The severity of a terrorist attack is successfully predicted with the accuracy of 89.98 percent using the neural network algorithm, along with others, as shown in Table 26.1.

When the proposed system output is compared with the existing available systems, it is noticed that the proposed system reveals some very critical information regarding the terrorist attacks that happened in the past. It also produces a significantly better result for the prediction of terrorist attacks compared to the already available systems that use the same data set. The existing model recommended by Sattar et al., 2021 [14] predicts with an accuracy of 98.12%; Pan, 2021 [13] proposed a system predicting the terrorist attacks within the range of accuracy between 97.16% and 96.82%; Olabanjo et al., 2021 [10] proposed a hybridized classical model, which

TABLE 26.1

Attack Severity Prediction Accuracy Matrix

Model	AUC	CA	F1	Precision	Recall
NenaïveNetwork	0.504	0.898	0.849	0.806	0.898
Naive Bayes	0.503	0.898	0.849	0.806	0.898
Logistic Regression	0.504	0.898	0.849	0.806	0.898
KNN	0.502	0.898	0.849	0.806	0.898
Random Forest	0.505	0.898	0.849	0.823	0.898
Decision Tree	0.504	0.898	0.849	0.821	0.898
CN2 Rule Inducer	0.504	0.898	0.849	0.821	0.898
SVM	0.502	0.501	0.595	0.816	0.501

TABLE 26.2

A Comparison of Intended Systems Work on GTD

Related Model Proposed By	Related Model Accuracy (%)	Proposed Work Accuracy (%)
Sattar et al., 2021[14]	98.12%	
Pan, 2021 [13]	96.82%–97.16%	98.20%
Olabanjo et al., 2021 [10]	97.81%	
Huamaní et al., 2020 [7]	75.45%–90.414%	
Singh et al., 2019 [15]	82%	
Agarwal et al., 2019 [1]	68%–82%	
Toure et al., 2016 [17]	96.3%	
Gao et al., 2019 [4]	94.8%	

gave an accuracy of 97.81%; and Huamaní et al., 2020 [7] can predict terrorist attack with an accuracy between 75.45% and 90.414%, which is significantly low compared to the proposed system. With the help of better pre-processing techniques and attribute selections, the proposed system can produce better prediction results with an accuracy of the 98.2 and 89.8 percentile for the prediction of attack and successfulness or severity of an attack, respectively.

Table 26.2 compares the proposed system with the existing systems that use different methods and practices on the GTD.

26.5   CONCLUSION

This chapter compared eight machine learning prediction algorithms for the proposed system, including RF, neural network, naive Bayes, logical regression, kNN, CN2 rule inducer, tree, and SVM. This experiment shows that RF, neural network, naive Bayes, kNN, and logical regression have the highest CA, up to 98.2%, followed by logical regression and CN2 rule inducer, which have the same classification accuracy but low AUC. Here, SVM produces the least significant result, with the accuracy of 7.9% and 50% for the AUC.

For the prediction of attack severity, RF produces the best results with the accuracy of 89.8% and AUC with 50.5%, while SVM again produces the worst accuracy for the prediction of attack severity with 50.10%.

With the help of data visualization and machine learning prediction algorithms, lots of shocking and vital information is disclosed, giving information related to attack patterns, weapons used, and most vulnerable locations and months and days. It has been observed that May 25th and 15th are the most vulnerable days in a calendar year. It also shows that every terrorist group has different geographical areas in which they operate. The topmost terrorist groups also operate in different geographical locations and never operate in other groups' areas. Other facts also reveal that the Taliban and Kurdistan Workers Party (PKK) have a perfect attack pattern that makes them almost unpredictable. Our prediction system also predicts future terrorist attacks, one of law enforcement agencies' basic but essential problems.

26.6   FUTURE WORK

Future work will focus on predicting group alliance possibilities. The attack type and location may also be predicted based on social media data like Facebook and Twitter. Local news agencies' reports, weather, and special occasion data may also help to predict the attacks with faster and better accuracy with the prediction of the severity of that attack.

Future work may also focus on developing a better prediction system by using better and enhanced prediction algorithms with smart pre-processing techniques.

REFERENCES

Agarwal, P., Sharma, M., & Chandra, S. (2019). Comparison of machine learning approaches in the prediction of terrorist attacks. 2019 12th Inte rnational Conference on Contemporary Computing (IC3), 1–7. https://doi.org/10.1109/IC3.2019.8844904.

Alhamdani, R. S., Abdullah, M. N., & Sattar, I. A. (2018). Recommender system for global terrorist database based on deep learning. International Journal of Machine Learning and Computing, 8(6), 6.

Bhatia, K., Chhabra, B., & Kumar, M. (2020). Data analysis of various terrorism activities using big data approaches on global terrorism database. 2020 6th International Conference on Parallel, Distributed and Grid Computing (PDGC), 137–140. https://doi.org/10.1109/PDGC50313.2020.9315784.

Gao, Y., Wang, X., Chen, Q., Guo, Y., Yang, Q., Yang, K., & Fang, T. (2019). Suspects prediction towards terrorist attacks based on machine learning. 2019 5th International Conferenc e on Big Data and Information Analytics (BigDIA), 126–131. https://doi.org/10.1109/BigDIA.2019. 8802726.

Hegde, L. V., Sreelakshmi, N., & Mahesh, K. (2016). Visual analytics of terrorism data. 2016 IEEE International Confer ence on Cloud Computing in Emerging Markets (CCEM), 90–94. https://doi.org/10.1109/CCEM.2016.024.

Huamaní, E. L., Mantari, A., & Roman-Gonzalez, A. (2020). Machine learning techniques to visualize and predict terrorist attacks worldwide using the global terrorism database. International Jou rnal of Advanced Computer Science and Applications, 11(4). https://doi.org/10.14569/IJACSA.2020.0110474.

Kolajo, T., & Daramola, O. (2017). Leveraging big data to combat terrorism in developing countries. 2017 Conference on Info rmation Communication Technology and Society (ICTAS), 1–6. https://doi.org/10.1109/ICTAS.2017.7920662.

Olabanjo, O. A., Aribisala, B. S., Mazzara, M., & Wusu, A. S. (2021). An ensemble machine learning model for the prediction of danger zones: Towards a gl obal counter-terrorism. Soft Computing Letters, 3, 100020. https://doi.org/10.1016/j.socl.2021.100020.

Ozgul, F., Erdem, Z., & Bowerman, C. (2009). Prediction of past unsolved terrorist attacks. 2009 IEEE International C onference on Intelligence and Security Informatics, 37–42. https://doi.org/10.1109/ISI.2009.5137268.

Pagan, J. V. (2010). Improving the classification of terrorist attacks a study on data pre-processing for mining the global terrorism database. 2010 2nd International C onference on Software Technology and Engineering, 5608902. https://doi.org/10.1109/ICSTE.2010.5608902.

Pan, X. (2021). Quantitative analysis and prediction of global terrorist attacks base d on machine learning. Scientific Programming, 2021, 1–15. https://doi.org/10.1155/2021/7890923.

Sattar, I. A., Alhamdani, R. S., & Abdulla, M. N. (2021). Design and implementation recommender system for Iraqi terrorist database based on deep learning. 2021 7th International Engineering Conference "Research & Innovation amid Global Pandemic" (IEC), 32–36. https://doi.org/10.1109/IEC52205.2021.9476083.

Singh, K., Chaudhary, A. S., & Kaur, P. (2019). A machine learning approach for enhancing defence against global terrorism. 2019 12th Inte rnational Conference on Contemporary Computing (IC3), 1–5. https://doi.org/10.1109/IC3.2019.8844947.

Spiliotopoulos, D., Vassilakis, C., & Margaris, D. (2019). Data-driven country safety monitoring terrorist attack prediction. Proceedings of the 2019 IEEE/ACM International Conference on A dvances in Social Networks Analysis and Mining, 1128–1135. https://doi.org/10.1145/3341161.3343527.

Toure, I., & Gangopadhyay, A. (2016). Real time big data analytics for predicting terrorist incidents. 2016 IEEE S ymposium on Technologies for Homeland Security (HST), 1–6. https://doi.org/10.1109/THS.2016.7568906.

Xia, T., & Gu, Y. (2019). Building terrorist knowledge graph from global terrorism database and Wikipedia. 2019 IEEE International Conferenc e on Intelligence and Security Informatics (ISI), 194–196. https://doi.org/10.1109/ISI.2019.8823450.

Zhenkai, L., Yimin, D., & Jinping, L. (2020). Analysis model of terrorist attacks based on big data. 2020 Chinese Control and Decision Conference (CCDC), 3622–3628. https://doi.org/10.1109/CCDC49329.2020.9164626.

Zijuan, L., & Shuai, D. (2017). Research on prediction method of terrorist attack based on random subspace. 2017 International Conference on Co mputer Systems, Electronics and Control (ICCSEC), 320–322. https://doi.org/10.1109/ICCSEC.2017.8446815.

27 Deep Learning Approach for Identifying Bird Species

B. Harsha Vardhan, T. Monish,
P. Srihitha Chowdary, S. Ravi Kishan,
and D. Suresh Babu

27.1 INTRODUCTION

In the study and research of birds, knowing the species of a bird is critical. We all like birdwatching in our free time, but our skills are insufficient to identify them. Even professionals such as ornithologists have difficulty in memorizing all of the species. Even if they remembered a few species, recognizing them is difficult due to the wide variety of sizes, shapes, and colors they come in.

As a result, our awareness of birds is insufficient to identify them. Birds are often recognized through a photo, audio, or video. Birds can be identified using processing technology that records audio or video. However, other sounds in the surroundings, such as insects and real-world noises, make processing such information more difficult.

As a result, rather than utilizing audio or video to identify birds, it is best to use a photo. Humans nowadays find it tough to collect and assemble various bird photos. Even if they are collected, recognizing them by referring to books is a far more complex and time-consuming task. Rather than depending on books, we created an interface based on a deep learning model that allows us to quickly identify species.

27.2 LITERATURE SURVEY

A. K. Reyes et al. [1] collected a dataset of Colombian bird audio recordings and performed audio feature extraction on it. In the process of feature extraction, they extracted the most useful information using MFCC, which stands for mel-frequency cepstral coefficients of an audio signal. This MFCC produced a set of 16 coefficients for each frame with a duration of 11.6 ms. For representing an audio recording, a vector is used. The vector is calculated by averaging all the frame coefficients. Finally, a matrix is built with all the vectors. To find the similarity between any two audio recordings, they used a distance function. There are many distance functions like Euclidean distance, histogram intersection, and cosine

DOI: 10.1201/9781003328414-27

similarity. They used the cosine similarity function. There are many methods in order to reduce the dimensionality of the vectors like MDS, which stands for multidimensional scaling; PCA, which stands for principal component analysis; and Isomap, which stands for isometric feature mapping, which is used for reducing the high dimensionality vectors. They used the PCA method.

A. Thakur et al. [2] proposed a multi-layer alternating sparse-dense framework in order to identify the species of a bird. They began by using short-time Fourier transform (STFT) to convert the input audio recording into something like a magnitude spectrogram having 512 FFT points on 20-ms frames with 50% overlap. With archetypal analysis (AA), 256 archetypes were learned for each class. Concatenating the previous and next frames around the present frame adds information to the current frame of the spectrogram. Following that, a class's spectrograms are translated into a super-frame-based matrix representation. They presented a technique AA to manage the data's outliers. To identify the audio clip, the bird vocalizations are split, and their associated super-frames are translated into the compressed super-frame format.

B. Chandu et al. [3] proposed a method that involves two stages. In the first stage, they collected all the sound recordings of different bird species and performed different pre-processing techniques on them. In general, all the audio signal recordings will be recorded by using a microphone that records other unnecessary frequencies along with the bird audio frequency. For this, the pre-processing technique used was pre-emphasis which reduces all the other unnecessary frequencies by using a filter. Because an audio signal is not a stationary signal, it has distinct qualities that can change with the duration of time. So, they first divided the recorded audio based on its length into a number of frames and then extracted the signal. The frame length was calculated by considering the total length of the audio signal and the sampling period used. After dividing into frames, they performed the silence removal operation for each frame using the thresholding function. If any signal falls under this threshold limit, it was considered background noise. Following the elimination of background sounds, the next stage was reconstruction, which was a process of integrating all of the frames acquired after the framing and silent removal processes. The next step was to create a spectrogram, which may be done by first translating the time domain data to frequency. For this operation, they used Fourier transforms. This process was repeated for all the audio recordings. They performed this operation in order to identify the species based on their recordings very easily since the spectrogram of one species is different from others. Next, they trained the neural network using these spectrograms. This method achieves 97% accuracy.

C. N. Silla et al. [4] proposed a method that involves several steps. The first one is feature extraction, the second one is model construction, and the third one is evaluating the performance of the model. They utilized the MARSYAS framework to extract features. They took three approaches for categorization. The first technique is flat classification, the second is a classifier with a local model per parent node, and the final one is a global-model hierarchical classification method. The flat classification approach is just like a multi-class classification approach and it can be used for the problems which are labeled up to the leaf nodes. To perform this multi-class classification, they used the classic naïve Bayes algorithm. In the local model hierarchical classification, they used the LCPN approach, which stands for local classifier per parent node approach. A multi-class classifier is used to train each non-leaf node in the class hierarchy in this technique. For testing this approach, they used a top-down approach. The naïve Bayes algorithm was utilized to conduct multi-classification in this second technique. In the third technique, global-model hierarchical classification, a single algorithm was employed to predict the class at any stage of the hierarchy. For this, they used GMNB, which stands for global model naïve Bayes algorithm. They compared these three approaches, and according to their experimental results, the global approach achieved higher accuracy than the flat and local model approaches. They performed these operations on 74 bird species only.

G. Sarasa et al. [5] employed NCD, which refers to normalized compression distance, as a similarity metric to identify bird species in this study. They tested the performance of NCD by using six compression algorithms which belonged to different compression families. Four of them are LZMA, LZMA2, PPMD, and Deflate provided by the 7z software. The CompLearn Toolkit includes two of them: Zlib and BZlib. They then employed the MQTC-based hierarchical clustering technique. Based on their findings, the normalized compression distance can be utilized as an alternate method for determining the species of a bird.

M. M. M. Sukri et al. [6] identified the bird species using ANN which stands for artificial neural network. First, they collected the bird sound records and performed pre-processing operations on them. They used the PSD method, which stands for power spectral density, in order to perform the pre-processing operation. Next, they trained the ANN model with the pre-processed data. The classifier in ANN is the MLP, which refers to multilayer perceptron. This MLP takes the chosen characteristics as input and produces a unique output for each bird species. To perform all these operations, they first recorded the raw data in mp3 format and converted it to a. wav file.

M. T. Lopes et al. [7] collected the audio recordings of 1619 song recordings of 75 bird species. For feature set operation, they compared different frameworks like MARSYAS, IOIHC, and Sound Ruler. Their results showed that the performance of the MARSYAS feature set was better than the other two feature sets. Based on this reason they used this MARSYAS framework to extract the features. Next, they split all the audio recordings according to pulses, which was a short sound interval with high amplitudes. To split the recordings into pulses, they used the audacity audio processing tool. They performed the experiments on the first database and second database according to three dimensions. One was the use of pulses and the second one was using classifiers. The third one was selecting the most frequent classes. They compared different classifiers like naïve Bayes, KNN, SMO-polynomial, MLP, and SMO-Pearson on the two databases. According to their testing results for the first database for classes 3, 5, and 8, the SMO with a polynomial kernel performed better. However, the MLP performed admirably in 12 and 20 classes. In both situations, the SMO method with the Pearson VI kernel function produced better results for the second database.

W. Tsai et al. [8] proposed a method that involves two stages. In the first stage, they perform classification whether it is a call or a song of a bird. In the second stage, the corresponding identifier will perform the operation. If a sound clip is identified as a call, a call identifier will handle it in the second stage; if it is classed as a song, a song classifier will handle it in the second stage. Both identifiers identify the species using auditory characteristics, timbre attributes, and pitch information. The sounds were transformed to MFCCs and evaluated using gaussian mixture models for timbre attributes. The timbre pattern of each bird species' cry or song varies. In the case of pitch characteristics, they transformed the sounds into MIDI note sequences and examined the dynamic shift of the notes.

27.3 METHODOLOGY

27.3.1 Dataset

A large collection of images of 315 different bird species are taken. For each type of bird species, a minimum of 110 images are taken in order to cover the species with different shapes, sizes, colors, and angles. In this way, all the species are organized into training and testing.

- There are 47,555 photos of 315 distinct bird species in the dataset.
- For training, 45,980 photos from 315 species were used.
- For testing 1575 photos from 315 species are used, each with 5 images.

27.3.2 Proposed Work

During the training of the model, all the images are loaded, resized into 224 by 224 pixels, and image augmentation to them in order to avoid the overfitting of the model. The image augmentation contains transformations like shear range, zoom range, and horizontal flip. Figure 27.1 represents the image augmentation of a bird.

- The shear range is used for slanting the image. We fix one axis and stretch the image at a certain angle.
- The zoom range is to zoom the image. If the value is less than 1.0, it will zoom in the image, and if the value is greater than 1.0, it will zoom out the image
- The horizontal flip is used to flip the image either horizontally or vertically.

27.3.3 Proposed Diagram

Figure 27.2 represents the diagrammatic representation of the process. First, load the dataset of different bird species and apply preprocessing techniques to them. We

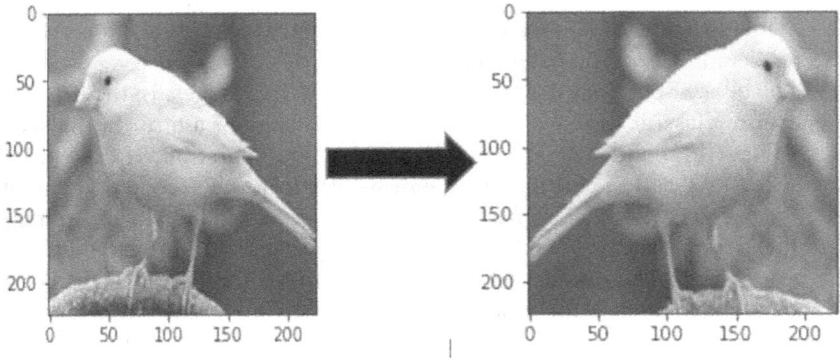

FIGURE 27.1   Image augmentation of a bird.

applied different transformations to them like rescaling, shear range, zoom range, and horizontal flip in order to avoid the model overfitting. After applying all the transformations to it, all the species are labeled as 0, 1, 2, . . ., which keeps track of mapping the bird to the corresponding species indices, where the indices are predicted by the model. The next step is to train the model. For training, a training dataset is used which consists of 45,980 images of 315 different bird species. We used the Adam optimizer and categorical cross-entropy for the loss function, and the number of epochs is 50. During training, it collects all the feature maps like the body, color, and size of the birds and finally saves the trained model. The next step is to develop an interface with the trained model using streamlit in order to provide convenience to the user to upload an image. When a bird is passed as an input by the user, the model will identify the bird from the image; extract the features from it; and identify the species based on its body, color, and size.

27.4   ARCHITECTURE

Figure 27.3 represents the architecture of the model. In this model, there are 16 layers which are 1 input layer, 13 conv2D layers, 5 Maxpooling2D layers, 1 flatten layer, and 1 dense layer. The input layer will take an image in 224 by 224 pixels and pass it to conv2D layers. The conv2D layer is a convolution layer that generates a tensor of outputs by winding a convolution kernel with layers input. The maxpooling2D layer will downsample the input window of size defined by pool size. The flatten layer converts the data into a one-dimensional array for input to the following layer. The dense layer will transmit all of the preceding layer's outputs to all of its neurons, with each neuron representing an output to the next layer.

27.5   RESULTS

Figure 27.4 represents the user interface. The basic design of the user interface is developed using streamlit. Streamlit is an open-source application framework written

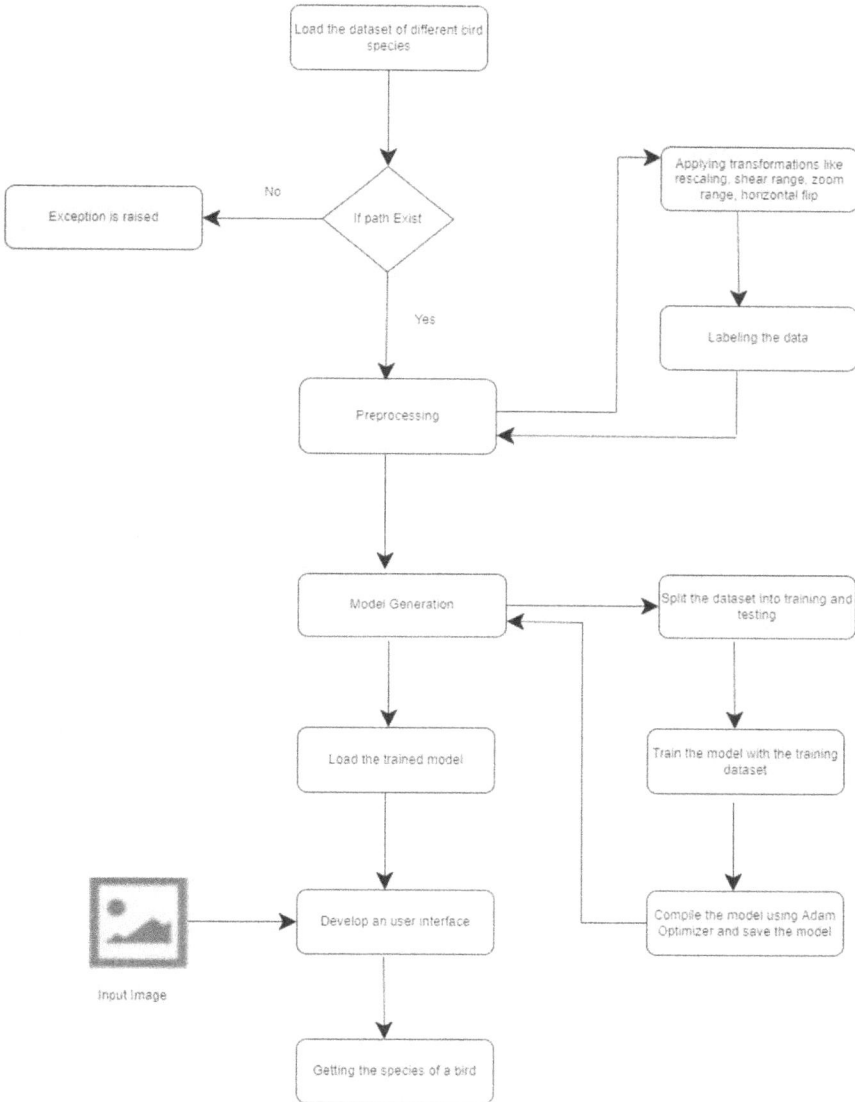

FIGURE 27.2 Flowchart.

in the Python language. It enables us to quickly develop web applications for data science and machine learning. Major Python libraries like sklearn, Keras, PyTorch, SymPy (latex), NumPy, pandas, and Matplotlib are all compatible with it.

Figure 27.5 represents the selection of the input image. There is a browse file option in the interface, which is shown in Figure 27.4, through which the user can select the required image. The format of the image should be in JPG or PNG file

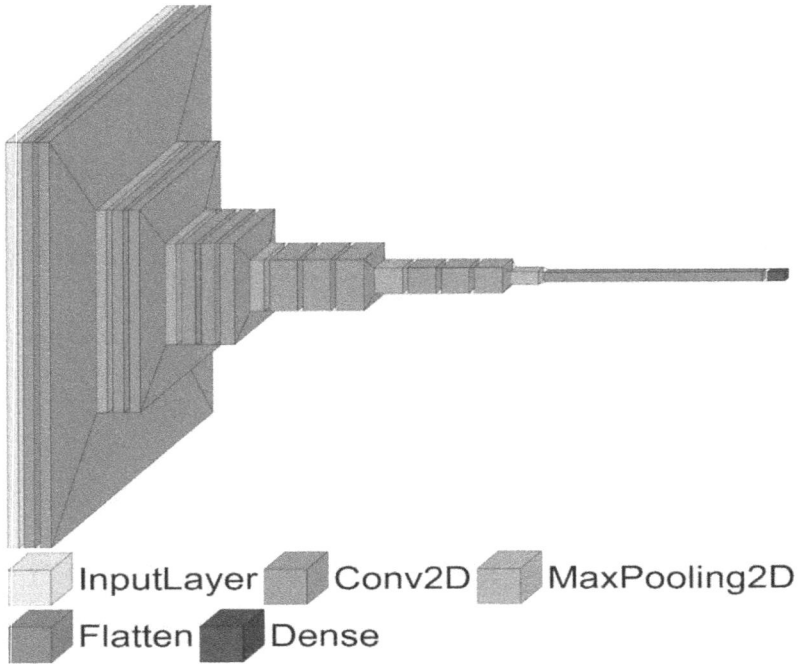

FIGURE 27.3 Architecture of the model.

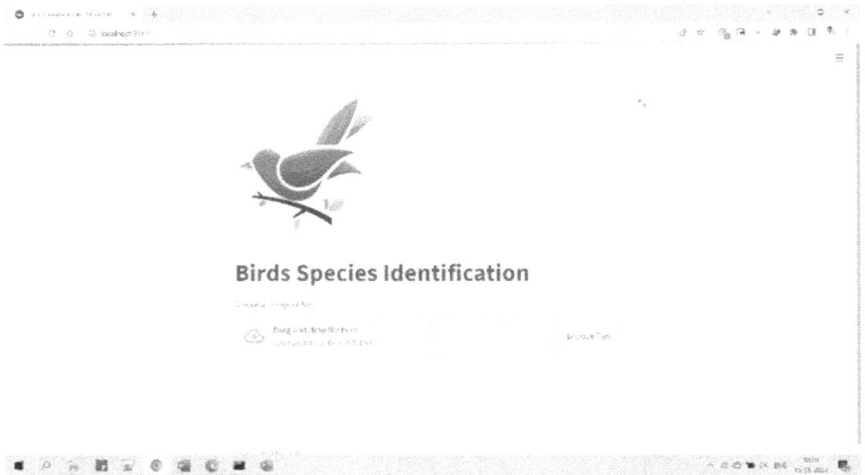

FIGURE 27.4 User interface.

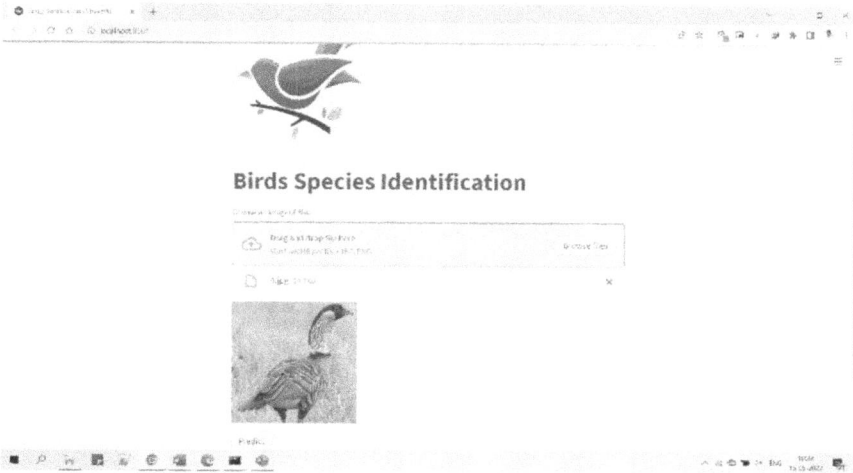

FIGURE 27.5   Input image.

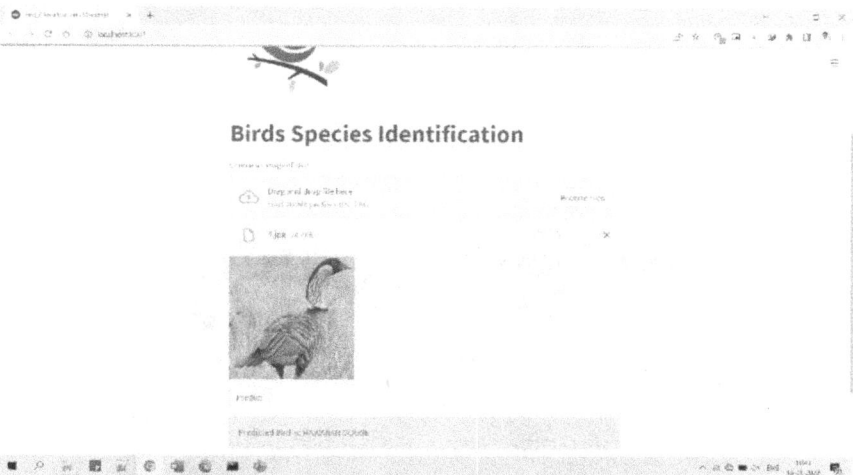

FIGURE 27.6   Species of a bird.

format. When the user selects the browse file option, it asks for an image to upload, which is shown in Figure 27.5.

Figure 27.6 represents the species of a bird. After the selection of the required image by the user, then the user can get the species of the given bird image with the predict button, which is shown in Figure 27.6. Figure 27.7 Represents the Confusion matrix without normalization. Figure 27.8 represents the confusion matrix with normalization. Taking the normalized values into consideration, the confusion matrix with normalization is plotted and is shown in Figure 27.8.

FIGURE 27.7   Confusion matrix without normalization.

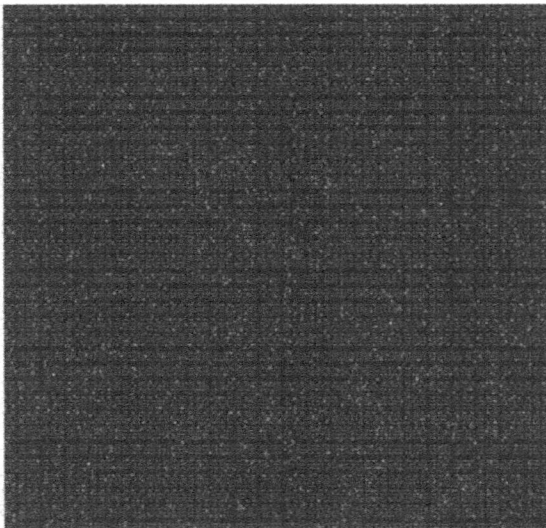

FIGURE 27.8   Confusion matrix with normalization.

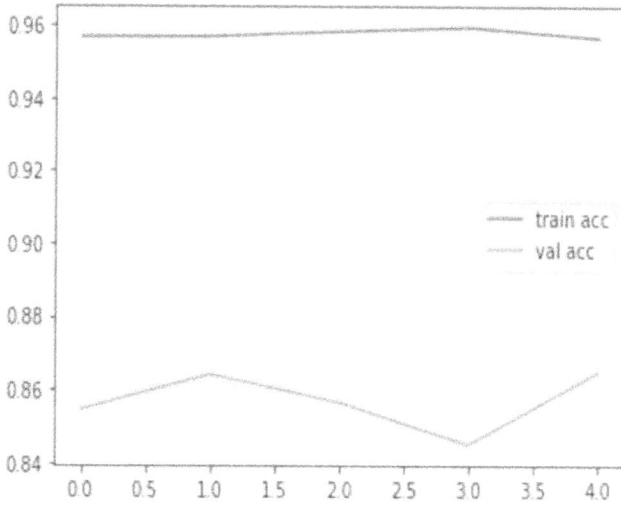

FIGURE 27.9 Accuracy graph.

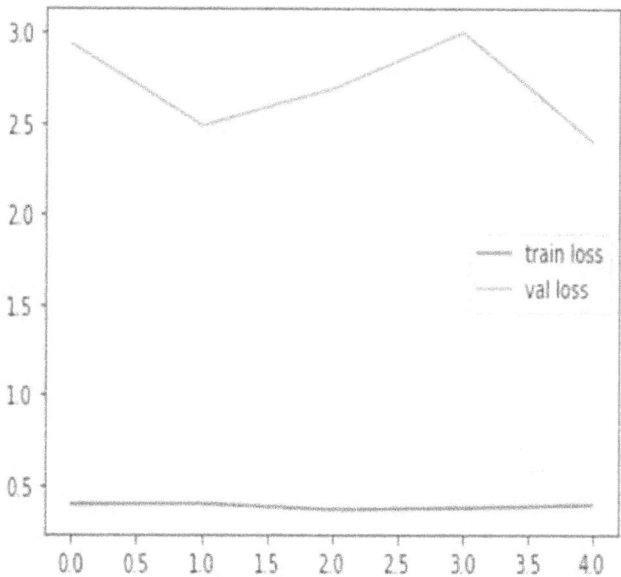

FIGURE 27.10 Loss graph.

Figure 27.9 represents the accuracy graph. We can conclude that this model achieves better accuracy compared to other models and gives better and more accurate results.

Figure 27.10 represents the loss graph. We can conclude that this model gets less loss during training.

FIGURE 27.11   ROC curve.

Figure 27.11 represents the receiver operation characteristics (ROC) curve. Each line in Figure 27.11 corresponds to each bird species in the dataset.

27.6 CONCLUSION AND FUTURE WORK

We used vgg16 architecture for training the model with 315 different species of 45,980 images in different angles, and a user interface was developed using the streamlit framework in order to make it convenient for the user to upload an image. With the trained model we can identify the species of a bird for the image given by the user. Based on the results, we can conclude that the model achieved better accuracy of 86% when compared to other models, and the loss rate is also minimum during training. In the future, we propose a method for recognizing numerous species at once.

REFERENCES

[1] A. K. Reyes and J. E. Camargo, "Visualization of Audio Records for Automatic Bird Species Identification," 2015 20th Symposium on Signal Processing, Images and Computer Vision (STSIVA), 2015, pp. 1–6, doi: 10.1109/STSIVA.2015.7330415.
[2] A. Thakur, V. Abrol, P. Sharma and P. Rajan, "Compressed Convex Spectral Embedding for Bird Species Classification," 2018 IEEE International Conference on Acoustics, Speech and Signal Processing (ICASSP), 2018, pp. 261–265, doi: 10.1109/ICASSP.2018.8461814.
[3] B. Chandu, A. Munikoti, K. S. Murthy, G. Murthy and C. Nagaraj, "Automated Bird Species Identification using Audio Signal Processing and Neural Networks," 2020 International Conference on Artificial Intelligence and Signal Processing (AISP), 2020, pp. 1–5, doi: 10.1109/AISP48273.2020.9073584.
[4] C. N. Silla and C. A. A. Kaestner, "Hierarchical Classification of Bird Species Using Their Audio Recorded Songs," 2013 IEEE International Conference on Systems, Man, and Cybernetics, 2013, pp. 1895–1900, doi: 10.1109/SMC.2013.326.

[5] G. Sarasa, A. Granados and F. B. Rodriguez, "An Approach of Algorithmic Clustering Based on String Compression to Identify Bird Songs Species in Xeno-Canto Database," 2017 3rd International Conference on Frontiers of Signal Processing (ICFSP), 2017, pp. 101–104, doi: 10.1109/ICFSP.2017.8097150.

[6] M. M. M. Sukri, U. Fadlilah, S. Saon, A. K. Mahamad, M. M. Som and A. Sidek, "Bird Sound Identification based on Artificial Neural Network," 2020 IEEE Student Conference on Research and Development (SCOReD), 2020, pp. 342–345, doi: 10.1109/SCOReD50371.2020.9250746.

[7] M. T. Lopes, L. L. Gioppo, T. T. Higushi, C. A. A. Kaestner, C. N. Silla Jr. and A. L. Koerich, "Automatic Bird Species Identification for Large Number of Species," 2011 IEEE International Symposium on Multimedia, 2011, pp. 117–122, doi: 10.1109/ISM.2011.27.

[8] W. Tsai, Y. Xu and W. Lin, "Bird Species Identification Based on Timbre and Pitch Features," 2013 IEEE International Conference on Multimedia and Expo (ICME), 2013, pp. 1–6, doi: 10.1109/ICME.2013.6607576.

28 AI in the Energy Sector

Ashwin Kumaar K, Abhishek Asodu Shetty,
Kirti Aija, and C. Naga Sai Kiran

28.1 INTRODUCTION

In recent times artificial intelligence (AI) has gained relevance in various sectors. The term is broad, and almost all sectors utilizes this advanced technology. AI becomes more and more important in the energy sectors because the potential that AI holds is enormous and the development that it can provide for the future design of the energy sector is astounding. The applications where AI can be utilized is almost infinite: smart grids, electricity trading, coupling etc. The digitalization is here, and along with it the energy sector is moving into digitalization. AI is versatile in that it can make the energy industry very efficient and evaluate the data for better performance of energy-sector digitization.

As electricity supplies more departments, sectors, and applications, the energy sector is becoming the most crucial pillar of the world's energy supply. From conventional energy to renewable energy, AI has a part in everything. AI satisfies the requirements like forecasting, coordination, digitization, power tracking, etc., to establish smooth operation in power grids and other energy sectors. AI is already transitioning the hurdles of the energy transmission in various domains, and it also concentrates on grid operation optimization, energy distribution, and demand-side management. AI applications in the energy sector have been successful and promising. The innovation, acceleration, and transition in energy sectors are becoming highly efficient and interconnected for a better future [1].

However, AI holds greater potential to provide more innovations to help with global improvement. Harnessing the power of AI for a global transition is certainly possible, provided industry collaboration for more innovations takes place.

28.2 TYPES OF SOLUTIONS CONTRIBUTING A SIGNIFICANT IMPACT ON THE AI-ENERGY SECTOR

28.2.1 Capacity Forecasting

Predict the real ability of the energy sector for total electricity generation. It's mostly based on ambience, energy plant conditions, rough and tough tests [1].

28.2.2 Plant Reliability

28.2.2.1 Predictive

Advances analytics is used to oversee the machines or equipment performance and predict errors, disruptions or disturbance, and failures along the lead time to track corrective plans [1].

 DOI: 10.1201/9781003328414-28

28.2.2.2 Values-Based

Find the best balance between the time and expense of repairs and the equivalent increase in plant performance using advanced analytics [1].

28.2.3 EFFICIENCY

The amount of electricity generated and produced for each unit of fuel consumption is optimized [1].

28.2.4 PERFORMANCE MANAGEMENT

Various aspects of plant operations on a day-to-day basis utilize a series of tools to help optimize performance [1].

28.3 AI FOR TRANSITIONING ENERGY PRINCIPLES

Figure 28.1 illustrates the application of artificial intelligence (AI) in transitioning energy principles. It showcases the integration of AI technologies, such as machine

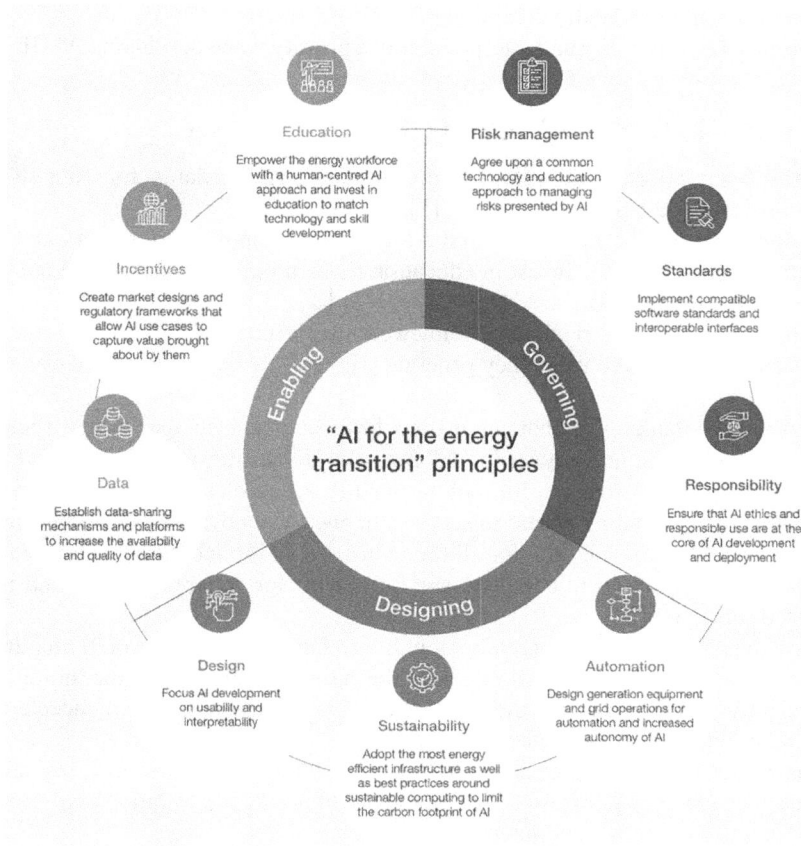

FIGURE 28.1   Artificial intelligence for transitioning energy principles.

learning and data analytics, to optimize energy systems, improve efficiency, and enable sustainable energy transitions. The figure provides a visual representation of the role of AI in revolutionizing the energy sector towards cleaner and more renewable sources.

28.4   GOVERNING THE USE OF AI

1. **Standards:** Use interoperable interfaces and comparable software standards [1].
2. **Risk management:** Agree on a shared technological and educational strategy for controlling the threats that AI presents [1].
3. **Responsibility:** Responsible use and AI ethics must be at the forefront of AI development and implementation [1].

28.5   DESIGNING AI THAT IS FIT FOR PURPOSE

1. **Automation:** Create grid operations and generation hardware for automation and more AI autonomy [1].
2. **Sustainability:** To reduce AI's carbon footprint, implement the infrastructure that is the most energy-efficient possible and follow best practices for environmentally friendly computing [1].
3. **Design:** Give usability and interpretability a priority when developing AI [1].

28.6   ENABLING THE DEPLOYMENT OF AI AT SCALE

1. Establish platforms, data-sharing procedures, and standards for data to improve its availability and quality [1].
2. **Education:** With a human-centred AI approach, empower consumers and the energy workforce. Invest in education to keep pace with advancements in technology and skill sets [1].
3. Create a market and regulatory frameworks with incentives to help AI use cases profit from the value they generate [1].

AI is no special bullet. Technology in any form can replace the energy sector. However, provided the urgency scale and complex global transmission, we have to be very specific about what we are looking for [2]. AI accelerates the transition of energy while also extending access to services in energy, innovation, affordable, and resilient service. Now is the time for all the industries and foundations to enable the AI-accessed energy future to complete and build trust for collaboration around the AI-enabled transition [2].

AI and the energy sector are greatly more interrelated than people could imagine. AI stands at the very centre of revolutionizing the entire energy sector in the future. AI can efficiently help to overcome the variable and unpredictable nature by accelerating and adapting conventional energy sources [2].

Texas power outages in February 2021 played a big role in the freezing temperatures. The electricity infrastructure in the world is starting to fail slowly due to weather conditions and climate intervals [2].

AI widely uses modern computing power to do computations and perform tasks faster than any normal human being. This will help and allow physical attributes to be interconnected for communication through the flow of vast amounts of data in real time [2].

There are two important fields where you can use AI methods effectively and sustainably which allows the energy sector to improve substantially in the current situations and support faster integration [2].

28.6.1 IMPROVING THE ACCURACY OF PREDICTION

Machine learning (ML) algorithms can identify patterns and insights from any raw data, which in turn predicts outcomes [2].

28.7 ANTICIPATE THE DEMAND FOR ENERGY

Effectively increasing the accuracy in very short time demand forecasting can help energy sectors improve the production line and decision-making, intensifying the dispatch methods' efficiency and lowering the required operating reserves. For consumers, AI can help track utility use and reduce bill costs through the optimization of variable systems [2].

28.8 FORECAST WEATHER CONDITIONS

Stabilizing the difference in sources of energy in the renewable portfolio is not easy, as the atmospheric conditions depend on it. Self-predicting weather models, large-scale satellite imagery, and real-time measurements can remarkably lower the energy generation price [2].

28.9 MAINTENANCE INFRASTRUCTURE PREDICTION

ML models that can predict the check-ups for the energy-sector infrastructure can improve their long-term performance and mitigate revenue loss and also gain the estimation cost for infrastructure redevelopment [2].

28.9.1 FULL AUTONOMY

The next step is to attain the output without human intervention, as the AI should act independently.

The growth of AI-based models has opened possibilities beyond the automation it is recommending, e.g. the Holy Grail company is on the verge of attaining autonomy over its sectors, particularly in the power grid area, which is one of the most complicated mechanism systems.

Day to day it is becoming more challenging to perform operations due to the advent of distributed energy. AI deep learning models possess the ability to automate and optimize the process. This could be the lead or a key to reducing grid congestion and integration, enabling fast recovery in the course of a natural disaster [2].

28.10   COMMON APPLICATIONS OF AI THAT ARE ENERGY RELATED

28.10.1   Smart Grid

Smart grids offer a revolutionary method for networks devoted to energy efficiency by utilizing multiple flows of data and electricity. The incorporation of cloud, AI, and digital technologies that facilitate self-regulation is the primary distinction between the typical networks. The collaboration between IBM's cloud-based analytics and London's National Grid is one of the most notable instances of the smart grid. Preventive and predictive maintenance, which are essential components of the grid's operation, are provided by the smart grid. Forecasting accuracy is boosted, resilience is strengthened, and security is improved using AI-powered smart grids [3].

28.10.2   Energy-Efficiency Programs

The Sustainable Development Goal of energy efficiency deserves serious attention. An AI-powered energy efficiency programme oversees energy consumption, creates smart forecasts, and regulates peak usage. The energy efficiency of a facility can be increased by an average of 10.2% to 40% using model-based predictive control. Up-to-date forecasts can be provided using predictive analytics and ML. The design and implementation of energy-efficiency initiatives within businesses, municipalities, and states are therefore based on these projections [3].

28.10.3   Smart Heaters

Due to their control over the entire heating system, smart heaters can be used in contemporary green solutions. Here, reasonably directing the electricity to allow it to divert any unused energy to specific locations is the key strategy [3]. The end-to-end signup process for smart heaters was developed to create a self-learning home system, and it will be given a legacy flow that was incomplete and missing payment capability [4].

A self-learning home system helps to reduce the cost of energy consumption, which in turn when implemented in a heating system, an element uses considerably high amounts of energy. By obtaining the usage data AI can help it use the heating coil in such a way that it is used only when it's required and the amount of water heated and temperature, etc., are controlled [3].

28.10.4   Digital Twins

The industrial energy complex was viewed as needing a digital twins framework to survive. One definition of a digital twin is "a multi-dimensional visual depiction of a process, facility, or actual thing." The study opportunities presented by these digital twins are greater than those of simulations since they function as real-time virtual models. Digital twins aid in the research of wind turbines and power-generation facilities in the energy and AI fields. To better service, experiment with, maintain, and optimize the energy network, whether it be conventional or renewable, a digital twin can be created using AI [4].

28.10.5 RENEWABLE ENERGY INTEGRATION

Large energy suppliers need to strike a balance between conventional and renewable energy sources. Now that ML and AI are being used in the energy sector, it is possible to forecast and predict the ideal conditions for precise renewable energy integration [4]. This manages the insertion of renewable energy sources into the current electrical system, to put it another way. There may be a role in spreading the energy outputs from wind and solar farms and forecasting their outputs to keep the system in balance [3].

AI in the energy sector assesses the current situation and assists in taking the necessary steps to realize the sector's full potential. Utilities are working to keep up with these new difficulties as worldwide demand rises. AI can be gradually incorporated into energy grids, renewable energy sources, and decentralized networks to optimize energy use and raise consumer satisfaction. Thus, AI in the energy sector can bring about sustainable practices, cut prices, and promote transparency [4].

Choosing the appropriate service provider with relevant experience is essential because 50% of oil and gas businesses intend to expand their investments in AI and ML. Some energy network projects may be out-of-date or adhere to particular technological requirements, necessitating a committed staff. Choosing a capable development vendor is essential for any project, including software for decentralized networks, smart grids, failure prediction systems, and digital twins [4].

28.11 AI IN ELECTRICITY TRADING

Power trading uses AI to make projections more accurate. With AI, it is easier to examine the massive amount of data in power trading in a systematic manner, such as weather data or historical data. Grid stability and supply security are both improved by better forecasts. AI can aid in and hasten the integration of renewable energy, particularly in the realm of forecasting. To improve projections in the energy sector, ML and neural networks are crucial [4].

28.12 ARTIFICIAL INTELLIGENCE IN VIRTUAL POWER PLANTS

In the virtual power plant, a lot of data is processed and predictions are made. The AI algorithms assist in coordinating numerous participants in the virtual power plant and produce better and better projections [4].

When it's required to plan which plant creates or consumes how much electricity when, for instance, this occurs. Live feed-in data, historical data, data from power trading centres, and weather forecasts are some of the data sources used as the basis for the analysis [4].

Some AI programs have already developed the level of intelligence necessary to engage in independent trading. This is what is referred to as automatic trading, algo trading, or algorithmic trading [4].

AI can also support automated monitoring and analysis of power market trades. As a result, deviations from the norm, like the misuse of market power, may be found and stopped more promptly and precisely [4]. The applications of AI in energy industry is shown in figure 28.2.

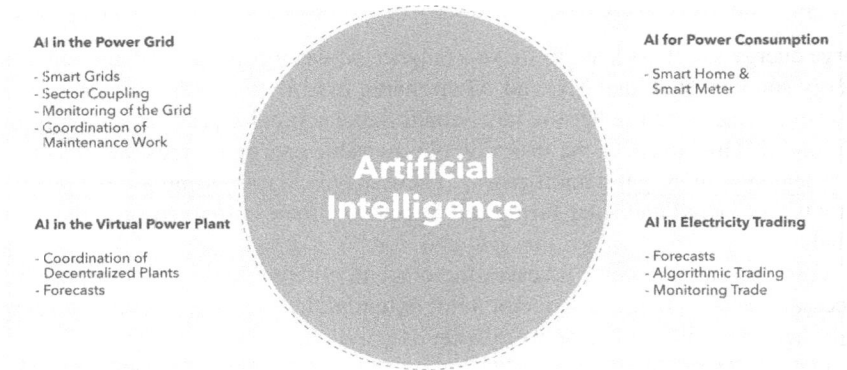

FIGURE 28.2   Artificial intelligence in the energy industry.

28.13   USING AI FOR POWER CONSUMPTION

Intelligently integrated consumers can help create a reliable and environmentally friendly electrical infrastructure. Although smart houses and smart metre solutions already exist, their adoption is still limited [5].

In a smart networked home, networked devices respond to electricity market pricing and adjust to household usage patterns to conserve energy and cut costs. Smart networked air conditioning systems are one instance. When electricity is readily available and reasonably priced, they increase their output in response to market prices. By examining user data, they can incorporate details about user preferences and periods into their computations [5].

28.14   ARTIFICIAL INTELLIGENCE IN THE ENERGY SECTOR: BLOCKS AND LIMITATIONS

The intelligence of any AI is only as good as its data. One of the major obstacles is this. Some of the biggest areas where the usage of AI is vulnerable are the areas of data security and protection. Intelligent digitally connected people reveal a lot about themselves, making the system more susceptible to cyberattacks. The number of cyberattacks on critical infrastructure increased from 2017 to 2018, according to the German Federal Office for Security (BSI) [5].

This vital infrastructure includes the sources of energy and the overall energy system. Because of this, cybersecurity will be even more crucial in the future to safeguard the heavily networked power system against intrusions and data theft [5].

Contrary to popular belief, AI can significantly aid in the defence against cyberattacks, not impair the power grid's security. Large amounts of data may be swiftly checked, helping to identify deviations. AI is also capable of making inferences from prior intrusions. In this field, ML has already had considerable success, as seen in the detection and protection of Trojans, for instance [5].

Many consumers have negative views of AI, particularly in the context of smart home devices. This makes sense given that the data of the most private zone is collected and discloses a lot about its users. According to studies, the main barrier to the adoption of smart metres is the uncertainty around the usage of personal data. These worries are well-founded because there is still no law on how to handle this private information, which is crucial for the development of the future electrical grid.

The energy consumption of AI itself is a different criticism against it. Large amounts of electricity are used in the processing of data. Analyzing how to construct data centres to be as energy-efficient and climate-neutral as feasible is vital when using AI to improve the energy system [6]. This conundrum might be resolved by placing data centres close to renewable energy production facilities, delaying power-intensive computing operations to times when there is plenty of power available, using more energy-efficient IT hardware, or writing software that uses the least amount of processing power possible [5].

A wide range of relevant application scenarios for AI in the energy sector is available to support the energy transition and provide a climate-friendly energy system. But it will be essential to safeguard user information and make AI use clear and understandable [5].

28.15 CONCLUSION

The potential for AI to speed up the global energy transition is significantly greater. Although AI applications in the energy sector have been creative and promising, adoption is still relatively low. With AI, there is a fantastic chance to hasten the development of the interconnected, highly efficient, and emission-free energy infrastructure that we will need for a better future. The intersection of AI and energy is a fantastic place to start for individuals wishing to alter the future of the energy sector. Technology innovation has fundamentally altered how we think about these two industries and how they interact. They is the perfect environment for creative thinkers to leave their imprint and have the potential to impact the world in ways we haven't even imagined [7].

REFERENCES

[1] "Artificial intelligence," *Next-kraftwerke.com*, 2018.https://www.next-kraftwerke.com/knowledge/artificial-intelligence (accessed Dec. 20, 2022).

[2] B. Boswell, S. Buckley, B. Elliott, M. Melero, and M. Smith, "An AI power play: Fueling the next wave of innovation in the energy sector," *McKinsey & Company*, May 12, 2022. https://www.mckinsey.com/capabilities/mckinsey-digital/how-we-help-clients/an-ai-power-play-fueling-the-next-wave-of-innovation-in-the-energy-sector (accessed Dec. 20, 2022).

[3] Espen Mehlum, D. Hischier, M. Caine, and World Economic Forum, "Here's how AI will accelerate the energy transition," *World Economic Forum*, Sep. 2021. https://www.weforum.org/agenda/2021/09/this-is-how-ai-will-accelerate-the-energy-transition/ (accessed Dec. 20, 2022).

[4] "3 ways AI is powering innovation in the energy sector," *Aimagazine.com*, May 23, 2022. https://aimagazine.com/technology/3-ways-ai-is-powering-innovation-in-the-energy-sector (accessed Dec. 20, 2022).

[5] "Artificial intelligence (AI) in the energy sector of the future | Informatec," *Informatec.com*, 2020. https://www.informatec.com/en/artificial-intelligence-ai-energy-sector-future (accessed Dec. 20, 2022).

[6] "Artificial intelligence in energy: Use cases and solutions," *N-ix.com*, Sep. 12, 2022. https://www.n-ix.com/artificial-intelligence-in-energy/ (accessed Dec. 20, 2022).

[7] A. Israel, "How AI will transform the energy sector," *Sifted*, Oct. 05, 2021. https://sifted.eu/articles/ai-energy-transform/ (accessed Dec. 20, 2022).

29 Artificial Intelligence in Fashion Design and IPRS

*Kaja Bantha Navas Raja Mohamed, Divya Batra,
Shivani, Vikor Molnar, and Sunny Raj*

29.1 INTRODUCTION

Artificial intelligence (AI) is the ability of robust computer-based technology to perform tasks and generate results that are usually done by humans. AI is playing major roles like advisory, purchase and trend prediction, online fashion retail, smart wearable products, mass customer interaction, electronics textiles, fashion, and lifestyle accessories [1]. The use of AIe alongside machine learning, deep learning, natural language processing, visual recognition, and data analytics reshaped the fashion industry. Automatic tagging and virtual dressing room applications in the fashion industry are interesting. Through machine vision techniques, we can eliminate manual tagging and make our fashion catalogue management process with visual analytics and AI in fashion. Customers can vary and validate their own images based on background, shape, texture and pose using generative adversarial network (GAN) and deep convolutional generative adversarial network (DCGAN) techniques. This chapter describes how the adoption of AI has contributed to fashion design. It is mainly divided into two parts. The first part briefly explains the role of AI in fashion design. The second part cites the World Intellectual Property Organization (WIPO) Hauge Design registration portal design IPRs in terms of AI. Lastly, the chapter discusses the limitations of this study and the breach of our own privacy using AI in our lives.

29.2 BACKGROUND AND DRIVING FORCES

In this world only humans have capability to use their brain with intelligence. Because of this intelligence, only humans have evolved to such an extent. With the help of this intelligence, humans have developed computers, smart phones, the Internet, etc. Humans have developed technology to such an extent that now they want to make a machine that can think, analyze, and react to situations like humans. This machine with advanced technology is called AI. The term was first coined decades ago in the year 1956 by John McCarty at the DARTMOUTH CONFERENCE at the science and engineering of making intelligent machine. AI is a technique by which you can make machines work and behave like humans. In the past AI has accomplished this by creating machines and robots that are being used in a wide range of fields

DOI: 10.1201/9781003328414-29

including healthcare, robotics, marketing, and business analytics. People often tend to think that AI, machine learning, and deep learning are same since they have common applications but they are different. AI is the science of getting machines to mimic human behaviors, but machine learning is the subset of AI that focuses on getting machines to make decisions by feeding them data. On other hand, deep learning is the subset of machine learning that uses the concept of neural networks to solve complex problems, so to sum it up AI, machine learning, and deep learning are interconnected fields. Machine learning and deep learning need AI to provide the set of algorithms and neural networks to solve data-driven problems; however, AI is not restricted to only machine learning and deep learning; it covers a vast domain of fields, including natural language processing, object detection, computer vision, robotics expert systems, and so on. AI can be structured along three evolutionary stages that are artificial narrow intelligence, artificial general intelligence, and artificial super intelligence. Artificial narrow intelligence, also known as weak AI, involves applying AI only to specific tasks, for example, Alexa, face verification in iPhone, and Autopilot feature of Tesla. It operates within a limited predefined range of functions; there's no genuine intelligence or no self-awareness despite being a sophisticated example of weak AI. Artificial general intelligence is also known as strong AI, and it involves machines that possess the ability to perform any intellectual task that a human being can. Machines don't possess the human sort of abilities. We have a robust processing unit which will perform high-level computations but it's not yet capable of thinking and reasoning like a human. There are many experts who doubt that AI will ever be possible. AI has an intelligent agent structure. Intelligent agent is an operation that takes a decision itself to put AI into action. It works on the basis of perception and action. It does this work on basis of three components, i.e. sensors, actuators, and effectors. Through sensors, AI observes changes in the outer environment, and with actuators it performs the role of controlling and moving a system.

There are five different types of AI are shown in Figure 29.1.

29.3 ARTIFICIAL INTELLIGENCE IN DESIGN

People Centered: Empathy with users inspires creativity when it is driven by design. Design-driven innovation arises from understanding an issue from the user's perspective and generating predictions about what would be relevant to her, rather than being driven by technological breakthroughs and what is achievable.

Abductive: Design takes a generative approach to solving problems. Rather relying exclusively on deductive and inductive reasoning (how things are) to create, design employs abductive reasoning, or the creation of hypotheses about how things may be. Abductions frequently result in the problem and the question that underlies the design being reframed.

Iterative: Abductions are iteratively changed and enhanced through rapid testing cycles. They enlist the help of the team and users, allowing them to share ideas, fail, learn, and develop until a satisfactory solution is found.

AI as a virtual assistant: Machines can't generate creativity like human minds, but what it can do is help in boosting creativity. It can help designers in doing normal

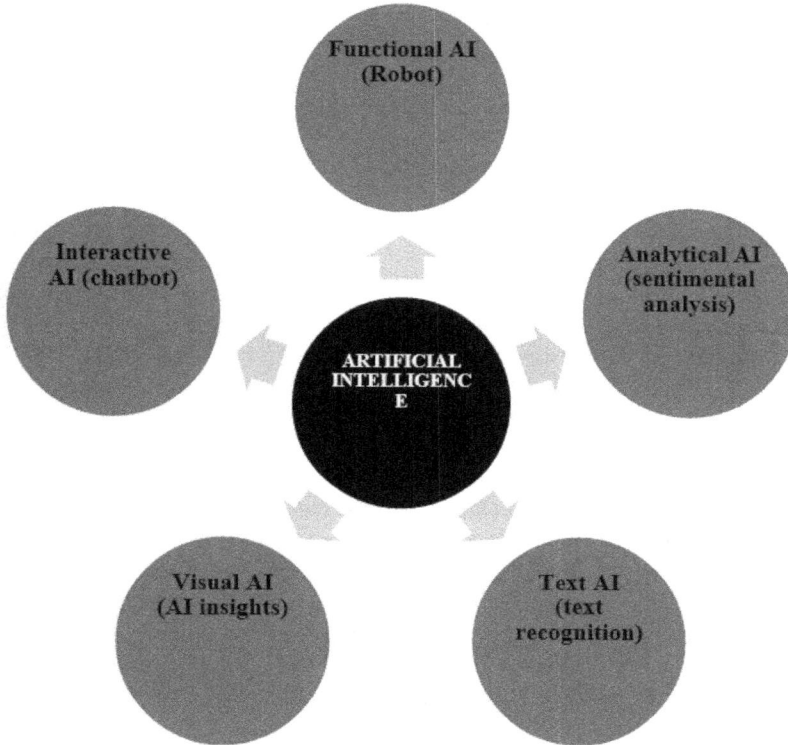

FIGURE 29.1 Different types of AI.

and time-consuming tasks. Because of this, the designer can utilize that saved time on idea building by focusing more on creativity and design aspects. AI acts like a personal assistant for designers; for example, the Sensei stitch of Adobe. It helps in creating better marketing experiences for the clients. It collapses the time between ideation and execution.

AI analyzes user behavior: AI can detect what our target market or client needs and wants. It can identify designs and platforms that users like with the help of algorithms. With the help of AI designers can create more user-oriented designs. It also suggests interesting tips and tricks for enhancing creativity. For example, Prisma Labs is a platform that helps users to express their feelings with the camera. Lensa is its Android version. It helps in editing photos for complex features easily like face retouch, eyebrow shading, etc.

AI creates multiple versions: AI can create multiple forms after recognizing a pattern. The algorithm extracts colors and patterns of a design and then creates thousands of variants, within the range of identified colors and patterns. It can be very helpful in logo and banner designing. It can be used to create various unique designs. It can give new ideas to designers when they are struggling with creativity. Various visual options can give customers a variety of choices and help them to make choices easily. For example, Nutella Unica, with an algorithm, created millions of combinations, and they sold millions of jars.

AI brings value to customers: Personalized user experience is the key to customer satisfaction. Platforms that make users feel loved and cared for attract and retain users. Look at Netflix. It recommends a series based on the user's history. It also tells you why it is recommending the series 'because you watched Lucifer'. It saves users from a lot of effort. Instead of going through hundreds of available options, the user finds the content they are interested in without any effort. It builds a strong bond between the brand and the user. In today's era everyone wants individuality and personalized experiences. The companies which provide this feature to their customers gain profits. It enhances customer satisfaction and loyalty. The best example for this is Netflix, as mentioned earlier.

29.4 ARTIFICIAL INTELLIGENCE IN FASHION

AI is transforming the fashion industry by playing a vital role in a number of important divisions. AI in fashion is revolutionizing the business in many ways, from design to manufacturing, logistics supply chain, and marketing [2]. Fashion brands are becoming increasingly interested in and concerned about AI and machine learning in order to improve user shopping experiences, increase sales system efficiency through intelligent automation, and optimize sales processes through predictive analytics and guided sales processes. Many AI technologies collect a large amount of consumer data that is used by companies to adapt to the market, identify trends, track competitors' activities via the Internet and social media, and improve consumer experiences and target consumer demands. However, consumers may not be aware of just how much information is being collected from the Internet, and fashion retailers and designers may be unaware of the privacy law issues that are implicated in the collection, processing, and use of big data and AI [3]. AI and big data have been extremely beneficial in allowing fashion companies to better address market demands, improve relationships with consumers, and use social media and other platforms to increase sales and provide a more interactive experience.

Some examples of AI in fashion applications are listed next:

AI in product development: AI helps in verifying if the company's competitors have successfully integrated their products. During the development cycle, product development teams sometimes use AI and machine learning techniques. It also helps in assisting in generating higher economies of scale, efficiency, and speed improvements throughout the product development process. AI assists with the three main aspects of product development: digital prototyping, product lifecycle management, and product profile management. It also provides enterprises with large-scale insights about their new product development [4, 5].

Inventory: AI has taken over the task of maintaining the accuracy of stock records. The good news is that companies now have access to a simplified solution that can cross-check data from an enterprise resource planning (ERP) system against customer trends data. This technology simplifies and improves inventory tracking, reducing the risk of raw material waste. This is exactly what AI provides to inventory management. AI in the supply chain has a plethora of applications [4–6]. Operational procurement using intelligent data and chatbots, supply chain planning to forecast demand and supply, warehouse management to optimize stock, faster and more accurate shipping to reduce lead times and transportation expenses, and optimal supplier selection using real-time data are all areas of impact in the supply chain and logistics [7].

In styling: Furthermore, the use of AI in fashion is helping each of us to choose the best outfits that suit our body types and fashion preferences. These AI-enabled clothing and costumes are adjusted not only for different situations and weather but also for the user's preferences and needs, body type, colors, and current fashion trends [8, 9].

Customer experience: The AI customer experience creates hyper-relevant digital ads. According to Joanna Coles, the former chief content officer of Hearst Magazines, "People hate advertising." This is because, according to Marc Pritchard, the chief brand officer at Procter & Gamble, ads are often irrelevant and sometimes "just silly, ridiculous or stupid." Ninety-one percent of people say advertisements are very invasive these days [10]. To deal with the challenge, brands are using AI to show relevant ads to the viewers. Machine learning is helping companies to predict what kind of advertisement viewers would like to watch based on their online behavior, profile, and audience segmentation [11].

Empowers personalized search: To enhance the online user experience companies are using AI. Consumers like to buy from the brands that provide them more personalized experience and individuality. As today's generation seeks their individuality in products and services, they are buying for their use. Many e- commerce platforms are categorizing and recommending the products and services based on user behavior. It provides feasibility to the customer to reach their choice of products, thus reducing their frustration in getting what they want.

24/7 customer service: For every brand to operate smoothly, there is need for providing customer service. But it is very difficult and expensive to provide it via human teams only. AI helps companies to provide such services via email, chat, messaging, SMS, and voice platforms. This takes queries of customers through their preferred channels. With human staff only, it will be hard to manage and would be costly to provide the omni channel experience that customers want.

Visual search: Visual search scans and detects user-input photos, similar to text-based search, and returns the most relevant search results. Customers may search for what they need without having to explain it, making online shopping more convenient and enjoyable. Users can capture screenshots of online outfits, recognize shoppable gear and accessories in the image, and then find the same outfit and shop for comparable fashions using AI-enabled apps.

Automated authentication: Forgeries related to fashion counterfeit items are also detected using computer vision and machine learning. Detecting fakes used to necessitate the use of professional traditions or the expert eye of other law enforcement professionals. AI systems can now detect counterfeit things that increasingly resemble the real thing [11]. Customs and border officials are using AI to help verify the authenticity of high-quality items that are commonly counterfeited, such as handbags and sunglasses. Ordinary buyers may fail to spot counterfeit items from a third-party seller while browsing vast online marketplaces. When a customer buys a product that appears to be genuine but isn't, it can leave a sour taste in their mouth and affects their perception of the brand.

29.5 FASHION DESIGN IPRS

Here we have discussed five different fashion intellectual property rights (IPRs) in terms of AI.

Clothing: Knight noted back in 2017 that Amazon's fashion design-specific AI initiative—a program that creates garment designs (by way of a tool called generative adversarial network) that can then be physically manufactured by humans. Unlike typical clothes, digital clothes are intangible and exist only on your computer or phone. People typically download or purchase digital clothes to practice their own digital design skills or to superimpose the clothes onto pictures to post on social media. One of the frontrunners in the digital clothes space is the design house The Fabricant, which uses three-dimensional AI technology to produce digital-only garments and often collaborates with brands, including Adidas, Under Armour, Puma, and Tommy Hilfiger. The Fabricant's contracts state that The Fabricant is the owner of their creations. Ralph Lauren's Polo Tech Shirt is a tee that combines the fashion label's preppy aesthetic with the ability to deliver live metrics (heart rate, breathing, and steps) from the shirt to your phone. Ralph Lauren's sportswear t-shirt is one that you would have worn anyway, but smarter. The Polo Tech shirt is a compression top, meaning it hugs and pulls at your body to promote circulation. Beneath that figure-hugging fabric is a large silver strip that goes around your lower ribcage. This is what delivers your biometrics to the attached module: The conductive silver fabric also connects sensors to detect your heart rate, breathing, and exertion. The shirt's material feels slightly stiffer along the band due to the snug fit of the shirt.

Eyewear: In 2020, sports and lifestyle company Bollé launched the Volt+, their first lens developed with AI. Researchers at EPIC, Bollé's state-of-the-art design and technology innovation lab in Lyon, France, evaluated 20 million lens formula combinations with AI. The algorithm assessed all the combinations, and the team eventually chose one high-contrast formula that provides the best possible color enhancement and depth perception for their active lifestyle customers. The 'anti-procrastination' smart glasses use AI to monitor what you look at all day [12]. A start-up named Auctify has what it claims is the solution: smart glasses that use AI to monitor what you're looking at and nudge you to pay attention [13]. Depending on your worldview, it's the product of your dreams or a productivity-hacking nightmare.

AI Wallet by Tag8: Tag8 Dolphin Wallet Tracker arguably is one of the most compact, high-range, and efficient Bluetooth trackers. It is a next-generation solution that helps you find your wallet when misplaced and also safeguards from theft, including theft of digital data stored in the cards carried in your wallet. The smart tracker comes with location tracking and anti-lost alarm system features [14]. The alarm rings on your smartphone and smart tracker as soon as your tracker goes out of range. The replaceable battery offers a battery life of over six months.

Amazon's Fashion Look replaced by Style by Alexa: Amazon's Echo Look came out in 2017 to help people get fashion and styling advice digitally [15]. But by 2020, this was replaced by Amazon's own Style by Alexa and StyleSnap application or webpage. StyleSnap is an AI-powered feature built into the Amazon app, and it's here to help you find looks you love quickly and easily [16]. All you have to do is take a photograph or screenshot of an outfit, upload it onto the Amazon app, and you'll be presented with items that look just like the ones in the picture. Sometimes, they're even the exact same. It's truly that easy.

Brand originality authentication AI: Dupe Killer by Deloitte: Dupe Killer is a new piece of technology that searches for design infringements using AI by learning the shape or configuration of a product and seeking out copies. This is different from detecting counterfeit goods, where the name is stolen and traded upon. Instead, Dupe Killer operates in a world where the only clues are visual. Counterfeits claim to be the brand, while design infringements lean on the brand's key features without ever mentioning the original product and are tricky to track and remedy. Some of the listed industrial design IPRs for fashion and accessory design are shown in Figure 29.2. The figure includes industrial design with world intellectual property rights reference numbers also. We have captured these industrial design images from the world intellectual property rights industrial design global database [17, 18].

29.6   LIMITATIONS

During the research it was observed that the products listed under WIPO are patented, but there are not enough products powered by AI which are patented in the world. Although the innovations are amazing, they come with some drawbacks that has restricted the inventor from gaining a patent for their technologically intelligent creation, for example, the VOLT+ sunglasses mentioned earlier. Also, there are AI products like some wallets being sold in the market from online and offline channels that have yet not received a patent or a trademark. Tag8 Wallet is one of the examples of the same. Another factor that limits AI in some way is the fact that AI is taking away our security from us, as it breaches our privacy by capturing all our data, not knowing what is personal and what is not [19, 20]. Having a smartphone itself is like giving away your privacy, for it collects every detail of our day and stores it where we can't reach and help ourselves from restraining any misuse of it.

29.7   CONCLUSION

AI has been available for two decades now; today what we see is the evolution of AI in machines and technology. But this isn't limited to here; it is being incorporated by every industry in this world because of the obvious benefits it has. There is much

D915344 - Artificial intelligence speaker

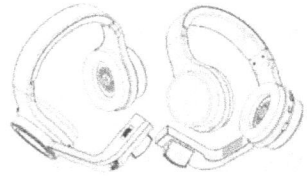

D914019 - Artificial intelligence headphones
with optical integrated glasses for augmented reality

D840396 - Wearable augmented reality (AR)
and virtual reality (VR) eyeglass communication
device with artificial intelligence (AI) data processing
apparatus

DM/078 537 - Pattern for artificial wicker-threads

1 5 7 8 6 0 4 - wallet tracker

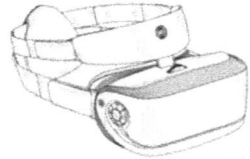

D818021 - Smart glass

A002016001504 - POLO SHIRT NET

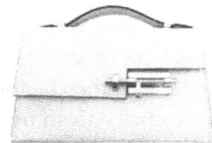

201730638859.X - Killer

FIGURE 29.2   Listed WIPO industrial designs towards artificial intelligence.

more to deep dive into to explore what more can with done with the help of AI. AI has conquered a good space in the fashion and design industry with complete rights or IPRs, especially during the last five to seven years.

REFERENCES

[1] Artificial Intelligence and the Fashion Industry, Professor Francesca Masciarelli, MariannaPupillo.

[2] Artificial Intelligence for Fashion: How AI is Revolutionizing the Fashion Industry by Leanne Luce.

[3] Artificial Intelligence for Fashion Industry in the Big Data Era By Sébastien Thomassey and Xianyi Zeng.

[4] https://www.igi-global.com/article/artificial-intelligence/211137.

[5] https://www.sciencedirect.com/science/article/pii/S014829632030583X.

[6] The Use of AI in Inventory Management - TFOT (thefutureofthings.com).

[7] https://builtin.com/artificial-intelligence/ai-in-supply-chain.

[8] How AI is Changing Fashion: Impact on the Industry with Use Cases, https://medium.com/vsinghbisen/how-ai-is-changing-fashion-impact-on-the-industry-with-use-cases-76f20fc5d93f.

[9] https://textilelearner.net/artificial-intelligence-in-fashion-industry.

[10] https://www.netomi.com/ai-customer-experience.

[11] https://hbr.org/2022/03/customer-experience-in-the-age-of-ai.

[12] https://www.theverge.com/2020/9/1/21404004/anti-procrastination-smart-glasses-productivity-boosting-auctify-specs-indiegogo.

[13] https://www.forbes.com/sites/bernardmarr/2021/06/21/ai-glasses-you-can-try-on-and-try-out-with-ar/?sh=4b22b4297e6d.

[14] https://www.croma.com/tag8-dolphin-wallet-tracker-rfid-protect-800021-black-/p/233155?utm_source=google&utm_medium=ps&utm_campaign=sok_pla_ssc-other_home_appliances&gclid=Cj0KCQjwxtSSBhDYARIsAEn0thRj9UvW49WK3X8W-mvqPKOMBCV9TYvc00T4G0Hm73ZSsY8l-N8Bi-8aAufUEALw_wcB.

[15] https://www.lutzker.com/ai-and-copyright-in-the-fashion-industry/.

[16] http://tesi.luiss.it/25378/1/212661_PUPILLO_MARIANNA.pdf.

[17] https://digitalcommons.uri.edu/cgi/viewcontent.cgi?article=1007&context=tmd_major_papers.

[18] https://www.amazon.com/stylesnap.

[19] https://www.insider.com/guides/style/amazon-stylesnap-review#click-the-camera-icon-in-the-upper-right-hand-corner-of-the-amazon-app-1.

[20] https://www.engadget.com/2016-03-18-ralph-lauren-polotech-review.html.

30 Artificial Intelligence in Education

A Critic on English Language Teaching

Dr. Harishree and Jegan Jayapal

30.1 INTRODUCTION

Artificial intelligence (AI) is an evolving field of study which plays a crucial role in post-pandemic education. The COVID-19 pandemic lockdown has made every field rely on web-based technology to sustain itself. Concurrently, the education field has also transferred its traditional teaching method and shifted to e-learning platforms. The students have been encouraged to self-learn with the guidance of the teachers through online networks. The usage of various information and communication technology (ICT) tools has paved the way for teachers to reach the students effectively in a lockdown situation. Because of the precipitated shift in the teaching-learning platforms, the scope of artificial intelligence in education (AIEd) has grown immensely. The term artificial intelligence has been "used to describe a collection of technologies that can solve problems and perform tasks to achieve defined objectives without explicit human guidance" (Schmidt & Strasser, 2022). Consequently, we can assure that AI can aid academicians in achieving a successful teaching-learning process post-pandemic. Besides, AI comprises multi-layered technologies which drive updated data and algorithms that can help education to transform according to the need of the hour.

Moreover, integrating AI-powered tutors has become a vital part of the future of education. Approval of courses to be offered by virtual universities by the governments (e.g., Virtual University of India) across the nations is a case in point for the conspicuous role of AI in education. Virtual universities can be online without a physical building, and universities and recognized institutes offer online degree programmes in addition to their physical programmes. In these virtual universities, students can ultimately earn their degrees through online learning, like the online Master of Business Administration at the University of Illinois Urbana-Champaign. India's National Educational Policy (NEP) 2020 has also encouraged more techno-based courses. As per recent modifications in the educational policy of India, all engineering graduate students must take at least one online course on massive open online course (MOOC) platforms during their course of study to earn their degree in India. Many foreign universities provide online diplomas for students who can learn the subject remotely (e.g., PG Diploma Business in Finance by the University of Essex). But most of these courses will be monitored by the course offering faculty, and each student of the course is evaluated individually by the

DOI: 10.1201/9781003328414-30

teacher with the help of AI systems. It is also economically viable for students worldwide to gain knowledge from the high-quality teachers in well-reputed universities.

According to (Ouyang & Jiao, 2021), AIEd has three paradigms: AI-directed, AI-supported, and AI-empowered. In all these paradigms, AI helps the learners in different roles as recipients, collaborators, and leaders. So, while using AI in teaching/learning, the role of the teacher and student should be considered. As a recipient, the learner can get the results of their task performed through AI; as a collaborator, the learner can use AI to improve their learning as a collaborator; and as a leader, the learner can use AI to lead others with the advancement in the AI while providing language learning tasks. But there were a few issues concerning facilitating the education to the target learners in each paradigm. For instance, AI could not identify the amount and kind of information it requires about the learners to deliver the assistance, the extent to which the learner's information should be integrated with AI systems, and the ways to address the complexity of AI systems with educational contexts. Therefore, frameworks have evolved in AIEd to help practitioners implement AI tools and techniques in classroom teaching according to the teaching-learning environment. There are a few notable frameworks discussed in this chapter to give insight into integrating AI in the teaching-learning process of language teaching and education. Moreover, the chapter also discusses the usage of AI in various aspects of teaching foreign languages in the present pandemic situation.

30.2 FRAMEWORKS OF ARTIFICIAL INTELLIGENCE IN EDUCATION

Since AI has obtained a staunch role in education, frameworks for understanding AIEd and integrating AI in education have been proposed by various theorists. For instance, Guan et al. (2020) have proposed three methods to implement AI in education, namely, learner-oriented AIEd, instructor-oriented AIEd, and institutional system–oriented AIEd. Similarly, (Langley, 2019) has proposed an integrative framework for AIEd with the following components:

1. Recognition and pattern matching
2. Decision-making and choice
3. Conceptual inference and reasoning
4. Execution and sequential control
5. Planning and problem solving
6. Integrated intelligent agents

Later, Hwang et al. (2020) developed the framework (Figure 30.1) evaluating the four roles of AI in education. The four roles of AI are intelligent tutor, intelligent tutee, intelligent learning tool or partner, and policymaking advisor. Firstly, the intelligent tutor has been working through intelligent tutoring systems (ITSs), adaptive/personalized learning systems and recommendation systems of AI to foster the students' education. For example, *AutoTutor* uses a dialogue-based tutoring system to instruct students on physics, computer literacy and critical thinking, and *ASSISTments* that give real-time feedback for students and data-driven reports for teachers. In addition, intelligent tutee has no special attention in the field of AI, but there are some examples like Microsoft *Tay* (a chatbot), which was shut down due to inappropriate comments. Third, an intelligent learning tool or partner helps the learners to collect and analyse data by focusing on the higher-order thinking skills (i.e., analyse, synthesis, and evaluate). Finally, a policymaking advisor can help to develop policymaking in the education field. The AI tools will help identify current trends and issues in educational settings; accordingly, new educational policies can be built and evaluated. This framework helps in understanding the role of AI in education by addressing the different roles and responsibilities in the field of education. The teaching, tools to teach, policymaking, and personally addressing the learners' progress are part of the education system. In addition, we can also identify that AI can help in developing our education system. It makes it vital for teachers and academicians to nurture AI in our educational curriculum.

Consequently, a common framework for AI in higher education has been proposed by Jantakun et al. (2021), using seven components to implement AI in higher education. The components are, namely, user interactive components of technology of AI, component and technology of AI, roles of AIEd, machine learning and deep learning, decision support system (DSS) modules (student, teaching and research modules), application of AI in education, and AI to enhance campus efficiencies. Subsequently, the eXplainable AI education (XAI-ED) framework was developed by Khosravi et al. (2022). The framework concentrates on human-computer interaction and the cognitive and learning sciences with six main aspects, as represented in Figure 30.2. This framework provides information on the basic concepts of AIEd. As the name of the framework suggests, it explains the models available in AI, approaches that can be used in AI, pitfalls of AI, the main stakeholders of AI, the main benefits of using this AI interface framework, and designing effective AI tools. Briefly, this framework considers both users, using present tools, analysis of the issues in using AI, and effective methods to develop new and advanced tools.

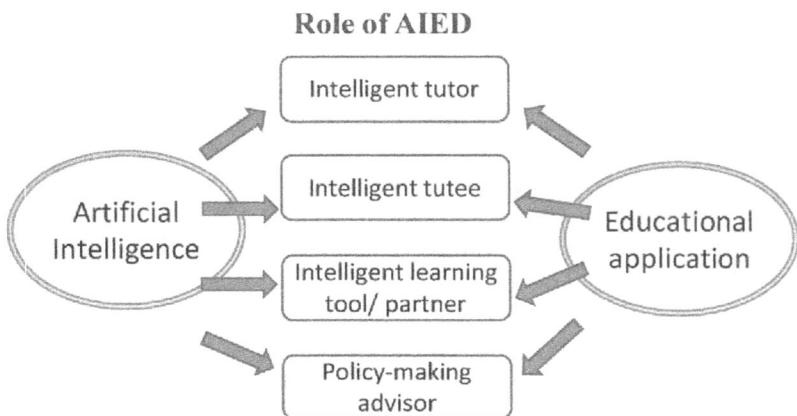

FIGURE 30.1   Framework of AIED (Hwang et al., 2020).

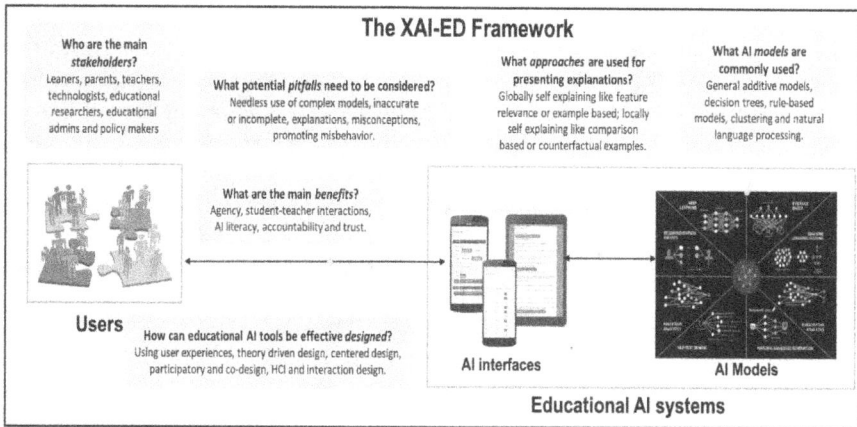

The XAI-ED Framework

Who are the main stakeholders?
Leaners, parents, teachers, technologists, educational researchers, educational admins and policy makers

What potential pitfalls need to be considered?
Needless use of complex models, inaccurate or incomplete, explanations, misconceptions, promoting misbehavior.

What approaches are used for presenting explanations?
Globally self explaining like feature relevance or example based; locally self explaining like comparison based or counterfactual examples.

What AI models are commonly used?
General additive models, decision trees, rule-based models, clustering and natural language processing.

What are the main benefits?
Agency, student-teacher interactions, AI literacy, accountability and trust.

Users

How can educational AI tools be effective designed?
Using user experiences, theory driven design, centered design, participatory and co-design, HCI and interaction design.

AI interfaces

AI Models

Educational AI systems

FIGURE 30.2   XAI-ED framework (Khosravi et al., 2022).

Thongprasit and Wannapiroon (2022) have proposed a framework for AI concentrating on the four types of AI technology functionality and three types of capabilities. The four types of functionalities are reactive machines, limited memory, theory of mind, and self-awareness, and the three capabilities are artificial narrow intelligence (ANI), artificial general intelligence (AGI), and artificial super intelligence (ASI).

These frameworks' central focus is understanding AI's role in education and the plausibility of integrating AI in education. Each framework provides different aspects to AIEd, and a deep analysis of these frameworks will provide suggestions to use AI tools according to the learners' needs. Additionally, it aids in adapting AI tools to the changing demands of the education field. It is necessary to identify the objective of teaching before considering AI tools to be integrated into the teaching-learning process of a language classroom. Different modules of AIEd also facilitate technology usage to succeed in the learning goals. Irrespective of these technological advancements and frameworks, AIEd is still evolving by addressing the various needs and problems in integrating AI as a part of classroom teaching. The XAI-ED framework provides a wholesome approach to AIEd to explore different issues while integrating AI in teaching and furnishing opportunities to improve the effectiveness of AIEd. In consonance with this, foreign language teaching, especially English language teaching, can also use AI tools to moderate learners' performance in the target language.

30.3   FOREIGN LANGUAGE TUTORING

AI provides educators with a wide range of opportunities to promote student learning. There were different powered tools of AI to make education interesting and motivating to the students. The use of AI in language teaching can be dated back to the early 2000s when different theories and tools were instilled to improve the learners' language skills. As AI plays different roles in education, it is significant in foreign language tutoring. AI-powered language learning focuses on natural language processing (NLP). NLP is a computer program that analyses and represents texts that naturally occur at one or more linguistic levels to attain human-like language processing. It is an area that combines both AI and linguistics. It can be categorized under computational linguistics, where formal analysis and language modelling will be processed, and NLP applications combine linguistics, computer science, and psychology. For instance, an NLP application like Intelligent Language Tutoring Systems (ILTS) concentrates on "lexical, morphological and syntactic aspects of language as well as aspects of meaning, discourse, and the relation of these to the extra-linguistic context" (Schmidt & Strasser, 2022).

Further, AI-powered language learning can be formulated under machine learning and deep learning, which uses NLP in formulating human-like language processing. Machine learning refers to computer systems that could learn and adapt using algorithms to learn on their own without human intervention. Deep learning works like image processing with the help of visual data sets. Various AI-powered applications have been developed in language tutoring systems in recent years using NLP, machine learning, and deep learning. AI writing tools like built-in spell checkers, Grammarly, chatbots, or virtual agents are some examples of AI-powered language tutoring systems.

European languages like French, German, and Spanish, and Asian languages like Japanese, Korean, Chinese, etc., hold major positions in foreign language tutoring. You can see a variety of online courses offered by major universities around the globe in these languages on online course websites like Coursera, EdX, Khan Academy, and so on. Nevertheless, many languages are used in modern technology, and different AI tools have been developed to learn different languages; English is still used in major global workplaces and technologies. It is difficult to succeed in the world of work without English language proficiency, as many non-native speakers still thrive on communicating with native English speakers professionally. Consequently, few AI systems help in teaching and learning the English language.

Accordingly, English language teaching encapsulates various methods and approaches like communicative language teaching, task-based language teaching, audio-visual methods, and so on. The aim of language teaching or learning should also be contemplated before using a particular method or approach. Similarly, AI systems should also be used according to the learning objectives and learners' needs for learning the language. According to (Yang et al., 2021), "AI research now focuses more on perception and human audio-visual literacy, which is the ability to see, hear, read, and write". This research can promote the students' motivation and effectiveness in learning a new language as it inspires exciting online tasks. Furthermore, there are a few algorithms like **BERT** (Bidirectional Encoder Representations from Transformers) and **GPT-2** (Generative Pre-Training) that help in promoting language teaching. So, AI tools and systems can help the teachers to provide individual attention to the learners' understanding and progress, which is vital for 21st-century learning. Accordingly, language tutoring will be effective with personalized tutoring, and AI can provide an adaptive learning environment that is discussed in more detail next.

30.4 PERSONALIZED TUTORING AND ADAPTIVE LEARNING

Personalized tutoring has often been viewed as an effective method to teach students, as it closely monitors the students at each stage of the teaching-learning process. Teachers can't provide individual attention to students in a classroom average of 50 to 60 students. Hence, the students' learning progress can be evaluated based on the average level of the whole class in the traditional classroom. AI helps teachers to evaluate an individual student's progress by providing various tools. It also helps the teachers plan additional learning techniques for low-average students to promote their understanding of the course. AI- empowered personalized tutoring would not replace the role of the teachers, but modifies the teachers' role as facilitators. AI helps in enhancing the abilities of teachers to tailor lessons for individuals. AI provides vast amounts of students' data to the teachers that can help them understand their learning progress and improvise their teaching techniques. For example, AI-powered algorithms like *Numerade* offer short-form videos to explain complex subjects of STEM (science, technology, engineering and mathematics) taught by highly skilled educators. It combines lesson plans with machine learning technology and videos of human teachers to provide a personal approach to all the students. Besides, in personalized tutoring, the learning objective, pedagogical approaches, and content to be taught will vary for a person according to the needs of the learners. There are different approaches to personalized tutoring, like adaptive learning and individualized learning, that will use AI algorithms to address the unique needs of the learners. Some examples of personalized tutoring and adaptive learning software are:

- SmartEd – software that uses games to make teaching-learning interesting and fun.
- Knewton's Alta – helps in personalized learning by assessing students' assignments.
  Knowji – an audio-visual app that works on spaced repetition algorithm, helping learners to progress in learning language learning.
- Fast ForWord – reading and language software that helps learners develop language and cognitive skills.
- Century Tech – helps in pointing out knowledge gaps.

AI and machine learning technology for adaptive learning have yet to be developed to meet all learners' needs.

Schmidt and Strasser (2022) have proposed a domain model, an evaluation model, and a learner model as three foundational pillars of the adaptation model for the machines to facilitate adaptive learning. According to the authors, a domain model contains information about the parts of the language that have been addressed in the exercises, which address all levels of grammatical concepts. Further, the evaluation model will assess the learners' performance individually, and the learner model will collect and update the learners' proficiency with respect to the domain model's content. Hence, this serves best to improve the learner's adaptive learning environment. In addition, Interact4School is a government-funded interdisciplinary project concentrating on Focus on Form (FOF). FOF is a language education approach where learners will be aware of the grammatical form of the language features they use for communicative purposes in the target language. In this project, a mixed-methods approach has been applied: language testing, classroom videography, questionnaires, interviews with learners and teachers, and conditions and effects of adaptive, individualized practice with intelligent feedback (Schmidt & Strasser, 2022). Framework for user modelling and adaptation (FUMA) is a student-based framework that applies machine learning in the courses in MOOC-like platforms to provide personalized tutoring to students. Hyper-personalization AI technology has been enabled by machine learning, where Carnegie Learning (that offers immediate guidance to the students and interacts with the learners closely) has evolved as an intelligent instructional program. Correspondingly, these AI tools and software facilitate language tutoring and learning in our current 21st-century technology-based education system. Further, adaptive learning can be assisted by employing question-answering AI tools as part of language teaching.

30.5 QUESTION ANSWERING

Question answering (QA) is a part of AI within the NLP and information retrieval fields. It is a building system that responds to the questions posed by humans using NLP. The fundamental reason for the development of QA is to improve man-machine interactions. For instance, we can take Apple's *Siri* (virtual assistant), Amazon's *Alexa* and Google's *Google Assistant,* which respond to human questions. According to (Sing et al., 2016), "Question Answering is a research area that combines research from different fields, with a common subject, which are Information Retrieval (IR), Information Extraction (IE) and Natural Language Processing (NLP)". In addition, QA systems can be classified into open-domain and closed-domain. The open domain uses everything from the World Wide Web (WWW), and the closed domain uses more domain-specific systems, which use NLP systems heavily.

The linguistic approach, statistical approach, and pattern matching approach are the three major approaches in the QA systems. The linguistic approach deals with NLP, the statistical approach deals with the World Wide Web database, and the pattern matching approach deals with text patterns with a computing approach. In 1964, ELIZA, an early NLP program, was created by Joseph Weizenbaum. It was one of the first chatbots which communicates with humans by the pattern matching approach. Following ELIZA, many little systems like Evi, QUORA, BING, StoQA, Wolfram Alpa, Answerbag, Blurtit, Kangavaril, and Ask. com have been developed. All the listed systems work based on AI QA systems by retrieving information or answers from the database based on the questions posted. In this line, these QA systems can help in collaborative learning and personalized tutoring of language teaching. By using these systems, the learner can post their questions and doubts to a large audience and get responses from intellectual users of the previously stated platforms worldwide.

Moreover, the learners can get different aspects of the subjects dealt with by different institutes. As in apps like *Siri* and *ALEXA*, QA systems have access to a large AI database, and they can be used to provide personalized responses according to the learner's previous learning history. Hence, QA systems can be very helpful in providing feedback and the status of the learning progress personally to all individual learners. In connection with the feedback, AI also provides tools and software for grading learners' performance automatically and provides individual feedback by simplifying the teachers' major workload.

30.6 AUTOMATED GRADING OF TESTS

Assessing and grading a large group of students is strenuous work, and giving individual feedback to each student in the large group is an overwhelming task for any teacher. Therefore, academicians are looking to implement AI in correcting the transcripts of students' tests. We can use Google forms, Microsoft forms, and a few other automatic grading tools to assess the students' performance. These tools can be used for the effective evaluation of multiple-choice questions (MCQs) but not short answers or long answers. It is difficult for AI tools to assess short answers, essays, and mathematical equations consisting of symbols. Research on automatic scoring tools has been in the education field since the early 1990s. (Ramesh & Sanampudi, 2022) have studied automated essay scoring systems using AI. The authors stated that pattern matching and statistical-based approaches had been used in the early 1990s to automatically score the essay-type answers, whereas in the last decade the regression-based approach has been implemented.

Further, the authors have identified that since 2014, the deep learning approach has been used in automatic grading systems. Some automated grading tools are also used in Ohio, Utah, and some other US schools to evaluate the students' assignments. Among them, the Utah assessment tool has developed a rubric with six characteristics: development of ideas, organization, style, word choice, sentence fluency, and conventions to assess the students' essays. There is also an Educational Testing Service (ETS) which automatically evaluates the students' script on the content-based approach. Akhtar (2015) has researched the "development and successful deployment of the Automated Grading and Feedback (AGAF) system". AGAF focuses on the different types of assignments and grading tools. For instance, a Microsoft Office assignment has been asked to be uploaded to the system, and a Survey Monkey survey and a blog have been used to assess the assignments. Moreover, a pattern matching approach has been used to evaluate the student's uploaded image file as an assignment. Likewise, Gradescope software works on machine learning and AI to assess and give feedback to individual learners' test scripts. It evaluates paper-based exams, programming projects, and even online homework.

In many developing countries, Google forms have been widely used for MCQs and quizzes. Kahoot and Quizlet are also some of the AI platforms used in e-class rooms to motivate students in online education. These platforms automatically grade the students' answers for objective-type and MCQ-type tests. Online education has a huge impact as the world is reviving a normal lifestyle. Teachers can also prefer automating grading systems in their courses to reduce their rigorous evaluation time and concentrate more on classroom assessment and pedagogical activities. But there are a few practical difficulties in integrating AI-based tutoring in language teaching/ education. To name a few difficulties, the technology is not available to all learners, and teachers are not trained to use AI, institutes are not equipped to implement

AI-based teaching, and so on. Hence, academicians should take the necessary steps to improve educational policies, and educational institutes should cater to the needs of upcoming AI-based education.

30.7 CONCLUSION

Many governments across the world still suggest that educational institutes use online modes in the classroom. AIEd has become accessible to everyone worldwide, but rural institutes still cannot use advanced technology above a certain level. Since educational institutes have already experimented with the implementation of AI in pedagogy, they can investigate the limitations and obstacles between the students and AIEd. This chapter throws light on the development of AI in the educational field through different frameworks and how AI tools can be implemented in teaching the English language. The education tools are not limited to language tutoring, but can be adapted to any field of study. As mentioned in the chapter, AI is still blooming, and different avenues to integrate AI into classroom teaching can be explored. Although AI provides many advantages, there are also some practical difficulties in real-life to imply AI in education. Hence, academicians and researchers should concentrate on possible solutions to overcome those difficulties. Eventually, AI can provide excellent support in academic institutes' teaching and learning processes. AI can assist teachers in promoting students' learning, but it is important to note that the pedagogical intervention of teachers is crucial in any learning environment.

REFERENCES

Akhtar, B. (2015). *An Automated Grading and Feedback System for a Computer Literary Course*. Appalachian State University, Boone, NC.

Guan, C., Mou, J., & Jiang, Z. (2020). Artificial intelligence innovation in education: A twenty-year data-driven historical analysis. *International Journal of Innovation Studies*, *4*(4), 134–147. https://doi.org/10.1016/j.ijis.2020.09.001.

Hwang, G.-J., Xie, H., Wah, B. W., & Gašević, D. (2020). Vision, challenges, roles and research issues of artificial intelligence in education. *Computers and Education: Artificial Intelligence*, *1*. https://doi.org/10.1016/j.caeai.2020.100001.

Jantakun, T., Jantakun, K., & Jantakoon, T. (2021). A common framework for artificial intelligence in higher education (AAI-HE model). *International Education Studies*, *14*(11). https://doi.org/10.5539/ies.v14n11p94.

Khosravi, H., Shum, S. B., Chen, G., Conati, C., Tsai, Y.-S., Kay, J., . . . Gašević, D. (2022). Explainable artificial intelligence in education. *Computers and Education: Artificial Intelligence*, *3*. https://doi.org/10.1016/j.caeai.2022.100074.

Langley, P. (2019). *An integrative framework for artificial intelligence education*. The Ninth AAAI Symposium on Educational Advances in Artificial Intelligence (EAAI-19), 9670–9677.

Ouyang, F., & Jiao, P. (2021). Artificial intelligence in education: The three paradigms. *Computers and Education: Artificial Intelligence*, *2*. https://doi.org/10.1016/j.caeai.2021.100020.

Ramesh, D., & Sanampudi, S. K. (2022). An automated essay scoring systems: A systematic literature review. *Artificial Intelligence Review*, *55*(3), 2495–2527. https://doi.org/10.1007/s10462-021-10068-2.

Schmidt, T., & Strasser, T. (2022). Artificial intelligence in foreign language learning and teaching: A CALL for intelligent practice. *Anglistik: International Journal of English Studies*, *33*(1), 165–184.

Sing, S., Das, N., Michael, R., & Tanwar, P. (2016). The question answering system using NLP and AI. *International Journal of Scientific & Engineering Research*, *7*(12), 55–60.

Thongprasit, J., & Wannapiroon, P. (2022). Framework of artificial intelligence learning platform for education. *International Education Studies*, *15*(1). https://doi.org/10.5539/ies.v15n1p76.

Yang, S. J. H., Ogata, H., Matsui, T., & Chen, N.-S. (2021). Human-centered artificial intelligence in education: Seeing the invisible through the visible. *Computers and Education: Artificial Intelligence*, 2. https://doi.org/10.1016/j.caeai.2021.100008.

31 Review of Learning Analytics Techniques and Limitations in Higher Education
A 21st-Century Paradigm

Tuhin Utsab Paul and Tanushree Biswas

31.1 INTRODUCTION

The present century is the age of analytics. Analytics has made inroads in the educational sector also. In academics, it is known as learning analytics. Learning analytics is the process of analyzing students and improving the educational performance based on the outcome of the analysis. These analytics also help educational institutes to improve their educational policies and move towards a more outcome-based pedagogy. Students around the world need advanced skills to succeed in the globalized, knowledge-based world of today. To succeed in this highly connected complex world scenario, students need certain 21st-century skills:

1. Collaboration
2. Skilled communication
3. Knowledge construction
4. Self-regulation
5. Real-world problem solving and innovation
6. Use of ICT in learning

Learning analytics will be an assessment and guiding tool for assessing the outcome of the skill-based education system and help in improving and planning policy for futuristic education. The National Education Policy 2020 by the Government of India also stressed the 21st-century skill-based education and outcome-based assessment. Higher education institutions must evaluate the learners' needs and also the requirement of Industry 4.0 and develop curriculum and pedagogy based on them. Learning analytics is one of the tools which help in assessing the outcome and performance of learners. The main purpose of learning analytics is to improve the outcome-based learning experience of the students and to improve the teaching pedagogy and teaching-learning environment for better productivity. Conventionally, questionnaires and surveys are used for identifying the strategy of the teaching-learning environment.

DOI: 10.1201/9781003328414-31

But in the 21st century, and with the effects of a global pandemic, information and communication technology (ICT)–based education had taken a front seat. Because of online teaching, in addition to survey-based analysis, computer-aided analysis can also be done to assess the learning outcome of students. This chapter focuses on the various learning analytics technologies that are currently in use and provides a detailed review of the various cutting-edge research studies in the field of learning analytics. Moreover, this chapter also discusses the gap in research in the area of learning analytics and how those gaps in research can lead the way forward for further research.

31.2 LEARNING ANALYTICS: NEED OF THE 21ST CENTURY

Since the last decade, the education sector has undergone a change with the introduction of the learning management system (LMS). The old structure of data archiving into files and records had been replaced with the LMS, which is an enterprise resource planning (ERP) designed for educational institutes. The classrooms and offices had moved towards a paperless functioning system. LMS is essentially software to manage the daily operational requirements of an institute such as attendance recording, marks recording, result publishing, etc. Although there is a scope of analysis in LMS, it is limited to a large extent due to the rigid functioning of the system. This gives rise to a new requirement of analyzing large amounts of data generated by the education sector irrespective of the institutional boundaries. The analysis of educational data across various institutes can give insight about student behavior, skill development, outcome analysis, etc. This requirement opens up a new field of research known as learning analytics. Learning analytics deals with the collection and analysis of data of learners on every tier of education.

Various analyses can be done on the continuously growing academic data. Analysis can be done to assess the performance of learners. Active and focused participation of learners can be analyzed in both traditional classroom setup and online teaching via video conferencing. Group discussion and collaboration of student activities can be analyzed. Continuous assessment of skill development of learners can be analyzed. Institution-level decision making, feedback systems, and pedagogy development can be guided by various outcomes of learning analytics. Students' success or failures can be analyzed, which can help in career counseling or lifelong learning.

Learning analysis can be done on two levels: course level and institutional level. Course-level analysis is based on the analysis of learner activity in a particular course and benefits both students and educators. On the other hand, institutional-level analysis helps in understanding student progression, outcome analysis, skill development, etc., which helps the institute to formulate policy for improvement of the teaching-learning system of the institute.

31.3 LEARNING ANALYTICS: TOOLS AND APPLICATION SOFTWARE

Learning analysis is done by various leading universities across the globe to monitor the progress of the students and to devise the academic plan. Universities either use off-the-shelf software or develop their own analytics software. To mention a few, the

University of California uses Moodog, the University of Michigan uses E2coach, and Purdue University uses Signals. Various software programs are available to give analytics on learning outcomes and can be integrated with the existing LMS. Google Analytics, Microsoft 360, Edmodo, and Blackboard are a few such software programs which have their own LMS or can be integrated with other LMS packages and give different analytics on the student progression. But all those software programs mostly depend on grading of assessments by educators, rubrics, and online evaluation. Learning analytics in a complete purview must address a broader aspect such as attention span of students, collaborative skills of learners, interactivity of class, skill development, communication, etc. For those, research needs to be undertaken in this direction and custom-made learning analytics tools need to be developed. Various tools for developing learning analytics software are SPSS, Weka, Rapidminer, web-based tools, Gephi visualization tool, image processing tools like Matlab, etc.

31.4 LITERATURE REVIEW

Research on learning analytics is based on various aspects such as the purpose of learning analytics, technologies used in learning analytics, various data collection methods, types of education, online, offline, or massive open online course (MOOC), etc. This chapter focuses on giving a detailed study of various state-of-the-art research papers in the various areas of learning analytics.

C.S. Elliot [1] worked at Arizona State University and collaborated with two other universities and six industry houses to develop a framework for designing curriculum based on industry requirements. The framework is known as JACME2T. This framework uses learning analytics techniques to design curriculum.

One of the first experiments on learning analytics was carried out by Khalid et.al. [2] where he analyzed student performance on an academic advisory domain using fuzzy logic, neural networks, and genetic algorithms. The system developed by Khalid is known as Intelligent Planning System (INPLANS).

Data mining was used by Carlos et. al. [3] to predict school dropout. The paper collected data from over 600 schools and used various classification algorithms to do a predictive analysis to predict school dropout.

A proactive intelligent intervention predictive algorithm was proposed by Usamah et. al. [4] to do academic analytics and improve students' performance. The paper considers performance of students in semester examinations and other data such as attendance data, etc., to do predictive analysis. This research uses neural networks and decision trees for predictive analysis.

Hsu-Chen et. al. (2012) [5] emphasized the fact that doing learning analytics with end semester questionnaire-based surveys is not fruitful because the students' learning needs cannot be addressed and the response is low. Moreover, analysis based on LMS data alone is not sufficient because learning happens not only in classrooms but in libraries, field work, and discussion. So, he proposed a framework for learning analytics where the Internet of Things (IOT) using sensors, RFID, etc., are used to track learners' activities in various places and do analytics on skill development and progression.

A wide variety of technology had been used to achieve the goals of learning analytics. The technologies include adaptive learning, artificial intelligence, predictive modeling, data mining, clustering, machine learning and neural networks. Various data mining technologies such as support vector analysis, clustering, or linear regression are used to analyze students' performance. Statistical methods such as Pearson spearman and Bray Curtis methods, or Euclidean squared methods are used for evaluating the performance of learners. With the growth of online teaching-learning environment, learning analytics is also used in online teaching. Mobile technology, social network analysis, IOT technologies, and computer vision techniques are used for learning analytics.

Carlota et. al. (2014) [6] explored how machine learning systems can be used in specific learning requirements, such as sequencing and performance prediction, and he proposed a minimally invasive algorithm for web services–based sequencing for integrating learning analytics with e-learning systems.

Since, in today's Internet-driven world, learners learn not only from books or classrooms but a lot of Internet and social networking sites, Tobias et. al. [7] used web analytics, which is an established field of study, in the area of learning analytics for collecting and analyzing learners' interaction on online learning platforms. This work uses Google Analytics with a MOOC platform to capture behavioral data about learners' activities. But this study had a limitation in terms of learner-specific metrics and a concern about data privacy of learners.

In the article by Ali et al. [8], researchers developed a learning analytics tool called LOCO-Analyst which analyzes the feedback from all different aspects of teaching-learning related to student activities in the LCMS. Instead of working on a predefined set of factors which could trigger the system and alarm the teacher regarding student performance, for example, in TeacherADVisor (TADV) and Student Inspector, this tool, on the other hand, specifically examines the feedback from different domains by tracking the students' activities and interactions among themselves as well as the teachers and broadly categorizes the analysis into local and global feedback.

On the other hand, eLAT, a learning analytics toolkit [9] developed by Dyckhoff et. al., can be integrated with any virtual learning environment (VLE) which can extract the data and further segregate it into different criteria like teacher-student profiles, lecture notes and resource materials, assignments and submission data, student activities like posting queries or responding or communicating in the VLE, etc. The researchers experimented with a prototype at RWTH Aachen University and conducted several iterative processes of developing the said prototype. The eLAT user interface provides an interactive panel where teachers could easily monitor the students' visibility in the VLE.

Some of the learning analytics tools have also developed focusing on the students posts, chats, or any communication over the discussion forum; for example, SNAPP [10]. Wise et. al. [11] discussed how learning analytics could also play a vital role specifically on online discussion platforms. The researchers emphasized the platform called Visual Discussion Forum, where the tool represents a tree structure of all the discussions and comments made in the platform. This helps the students in having a clear graphical representation of the topic of conversation. The researchers further worked on the extracted analysis, developing metrics on the various log-generated data like number and length of any post created or deleted, time of posting, etc., and tried to investigate the active discussion participation of students on online forums throughout a semester system of a course using both embedded and extracted learning analytics.

And now due to the rise in the trend in learning analytics, educational institutions are stressing the fact there should be a standard for e-learning. Different learning analytics tools could be seen in the market where each analytic tool focuses on certain aspects of analytics. As seen earlier, some focus on feedback systems [8] or some use various discussion forums [10, 11] to gather information and present a meaningful interpretation which can predict student performance. In Ref. [12], one can observe that the researchers discussed the importance of e-learning standards and provided a comparative analysis of two common e-learning standards: the IEEE Standard for Learning Technology and The Experience API. The IEEE Standard for Learning Technology is a complex rigid model, whereas as per the researcher The Experience application programming interface (API) is a more suitable model for the learning analytics tools. One of the main advantages of The Experience API is the flexibility of the model to transfer and analyze student-relevant data over any LMS.

To understand the complexity of the learning analytics tools and their impact on all the spheres of the education system, including teachers, learners (students), and the institutions as a whole, researchers Drachsler and Greller [13] conducted a survey through various learning analytics groups and platforms. Their main focus was on five different angles of observing the growth of learning analytics, which included "who shall be affected majorly by the impact of the LA?", "what are the roadblocks in the evolution of LA", "what should be the main areas LA tools could [be] focused upon", and "how and what should be the methodology for the Data Mining?" The results as per the survey were contrasting as compared to the general trend of learning analytics.

On the other hand, the higher education system all over the world has been incorporating learning analytics, especially in their research domains. But the present scenario of data collection and usage is hampering the effectiveness of the LMS. The researchers in [14] provided a case study of a reputed university which has invested in learning analytics and implemented an LMS throughout its system, but the numbers generated are not able to justify the students' progression or performance or any teaching efficiency. The paper stressed the limitations and constraints in implementing a systematic LMS in the university and suggested a better planning and adoption of various policies in the integration of technological innovations into the teaching-learning process.

The MOOCs is also a new and increasing trend in higher education but the percentage of students leaving the course is also high in this indirect teaching-learning environment. The learning analytics could play a vital role in bridging the gap between the learner and the informal non-traditional teaching environment. The researchers in [15] incorporated the idea of the marketing funnel into the system of MOOC where learning analytics can analyze the pattern and progress of each learner from the beginning of the course to the final assessment and evaluation process. With careful analysis and evaluation, learning analytics can easily monitor the progress of the learner and could detect the possible drop-outs as well. Since not much research has been done in the field of implementation of learning analytics in the MOOC, there is scope in exploring and validating the concept.

31.5 GAPS IN RESEARCH

As we have seen, learning analytics could be very useful in understanding the learning characteristics of the students and can help the teachers as well as the students at various different levels. As we have seen earlier, learning analytics can help the teacher identify student progress by looking into the aspects like "how many times the student is opening a file, or how many times the student has responded or post queries, or how much time taken by student to submit any assignment in the LMS".

In all the previous research, we can see the performance of a student in a particular course could be predicted by looking into the data for assessment, analysis, and feedback generated through any LMS.

Along with it, student behavior in a physical class could also be monitored and analyzed. This perspective could further add a few characteristics like hand, eye, head orientation, and other body gestures of a student during a class. For example, when a teacher is teaching a concept, if the student is not paying attention by continuously looking here and there, this insight could help the teacher in better understanding the student behavior toward the lecture session and also self-evaluate the effectiveness of the course content.

Bohong Yang et. al. [16] focused on the case study of students' behavior during a class and analyzed it based on numerous parameters. The recordings of the physical classroom were thoroughly studied, and the student's concentration level was analyzed by looking into the patterns like whether the student is looking at the board, or facing the teacher, or continuously looking up and down to write notes. Similarly, they also examined how a teacher's audio data, and modulation can affect the concentration, along with content of the class lecture and teaching style.

31.6 CONCLUSION

The education system has been completely transformed due to the COVID-19 pandemic, and as a result, teaching pedagogy has changed dramatically, which has paved the way for the new reforms in the system, especially the concept of the e-classroom.

Since the teaching has shifted from the physical classroom to remotely and on digital platforms, the challenge to keep the concentration level of all the students during the whole online session by the teacher has become a daunting task. The lack of physical classroom teaching is affecting the students as well to focus on every topic. This could affect the quality and the effectiveness of the teaching.

If one can use the concept of learning analytics on the students' behavior during the online classes, the teacher could be able to examine the behavior, which in turn could help in improving the teaching efficiency.

Computer vision techniques can be used to analyze the student behavior pattern throughout a series of online classes with the help of high-definition webcam along with the use of headphones and microphones. Using a stable bandwidth as well as a stable online platform for teaching, one could analyze the attention level of a student on the topic of discussion in the class. During online class, by recording the screen, eye, and head movements of students pattern recognition techniques can be analyzed and the duration of concentration looking at the screen can be captured. Hence, an estimate of focused and concentrated learning phase can be determined. Techniques such as data mining and statistical methods like regression analysis, regression trees, correlation matrices, Pearson Spearman, Bray-Curtis methods, and various clustering methods and other visualization techniques could assess student performance. This shall give a concise picture of a teaching characteristic of a teacher, as well the students' concentration level during the class. The teachers thus could identify students who need guidance and in a way can modify certain teaching styles in their online sessions. Those analyses can also help in modifying the teaching pedagogy to help make the class more interesting and helpful to the learners.

REFERENCES

[1] Charles S. Elliott, "JACME2T: An industry - academic consortia to enhance continuing engineering education", FIE Conference, 1998.

[2] Khalid Isa, Shamsul Mohamad and Zarina Tukiran, "Development of INPLANS: An analysis on students' performance using neuro-fuzzy", Symposium on Information Technology, vol 3, pp. 1–7, 2008.

[3] Carlos Márquez-Vera, Cristóbal Romero Morales and Sebastián Ventura Soto, "Predicting school failure and dropout by using data mining techniques", IEEERITA, vol. 8, pp. 7–14, 2013.

[4] Usamah bin Mat, Norlida Buniyamin, Pauziah Mohd Arsad and Rosni Abu Kassim, "An overview of using academic analytics to predict and improve students' achievement: A proposed proactive intelligent intervention", IEEE Conference on Engineering Education (ICEED), pp. 126–130, 2013.

[5] Hsu-Chen Cheng and Wen-Wei Liao, "Establishing an lifelong learning environment using IoT and learning analytics", ICACT, pp. 1178–1183, 2012.

[6] Carlotta Schatten, Martin Wistuba, Lars Schmidt Thieme Sergio and Gutierr´ez-Santos, "Minimal invasive integration of learning analytics services in intelligent tutoring systems", ICALT, pp. 746–748, 2014.

[7] T. Rohloff, S. Oldag, J. Renz and C. Meinel, "Utilizing web analytics in the context of learning analytics for large-scale online learning," 2019 IEEE Global Engineering Education Conference (EDUCON), Dubai, United Arab Emirates, pp. 296–305, 2019.

[8] Liaqat Ali, Marek Hatala, Dragan Gašević and Jelena Jovanović, "A qualitative evaluation of evolution of a learning analytics tool", *Computers & Education*, Vol. 58, No. 1, pp. 470–489, 2012.

[9] Anna Lea Dyckhoff, Dennis Zielke, Mareike Bültmann, Mohamed Amine Chatti and Ulrik Schroeder, "Design and implementation of a learning analytics toolkit for teachers", *Journal of Educational Technology & Society*, Vol. 15, No. 3, pp. 58–76, 2012.

[10] S. Dawson, A. Bakharia and E. Heathcote, "SNAPP: Realising the affordances of real-time SNA within networked learning environments", *Learning*, pp. 125–133, 2010.

[11] Alyssa Friend Wise, Yuting Zhao and Simone Nicole Hausknecht, "Learning analytics for online discussions: A pedagogical model for intervention with embedded and extracted analytics", Proceedings of the Third International Conference on Learning Analytics and Knowledge, pp. 48–56, 2013.

[12] A. del Blanco, A. Serrano, M. Freire, I. Martínez-Ortiz and B. Fernández-Manjón, "E-Learning standards and learning analytics. Can data collection be improved by using standard data models", Global Engineering Education Conference (EDUCON) IEEE, pp. 1255–1261, 2013.

[13] Hendrik Drachsler and Wolfgang Greller, "The pulse of learning analytics understandings and expectations from the stakeholders", Proceedings of the 2nd international conference on learning analytics and knowledge, pp. 120–129, 2012.

[14] Leah P. Macfadyen and Shane Dawson, "Numbers are not enough. Why e-learning analytics failed to inform an institutional strategic plan", *Journal of Educational Technology & Society*, Vol. 15, No. 3, pp. 149–163, 2012.

[15] Doug Clow, "MOOCs and the funnel of participation", Proceedings of the 3rd International Conference on Learning Analytics and Knowledge, pp. 185–189, 2013.

[16] Bohong Yang, Zeping Yao, Hong Lu, Yaqian Zhou and Jinkai Xu, "In-classroom learning analytics based on student behavior, topic and teaching characteristic mining", *Pattern Recognition Letters*, Vol. 129, 2019.

Index

For Product Safety Concerns and Information please contact our EU
representative GPSR@taylorandfrancis.com
Taylor & Francis Verlag GmbH, Kaufingerstraße 24, 80331 München, Germany

www.ingramcontent.com/pod-product-compliance
Lightning Source LLC
Chambersburg PA
CBHW060755220326
41598CB00022B/2446